An Introduction to Modal Logic

An Introduction to Modal Logic

G. E. HUGHES

Professor of Philosophy,
Victoria University of Wellington

M. J. CRESSWELL

Professor of Philosophy,
Victoria University of Wellington

LONDON

METHUEN AND CO LTD

First published 1968
by Methuen and Co Ltd
11 New Fetter Lane, London EC4P 4EE
Reprinted with corrections 1972

First published as a University Paperback 1972
Reprinted twice
Reprinted 1977

Printed in Great Britain by
Fletcher & Son Ltd
Norwich

ISBN 0 416 29460 X

Distributed in the USA by
HARPER & ROW PUBLISHERS, INC.
BARNES & NOBLE IMPORT DIVISION

Contents

Preface

Modal logic can be described briefly as the logic of necessity and possibility, of 'must be' and 'may be'.

We had two main aims in writing this book. One was to explain in detail what modal logic is and how to do it; the other was to give a picture of the whole subject at the present stage of its development. The first of these aims dominates Part I, and to a lesser extent Part II; the second dominates Part III. Part I could be used on its own as a text-book for an introductory course of instruction on the basic theory and techniques of modal logic.

We have tried to make the book self-contained by including at the appropriate points summaries of all the non-modal logic we use in the exposition of the modal systems. It could therefore be tackled by someone who had not studied any logic at all before. To get the most out of it, however, such a reader would be well advised to buy himself another book on logic as well and to learn something more about the Propositional Calculus and the Lower Predicate Calculus than we have been able to tell him here.

We have excluded ancient and mediaeval modal logic from our survey and confined our attention to the modern development of the subject which had its effective beginning in the work of C. I. Lewis some fifty years ago. We have also not attempted to deal with deontic, epistemic or certain other closely related logics, which would fall under the heading of modal logic in a wider sense of the term than ours. Nor have we discussed many-valued logics, though these have sometimes been grouped with modal logics. But within the limits thus indicated we have tried to be as comprehensive as we could in a work of this length; and we have given full and detailed references to the literature throughout.

It should be clearly understood, however, that this is a book on formal logic. Modal logic does in fact raise a considerable number of philosophical problems, and in dealing with the interpretation of modal systems we have not been able to avoid them altogether. But we have confined the philosophical remarks we do venture to make to those which seem necessary in order to show the point of the formal work itself. Some of the problems raised by modal logic seem to us to be among the most important and fundamental in philosophy, but it would require a separate book, and a very different one from ours, to discuss them adequately. In our view there is also a link of a different kind between philosophy and modal logic, in that modal logic can be used to clarify a number of philosophical problems themselves; and we should like to think that what we have written might encourage some philosophers to make a serious study of modal logic with this end in view.

The notation we have used for modal formulae is only one of many in current use. In the text we have translated other writers' notations into our own throughout; but in Appendix 4 we have indicated how theirs differ from ours, and the reader who consults the originals should then have little difficulty in following the formulae as he finds them there. We have also adopted the convention that a formula or symbol may be treated as its own name and hence does not need to be placed within inverted commas when it is being spoken about.

We wish to thank the Publications Committee of the Victoria University of Wellington for a grant towards the cost of producing this book, and to express our gratitude to Mrs Helen Fleming, Mrs Maureen Mowat and Mrs Isbel Raudnic for their patience and devotion in typing a manuscript which was even more difficult to decipher than the printed appearance of the book might suggest.

G. E. HUGHES
M. J. CRESSWELL

Wellington, New Zealand

NOTE ON REFERENCES

All references to the literature are given in footnotes, and in abbreviated form, by name of author and date of publication. Full details will be found in the Bibliography on pp. 356–371, which is arranged in alphabetical order of authors and chronologically under each author's name.

NOTE TO THE SECOND PRINTING

The reprinting of this book has enabled us to correct a number of errors, but otherwise the text remains unaltered. The most important change is the re-casting of the completeness proof for S4 on pp. 112–15, which was defective in the original version.

There have been considerable developments in modal logic even in the two or three years since the book was first published. Most of these have appeared in articles in the journals listed at the beginning of the Bibliography on p. 356.

G.E.H.
M.J.C.
April, 1971

PART ONE

Modal Propositional Logic

CHAPTER ONE

Non-modal Propositional Calculus

Subsequent chapters of this book will presuppose a familiarity with at least the rudiments of the 'ordinary' (two-valued) Propositional Calculus. When we use the expression 'Propositional Calculus' (or the abbreviation 'PC') *simpliciter*, it is to this non-modal system of logic that we shall be referring. The present chapter outlines, in a very summary fashion, those elements of PC which we shall take for granted in what follows, and at the same time explains some of the terminology which we shall use throughout the book. It is not intended to act as an introduction to PC for its own sake: there are many readily-available books which do that[1], and we are not trying to add to their number.

When we construct a logical system we do two distinguishable things: (a) we set up a system of symbols, with rules for combining them into formulae and manipulating these formulae in various ways; (b) we give an interpretation, or attach a meaning, to these symbols and formulae.

If we confine ourselves to doing (a), we are said to have an *uninterpreted* system. If we do both (a) and (b), we are said to have an *interpreted* system.

This distinction is important. There are some things which we can prove about a logical system by considering it purely as an

[1] Most current logic textbooks give an account of PC in more or less detail. A book which uses almost the same terminology as ours and develops PC in a very similar (though of course much fuller) fashion is Hughes & Londey [1965]. For a more detailed account (and some historical notes) *vide* Church [1956]. Some other suitable introductions are found in Quine [1940], [1950], Ambrose & Lazerowitz [1948], Prior [1955] and Faris [1962]. Terminology and notation vary somewhat but this should not confuse the careful reader.

uninterpreted system; there are other things we cannot prove about it unless we take into account its interpretation as well. Moreover, in general a system can be given a number of different interpretations, and it is sometimes possible to demonstrate that a system has certain features by considering some other interpretation than the one it was originally intended to have.

We shall bear this distinction in mind in the brief exposition of PC which follows, and throughout the rest of the book as well. What this comes to in practice is that although we select a certain system of symbols and rules because we think it is capable of being given a certain interpretation, we take care to be able to describe that system without making any reference to the intended interpretation, or indeed to any other possible interpretation. And although we shall often find it convenient to call various symbols and rules by names which suggest the interpretation we have mainly in mind, these names can where appropriate be regarded simply as neutral labels.

Primitive symbols and formation rules of PC

We take as *primitive* (or undefined) symbols of PC the following:

A set of *letters*: p, q, r, . . . (with or without numerical subscripts). We suppose ourselves to have an unlimited number of these.

The following four symbols: $\sim$, $\vee$, (,).

Any symbol in the above list, or any sequence of such symbols, we call an *expression*. An expression is either a *formula* – more exactly a *well-formed formula* (wff) – or else it is not. We are concerned only with expressions which are wffs. The following *formation rules* of PC specify which expressions are to count as wffs:

FR1 A letter standing alone is a wff.

FR2 If α is a wff, so is $\sim\alpha$.

FR3 If α and β are wffs, so is $(\alpha \vee \beta)$.

In these rules the symbols α and β are used to stand indifferently for any expressions. Thus the meaning of FR2 is: the result of prefixing $\sim$ to any wff is itself a wff. Symbols used as α and β are used here are known as *meta-logical* variables. They are not among the symbols of the system (PC in this case), but are used in talking about the system.

Examples of wffs are: p, $\sim q$, $\sim\sim\sim q$, $(p \vee \sim q)$, $((p \vee r) \vee$

$\sim(q \vee \sim(\sim r \vee p)))$. For convenience, however, we allow ourselves to omit the outermost brackets round any complete wff (though not any subordinate part thereof). No ambiguity in interpretation or unclarity about what is permitted by the rules will result from this notational simplification.

Interpretation

We interpret the letters as variables whose values are *propositions*. We shall usually call them *propositional variables*. We assume that the reader is familiar with the notion of a proposition, and shall not enter into the philosophical issues which this notion raises. Rough synonyms of 'proposition' are 'statement' and 'assertion', where these words are used to refer to *what is stated or asserted*, not to the *act of stating or asserting*. Every proposition is either true or false, and no proposition is both true and false. (Hence if something is neither true nor false, or is capable of being both true and false, it is not to count as a proposition in the present context.) Truth and falsity are said to be the *truth-values* of propositions.

Now it is possible to form more complex propositions out of simpler ones. E.g., out of the proposition that Brutus killed Caesar we can form the proposition that it is not the case that Brutus killed Caesar. This is a proposition which is true if the original proposition is false, and false otherwise. In general, putting 'it is not the case that' in front of a sentence will result in a sentence which expresses a proposition which is true if the original sentence expresses one which is false, and a false proposition if it does not.

Similarly, from the proposition that Brutus killed Caesar and the proposition that Cassius killed Caesar we may form the proposition that either Brutus killed Caesar or Cassius killed Caesar. This proposition will be true iff[2] at least one of the original propositions is true, and therefore false iff both of these are false.

'It is not the case that' and 'either . . . or . . .', when used in the way we have just described, may be said to be *proposition-forming operators* on propositions, because they make new propositions out of old ones. The propositions on which such

[2] 'Iff' is a convenient abbreviation, which we shall regularly use, for 'if and only if'.

an operator operates are called its *arguments*. If an operator requires only a single argument, as 'it is not the case that' does, it is said to be *monadic*; if, like 'either . . . or . . .', it requires two, it is said to be *dyadic*.

Our explanation of these operators, 'it is not the case that' and 'either . . . or . . .', showed that the truth-value of a proposition formed by means of either of them depends in every case only on the truth-value of the operator's argument or arguments. In other words, whenever we are given the truth-value of the argument or arguments, we can deduce the truth-value of the complex proposition. An operator which has this property is said to be a *truth-functional* operator, and the propositions it forms are said to be *truth-functions* of its arguments. Not all proposition-forming operators are of this kind. For example, given merely the truth or falsity of the proposition that Brutus killed Caesar we cannot deduce the truth or falsity of the proposition that Napoleon believed that Brutus killed Caesar; and given merely that two propositions are both true we cannot deduce from this either the truth or the falsity of the proposition that the first follows logically from the second (though if we are given that one proposition is false and another true, we *can* deduce from this that it is false that the first follows logically from the second). Hence although 'Napoleon believed that' and 'follows logically from' are proposition-forming operators on propositions (monadic and dyadic respectively), they are not truth-functional operators.

We interpret $\sim$ and $\vee$ as 'it is not the case that' and 'either . . . or . . .' respectively, in the senses we have explained, and we usually read them simply as 'not' and 'or'. $\sim$ so interpreted is called the *negation sign*; $\sim p$ is said to be the *negation* of p. Using 1 and 0 for the truth-values truth and falsity respectively, we can express the meaning we attach to $\sim$ in the following *basic truth-table for negation*:

	$\sim$
1	0
0	1

Here the left-hand column tabulates the possible truth-values of a given proposition, and the right-hand column sets down the corresponding truth-values of the negation of that proposition. When interpreted in the way we have described, ∨ is known as the *disjunction sign* and its arguments are called *disjuncts*; $p \vee q$ is said to be the *disjunction* of p and q. The basic truth-table for disjunction is:

∨	1	0
1	1	1
0	1	0

The possible truth-values of the first disjunct are tabulated in the leftmost vertical column and those of the second in the topmost horizontal row. The truth-value of their disjunction is found by reading across and down.

These basic truth-tables bring out clearly the truth-functional nature of the operators[3]. In fact, not merely ∼ and ∨, but all operators in PC, are truth-functional and for this reason PC is sometimes called the theory of truth-functions. We said earlier that we interpret p, q, r, . . . as variables whose values are propositions; but in view of the fact that the only feature of the arguments of the operators which is relevant to the truth-value of the complex propositions they form is their truth-value, it is equally satisfactory from a formal point of view to regard the variables as having as their range of values, not the whole infinite set of propositions, but simply the two truth-values 1 and 0.

Further operators

A number of other operators can be defined in terms of the primitive ones. From the point of view of the uninterpreted system, the advantage of using these new operators is that they enable us to write many wffs more succinctly: i.e. from this point of view the definitions provide useful notational abbreviations. From the point of view of the interpreted system, they have the

[3] Strictly we should at this point distinguish between the *symbol* ∼ of PC and the *proposition-forming operator* 'it is not the case that', which is not a symbol of any logical system; but when we have in mind the intended interpretation of ∼ as 'it is not the case that' it should not cause confusion if we refer to the symbol also as an operator (and similarly with ∨ and the other symbols we shall introduce).

additional advantage of making the meaning of many wffs much easier to grasp.

We introduce three new operators, ., $\supset$ and $\equiv$, though it would be possible to have several others as well. The definitions are:

$$[\text{Def}\ .]\quad (\alpha \,.\, \beta) =_{\text{Df}} {\sim}({\sim}\alpha \vee {\sim}\beta)$$
$$[\text{Def} \supset]\quad (\alpha \supset \beta) =_{\text{Df}} ({\sim}\alpha \vee \beta)$$
$$[\text{Def} \equiv]\quad (\alpha \equiv \beta) =_{\text{Df}} ((\alpha \supset \beta) \,.\, (\beta \supset \alpha))$$

In these definitions α and β represent any wffs of PC. The symbol '$=_{\text{Df}}$' is read as 'is defined as'. The meaning of the first definition is that whenever we have a wff of the form ${\sim}({\sim}\text{—} \vee {\sim}\text{—})$, where the blanks are filled by any wffs we please, we can replace this wff by an expression which consists of the wff which filled the first blank followed by a dot followed by the wff which filled the second blank, the whole being enclosed in brackets. Analogous explanations apply to the two other definitions. Similarly, we can expand any expression of the form on the left into the corresponding expression of the form on the right.

Expressions which can be transformed, by applying definitions, into wffs as specified by the original formation rules, are themselves to count as wffs. When a wff contains no symbols except primitive ones it is said to be written in *primitive notation*. The definitions enable us to write all wffs in primitive notation if we wish to do so.

Interpretation of ., $\supset$ and $\equiv$

The interpretation we have already given to $\sim$ and $\vee$ will determine the interpretation we give to the operators defined in terms of them. Thus, we can calculate the truth-values of $p \,.\, q$ for all possible truth-values of p and q by calculating the appropriate truth-values of the wff of which it is an abbreviation, viz. ${\sim}({\sim}p \vee {\sim}q)$, and the basic truth-tables for $\sim$ and $\vee$ enable us to do this. It turns out that $p \,.\, q$ will be true when both p and q are true, but false in all other cases. The basic truth-table for . will therefore be:

.	1	0
1	1	0
0	0	0

When . is so interpreted, it is called the *conjunction sign*; it may be read as 'and'. A proposition formed with . is known as a *conjunction*, and the arguments are called *conjuncts*.

Similar considerations give the following basic truth-table for $\supset$:

$\supset$	1	0
1	1	0
0	1	1

I.e. a proposition formed with $\supset$ is false when the first argument is true and the second false, but true in all other cases. When so interpreted, $\supset$ is known as the (*material*) *implication sign*. It may be read as '(materially) implies' or as 'if [the first argument], then [the second argument]'. The first argument is known as the *antecedent*, the second as the *consequent*. The precise relation of material implication to the various uses of the word 'if' in English raises complex questions into which we shall not enter here. It may plausibly be claimed, however, that material implication represents the truth-functional component in the meaning of 'if' in at least a great many of its standard uses.

The basic truth-table for $\equiv$ works out as:

$\equiv$	1	0
1	1	0
0	0	1

I.e. a proposition formed with $\equiv$ is true when both arguments have the same truth-value, false when they have different truth-values. When so interpreted, $\equiv$ is known as the (*material*) *equivalence sign*. It may be read as 'is (materially) equivalent to', or as 'if and only if'.

Clearly these new operators, like the primitive ones, are truth-functional.

(We could have chosen other operators than $\sim$ and $\vee$ as primitive. Some authors, for example, take $\sim$ and . as primitive and define $\vee$ in terms of these. But whatever primitives we use, provided that all the operators can consistently be given the basic truth-tables listed above, the system of PC so obtained will be exactly equivalent to the one we have set down here.)

Validity

If we regard the variables, p, q, r, . . . as taking the whole range of propositions as their values, we can say that a wff of PC becomes a proposition when all its variables are replaced by propositions. A wff is said to be *valid* iff the result of *every* such replacement is a true proposition. (It is assumed that the replacement is carried out uniformly, i.e. that two or more occurrences of the same variable are always replaced by the same proposition.) If, however, we speak instead of the variables taking simply the two truth-values 1 and 0 as their values, we shall say that a wff is valid iff it always has the value 1, no matter what truth-values are (uniformly) assigned to its variables. We shall normally choose to speak in this second way; since all the operators in PC are truth-functional, exactly the same formulae will turn out to be valid in each case. Simple examples of valid wffs are $p \vee {\sim}p$ and $(p \,.\, q) \supset p$. (A valid wff of PC is often called a *tautology* or a *PC-tautology*.)

A wff is said to be *unsatisfiable* iff it always has the value 0, no matter what truth-values are (uniformly) assigned to its variables. A simple example of an unsatisfiable wff is $p \,.\, {\sim}p$. Many wffs, such as $p \supset q$, are of course neither valid nor unsatisfiable.

The basic notion in the above account of validity is that of assigning a value to each variable in a formula and being able in consequence to calculate the value of the whole formula. We shall say that V is a *value-assignment* to a set of variables when to each variable in the set V assigns either the value 1 or the value 0. We write '$\mathrm{V}(p) = 1$' for 'V assigns the value 1 to p', '$\mathrm{V}(q) = 0$' for 'V assigns the value 0 to q', and so on. We shall say that V is a *PC value-assignment* when in addition to satisfying the above condition it also satisfies two other conditions which reflect the basic truth-tables for $\sim$ and $\vee$. We can state explicitly the conditions for V's being a PC value-assignment as follows:

1. For every propositional variable, p_i (in a given set), either $\mathrm{V}(p_i) = 1$ or $\mathrm{V}(p_i) = 0$ but not both.

2. [V$\sim$]. For any wff, α, $\mathrm{V}({\sim}\alpha) = 1$ if $\mathrm{V}(\alpha) = 0$; otherwise $\mathrm{V}({\sim}\alpha) = 0$.

3. [V$\vee$]. For any wffs, α and β, $\mathrm{V}(\alpha \vee \beta) = 1$ if either $\mathrm{V}(\alpha) = 1$ or $\mathrm{V}(\beta) = 1$; otherwise $\mathrm{V}(\alpha \vee \beta) = 0$.

A PC value-assignment can satisfy conditions 2 and 3 in only one way, but it can satisfy condition 1 in any of 2^n ways, where n is the number of variables under consideration. With respect to n variables, therefore, there will be 2^n distinct PC value-assignments, which we can label $V_1, \ldots, V_{2^n}$ respectively; and each of these can be defined exactly by specifying $V(p_i)$ as 1 or 0 for each variable p_i in the set under consideration. Now every wff can be expressed in terms of variables, $\sim$ and $\vee$ alone; so once we have defined a particular value-assignment, V, in this way, we can calculate $V(\alpha)$ as 1 or 0 where α is any wff all of whose variables occur in the set. If $V(\alpha) = 1$, we shall say that V is a *verifying* value-assignment for α (or that V *verifies* α), and if $V(\alpha) = 0$ we shall say that V is a *falsifying* value-assignment for α (or that V *falsifies* α).

We can now define validity in PC (or PC-validity, as we shall usually call it) more precisely by saying that a wff, α, is PC-valid iff, for every PC value-assignment, V, with respect to the variables occurring in α, $V(\alpha) = 1$ (i.e. iff every such assignment verifies α).

In virtue of the definitions of ., $\supset$ and $\equiv$ we can easily work out consequential value-assignment conditions for these operators, parallel to [V$\sim$] and [V$\vee$]. These will be:

4. [V.]. For any wffs, α and β, $V(\alpha \,.\, \beta) = 1$ if $V(\alpha) = 1$ and $V(\beta) = 1$; otherwise $V(\alpha \,.\, \beta) = 0$.
5. [V$\supset$]. For any wffs, α and β, $V(\alpha \supset \beta) = 1$ if either $V(\alpha) = 0$ or $V(\beta) = 1$; otherwise $V(\alpha \supset \beta) = 0$.
6. [V$\equiv$]. For any wffs, α and β, $V(\alpha \equiv \beta) = 1$ if $V(\alpha)$ is the same as $V(\beta)$; otherwise $V(\alpha \equiv \beta) = 0$.

Since all wffs can be written in primitive notation, these rules are strictly unnecessary; they are however useful to have in practice.

Testing for validity: (i) the truth-table method

In what is known as the truth-table method of testing a PC formula, α, for validity, all possible PC value-assignments with respect to the variables in α are tabulated, and for each such value-assignment, V, the rules we have just stated (or, what comes to the same thing, the basic truth-tables for the operators) are used to calculate $V(\alpha)$ as 1 or 0. The result is a column of 1's and/or 0's, each of which represents V(α) for some value-assignment, V. This column is known as the *truth-table* of the wff. If and only if it consists entirely of 1's, the wff is valid.

An example should make the procedure clear. Let α be $((p \supset q) \cdot r) \supset ((\sim r \vee p) \supset q)$. Here we have three distinct variables, and therefore eight PC value-assignments, $V_1, \ldots, V_8$. The construction of the truth-table proceeds as follows:

	p	q	r	$((p \supset q)$	$\cdot\, r)$	$\supset$	$((\sim r$	$\vee\, p)$	$\supset q)$
(V_1)	1	1	1	1	1	1	0	1	1
(V_2)	1	1	0	1	0	1	1	1	1
(V_3)	1	0	1	0	0	1	0	1	0
(V_4)	1	0	0	0	0	1	1	1	0
(V_5)	0	1	1	1	1	1	0	0	1
(V_6)	0	1	0	1	0	1	1	1	1
(V_7)	0	0	1	1	1	1	0	0	1
(V_8)	0	0	0	1	0	1	1	1	0
				(1)	(2)	(6)	(3)	(4)	(5)

The complete list of value-assignments is set down to the left of the vertical line. The columns to the right are numbered in the order in which they are obtained. Thus column (1), for $p \supset q$, is obtained from the columns under p and q by [V$\supset$]; column (2) is obtained from (1) and the column under r, by [V.]; . . . until finally column (6), the truth-table for the whole wff, is obtained from (2) and (5). The first 1 in (6) means that $V_1(\alpha) = 1$, the second that $V_2(\alpha) = 1$, and so on. Since (6) consists entirely of 1's, $V(\alpha) = 1$ where V is any of $V_1, \ldots, V_8$, and hence α is PC-valid.

Testing for validity: (ii) the Reductio method

A formula can usually be tested more expeditiously by trying to find a falsifying value-assignment for it. The *Reductio* method enables us to find such a value-assignment if there is one.

We begin by supposing that there is some value-assignment, V, for which $V(\alpha) = 0$. We express this supposition by writing 0 under the main operator of α. (In general, if β is any well-formed part of α, we write 1 (or 0) under the main operator of β to signify that $V(\beta) = 1$ (or 0).) From this supposition certain consequences follow, by the basic truth-tables, about the values which must be assigned to certain well-formed parts of α; e.g. if α is of the form $\beta \supset \gamma$, we can have $V(\alpha) = 0$ only if we have $V(\beta) = 1$ and $V(\gamma) = 0$. From these new values certain other consequences follow in the same way, and so on, until finally we either (i) reach a consistent value-assignment to all the variables in α (in which case α is invalid), or (ii) find that we cannot reach such a consistent value-assignment (in which case α is valid).

As an example, let α be the formula we used to illustrate the truth-table method, viz. $((p \supset q) \cdot r) \supset ((\sim r \vee p) \supset q)$. We set out the whole working immediately and then explain it.

$$\begin{array}{cccccccccccccccccc}
 & 6 & 3 & 5 & & 2 & 3 & & 1 & & 8 & 9 & 4 & 7 & & 2 & 4 & \\
((& p & \supset & q &) & \cdot & r &) & \supset & ((& \sim & r & \vee & p &) & \supset & q &) \\
 & 0 & 1 & 0 & & 1 & \underline{1} & & 0 & & 1 & \underline{0} & 1 & 0 & & 0 & 0 &
\end{array}$$

The numerals above the formula indicate the order of the steps. Step 1 is the initial assignment of 0 to α. Since α is of the form $\beta \supset \gamma$, if $V(\alpha) = 0$, we must have $V(\beta) = 1$ and $V(\gamma) = 0$ (step 2). The 1's at step 3 are required by [V.] since β is a conjunction and $V(\beta) = 1$. The remaining steps should now be clear. We finally reach the conclusion (indicated by underlining) that if we are to have $V(\alpha) = 0$ we must have both $V(r) = 1$ and $V(r) = 0$, which violates condition 1 for a PC value-assignment. Hence we can never have $V(\alpha) = 0$, and α is therefore valid.

Other cases are sometimes not so simple. Suppose that α is the converse of the previous formula, viz. $((\sim r \vee p) \supset q) \supset ((p \supset q) \cdot r)$. Steps 1 and 2 can proceed as before, but the values at step 2 do not determine further values uniquely. We can however list exhaustively the alternatives left open at step 3 by the assumption that $V((\sim r \vee p) \supset q) = 1$, as follows:

$$\begin{array}{lcccccccccccccccccc}
 & & & & 3 & & & 2 & 3 & & 1 & & & & & & 2 & & \\
 & ((& \sim & r & \vee & p &) & \supset & q &) & \supset & ((& p & \supset & q &) & \cdot & r &) \\
\text{(a)} & & & & 1 & & & 1 & 1 & & 0 & & & & & & 0 & & \\
\text{(b)} & & & & 0 & & & 1 & 1 & & 0 & & & & & & 0 & & \\
\text{(c)} & & & & 0 & & & 1 & 0 & & 0 & & & & & & 0 & &
\end{array}$$

(a), (b) and (c) represent all the value-assignments to $(\sim r \vee p)$ and q which are compatible with $V((\sim r \vee p) \supset q) = 1$. If each of these leads us to an inconsistency, α is valid; if even one of them is compatible with a consistent assignment to the variables, α is not valid. In fact (b) and (c) both lead to inconsistencies; but (a) does not – it is compatible with $V(q) = 1$, $V(r) = 0$ and $V(p) = 1$ or 0. Hence the whole formula is not valid.

Provided we consider in this way all alternative value-assignments as the need arises, we can test the validity of any wff of PC whatever by the Reductio method. We shall make considerable use of this method in Chapters 5 and 6.

Each of the two methods we have described gives us an effective (i.e. mechanical and finite) procedure for deciding of any given wff of PC whether it is valid or not. Another way of expressing this is by saying that each method gives us a *decision procedure* for PC[4].

Substitution of equivalents

It is clear from the basic truth-table for $\equiv$ that iff a wff $(\gamma \equiv \delta)$ is valid, γ and δ have identical truth-tables. Hence if α is any wff which contains γ, and we obtain β from α simply by replacing γ by δ, then α and β will be equivalent, in the sense that $(\alpha \equiv \beta)$ is valid. When we make such a replacement we shall say that we perform an *equivalence transformation* (of α into β).

We list here some valid equivalences which are especially useful in making such transformations. In most cases we mention a name by which the equivalence is commonly known, and an abbreviation by which it will usually be referred to in this book.

PC1 $(p \,.\, q) \equiv {\sim}({\sim}p \vee {\sim}q)$ } [De Morgan Laws – DeM]
PC2 $(p \vee q) \equiv {\sim}({\sim}p \,.\, {\sim}q)$ }
PC3 $p \equiv {\sim}{\sim}p$ [Law of Double Negation – DN]

[4] Alternatively we can say that each method is a solution of the *decision problem* for the class of valid wffs of PC. Every class of objects has a decision problem (i.e. the problem of finding an effective method for deciding of any arbitrary object whether it is a member of that class or not) but we shall only be interested in the decision problem for classes of formulae. For some classes (such as the class of valid wffs of PC) the decision problem is solvable (i.e. a decision procedure can be found), but for others it is not. If the decision problem for the class of valid wffs of a certain logical system is solvable, that system is often said to be a *decidable* one; if it is not solvable, the system is said to be *undecidable*.

PC4 $(p \vee q) \equiv (q \vee p)$
PC5 $(p \cdot q) \equiv (q \cdot p)$ } [Commutative Laws – Comm]

PC6 $((p \vee q) \vee r) \equiv (p \vee (q \vee r))$
PC7 $((p \cdot q) \cdot r) \equiv (p \cdot (q \cdot r))$ } [Associative Laws – Assoc]

PC8 $p \equiv (p \vee p)$

PC9 $p \equiv (p \cdot p)$

PC10 $(p \supset q) \equiv ({\sim}q \supset {\sim}p)$ [Law of Transposition – Transp]

PC11 $(p \cdot (q \vee r)) \equiv ((p \cdot q) \vee (p \cdot r))$
PC12 $(p \vee (q \cdot r)) \equiv ((p \vee q) \cdot (p \vee r))$ } [Distributive Laws – Distrib]

Any substitution-instance of a valid wff (i.e. any wff formed from a valid wff by uniformly replacing one or more variables by wffs) is itself valid. We can therefore use PC10 not merely to replace $p \supset q$ by ${\sim}q \supset {\sim}p$ in any formula but to switch round the antecedent and consequent of *any* implication, negating each as we do so; and analogously with other valid equivalences.

Repeated applications of the Associative Laws enable us to re-group the disjuncts (or conjuncts) in any purely disjunctive (or conjunctive) wff, or in any substitution-instance of such a wff, in any way we please. In view of this, it is convenient to dispense with interior bracketing in such wffs. We can then write, e.g., $p \vee q \vee r \vee s$, which will be interpreted to mean that at least one of the propositions p, q, r and s is true, and $p \cdot q \cdot r \cdot s$, which will mean that p, q, r and s are all true. Our formation rules do not at present permit such expressions, so we license them by the definitions:

$$(\alpha \vee \beta \vee \gamma) =_{\text{Df}} ((\alpha \vee \beta) \vee \gamma)$$

$$(\alpha \cdot \beta \cdot \gamma) =_{\text{Df}} ((\alpha \cdot \beta) \cdot \gamma)$$

Repeated applications of Comm (together with Assoc if necessary) enable us to re-arrange disjuncts or conjuncts in any order.

In virtue of PC8 and PC9, any wff is equivalent to the disjunction (or conjunction) of itself and itself. In view of this we shall when convenient speak of any (non-disjunctive) wff as a disjunction with one argument; such a disjunction we call a *degenerate* disjunction. We shall speak analogously of degenerate conjunctions.

Conjunctive normal form

A wff is said to be in *conjunctive normal form* (CNF) if it is a conjunction (possibly degenerate), each conjunct in which is a disjunction (again possibly degenerate), the disjuncts in which are of certain specified forms.

We have defined 'CNF' in a general way first, so as to make the definition applicable to other systems than PC. In the specific type of CNF which is relevant to PC – which we shall call PC–CNF – every disjunct must be either a propositional variable or the negation of a propositional variable. Thus the following wffs are in PC–CNF:

$$(1)\ p; \quad (2)\ p \,.\, (q \vee p); \quad (3)\ p \vee q \vee r;$$

$$(4)\ (p \vee {\sim}p \vee q) \,.\, (q \vee r \vee {\sim}r) \,.\, (p \vee r \vee {\sim}r)$$

Wffs in PC–CNF have this important property: they are valid iff every conjunct contains among its disjuncts some unnegated variable and also the negation of that variable. Thus the last of the examples given above is valid, but the others are not.

By using equivalences drawn from the list in the previous section, we can transform any wff of PC, α, into an equivalent wff, α', which is in PC–CNF[5]. α is then said to be *reduced to CNF*. Since α is equivalent to α', it will be valid iff α' is valid. Since we have just stated a mechanical test for validity of wffs in PC–CNF, reduction to CNF gives us another decision procedure for PC.

Axiomatization of PC

An *axiomatic basis* for a logical system consists of (a) a list of *primitive symbols*, together with any definitions that may be thought convenient; (b) a set of *formation rules*, specifying which formulae are to count as wffs; (c) a selected set of wffs, known as *axioms*; and (d) a set of *transformation rules*, licensing various operations on the axioms, and also (normally) on wffs obtained by previous applications of the transformation rules. The wffs we obtain by applying the transformation rules are known as *theorems*. A wff which is either an axiom or a theorem of a given system is often called a *thesis* of that system.

[5] For a proof of this *vide*, e.g., Hughes and Londey [1965], pp. 365–7.

We take care to avoid all reference to interpretation in stating formation and transformation rules, and indeed in specifying the whole system; this is often a matter of considerable importance when we come to demonstrate that an axiomatic system has certain properties. When, however, we have an interpretation in mind, we are usually interested in constructing a deductive system of *valid* wffs. We therefore select as axioms wffs which under our intended interpretation are valid ones, and we also see to it that the transformation rules are such that when they are applied to valid wffs the theorems they yield are always valid too. Such transformation rules are said to be *validity-preserving.*

The system PM

Probably the best-known axiomatization of PC derives from Whitehead and Russell's *Principia Mathematica*[6]. We call it *the System PM*.

The primitive symbols, definitions and formation rules are as given earlier in this chapter.

There are four[7] axioms:

A1 $(p \vee p) \supset p$
A2 $q \supset (p \vee q)$
A3 $(p \vee q) \supset (q \vee p)$
A4 $(q \supset r) \supset ((p \vee q) \supset (p \vee r))$

There are two primitive (i.e. initially given) transformation rules:

TR1 *The Rule of (Uniform) Substitution*: The result of uniformly replacing any variable in a thesis by any wff is itself a thesis[8].

[6] Whitehead and Russell [1910].

[7] The original system as given in *Principia Mathematica* contained a fifth axiom as well, viz. $(p \vee (q \vee r)) \supset (q \vee (p \vee r))$; but this was later shown to be unnecessary. For a historical sketch of the development of PC *vide* Church [1956] pp. 155–166.

[8] This rule reflects the principle we stated earlier, that every substitution-instance of a valid wff is itself valid. Thus if $(p \vee p) \supset p$ (A1) is valid, so is $(q \vee q) \supset q$, or $((p \,.\, q) \vee (p \,.\, q)) \supset (p \,.\, q)$, or in fact any wff of the form $(\alpha \vee \alpha) \supset \alpha$, where α is any wff. By TR1, if $(p \vee p) \supset p$ is a thesis, so is any wff of the form $(\alpha \vee \alpha) \supset \alpha$; and the same will hold for any other

TR2 *The Rule of Modus Ponens* (or *Detachment*): If α and $(\alpha \supset \beta)$ are theses, so is β.

We list here some theorems of PM which will be used in the immediately following chapters. Since the equivalences numbered PC1–PC12 on pp. 14f are also theorems of PM we number our present list in sequence with them.

PC13 $p \equiv p$ [Law of Identity – Id]
PC14 $p \supset (p \vee q)$
PC15 $(p \,.\, q) \supset p$
PC16 $(p \,.\, q) \supset q$
PC17 $p \supset (q \supset p)$
PC18 $\sim p \supset (p \supset q)$
PC19 $p \supset (q \supset (p \,.\, q))$ [Law of Adjunction – Adj]
PC20 $(p \supset (q \supset r)) \equiv (q \supset (p \supset r))$ [Law of Permutation – Perm]
PC21 $(p \supset q) \supset ((q \supset r) \supset (p \supset r))$
PC22 $(q \supset r) \supset ((p \supset q) \supset (p \supset r))$ } [Laws of Syllogism – Syll]
PC23 $(p \supset (q \supset r)) \supset ((p \,.\, q) \supset r)$ [Law of Importation – Imp]
PC24 $((p \,.\, q) \supset r) \supset (p \supset (q \supset r))$ [Law of Exportation – Exp]
PC25 $(p \supset q) \supset ((p \supset r) \supset (p \supset (q \,.\, r)))$ [Law of Composition – Comp]
PC26 $(p \supset q) \supset ((r \supset s) \supset ((p \,.\, r) \supset (q \,.\, s)))$
PC27 $(p \supset r) \supset ((q \supset r) \supset ((p \vee q) \supset r))$
PC28 $(\sim p \equiv q) \equiv (p \equiv \sim q)$
PC29 $(p \equiv q) \equiv (\sim p \equiv \sim q)$
PC30 $(\sim p \supset p) \equiv p$
PC31 $((q \supset p) \,.\, (\sim q \supset p)) \equiv p$
PC32 $((p \supset q) \,.\, (p \supset \sim q)) \equiv \sim p$

axiom or theorem. In some axiomatizations of PC, A1 is replaced by a rule to the effect that any wff of the form $(\alpha \vee \alpha) \supset \alpha$ is a thesis, A2 by the rule that any wff of the form $\beta \supset (\alpha \vee \beta)$ is a thesis, and so on. Such a rule is called an *axiom-schema* (*vide* Church [1956], pp. 148–151). If all the axioms are written as schemata, TR1 becomes unnecessary and TR2 can be the only transformation rule. Each way of setting out the basis yields exactly the same formulae as theses, and as far as PC is concerned there is little to choose between them. Throughout Part I we shall state all axioms in terms of particular variables and have a rule of substitution, even when the systems we discuss were originally set out by means of axiom-schemata.

Consistency

In what, having regard to the standard interpretation of PC, is probably the most natural sense of the word, an axiomatic system is said to be *consistent* iff no thesis is the negation of any other thesis. (Stated without reference to interpretation, the condition is that for no wff, α, are both α and $\sim\alpha$ theses [9].)

PM is consistent in this sense. The simplest way of proving this is to show (a) that every axiom is valid; (b) that the transformation rules are validity-preserving[10]; and (c) that if a wff, α, is valid, then $\sim\alpha$ is not valid.

(At least two other senses have been given by logicians to the word 'consistent' as applied to axiomatic systems. In one of these a system is said to be consistent iff no wff consisting of a single propositional variable is a thesis. In the other, a system is said to be consistent iff not every wff is a thesis. PM can easily be shown to be consistent in these senses also.)

Completeness

We distinguish between the *weak completeness* and the *strong completeness* of an axiomatic system.

To say that an axiomatic system is weakly complete is to say that every valid wff of the system is derivable as a thesis.

[9] More generally, where $*$ is any operator (or complex of operators) we can say that a system is consistent with respect to $*$ iff for no wff, α, are both α and $*\alpha$ provable (cf. Church [1956], p. 108). But in PC, $\sim$ is the only operator which makes this a pointful definition.

[10] A proof of this kind is applicable because any theorem of PM will be the result of taking a number of initial objects (the axioms) and applying to them a finite number of determinate operations (the transformation rules). In essence the proof takes the form of showing (a) that a certain property (in this case validity) holds of a set of initial objects, and (b) that if the property holds up to a given stage it holds at the next. A proof of this kind is a proof by (mathematical) *induction*, or more simply an *inductive* proof. In the present case it is an *induction on the proof in PM* of formulae. Another form an inductive proof might take would be that of showing (a) that a certain property holds for all PC variables and then showing (b) (i) that if the property holds for any wff α it holds also for $\sim\alpha$, and (b) (ii) that if it holds for any wffs α and β then it holds for $(\alpha \vee \beta)$. This would prove that the property holds for any PC wff whatever. We call such a proof an *induction on the construction of PC wffs*. We shall from time to time have occasion to use inductive proofs in these forms and in others.

PM is weakly complete. It would take too long to prove this here, but the most usual method of doing so is by showing (a) that every valid wff in PC–CNF can be derived as a theorem, and (b) that if a wff in PC–CNF is derivable as a theorem, so is any wff which can be reduced to it by certain equivalence transformations which are sufficient to reduce any wff to PC–CNF. The proof of (b) involves (i) deriving as theorems most of the equivalences listed earlier in this chapter and also (ii) showing that PM possesses one important feature which is worth while calling attention to separately. This is that the following rule of Substitution of (Proved) Equivalents holds in it:

If α is a thesis and β differs from α only in having some wff, δ, at one or more places where α has a wff, γ, then if $(\gamma \equiv \delta)$ is a thesis, β is a thesis.

To say that two wffs, γ and δ, are proved equivalents is simply to say that $(\gamma \equiv \delta)$ is a thesis. What the rule states is that if we replace any well-formed part of a thesis by some proved equivalent of that part, the result is also a thesis. In contrast to TR1, this rule does not require that the permitted replacements be made uniformly.

This is not one of the primitive transformation rules of PM, but it can be established as a *derived transformation rule*; i.e. it can be shown in a quite general way that any theorem which can be obtained by using it could also have been obtained, though more circuitously, by using only the rules of Substitution and Modus Ponens.

To say that an axiomatic system is strongly complete is to say that it cannot have any more theses than it has without falling into inconsistency. More exactly, it is to say that if any wff whatever which is not derivable from the basis were added to the axioms, the system would no longer be consistent. Since 'strong completeness' is defined in terms of consistency, there will be three senses of 'strongly complete' corresponding to the three senses of 'consistent' which we mentioned earlier.

PM is strongly complete in all three senses[11]. Most systems discussed in this book, however, are not strongly complete, and

[11] And more generally any weakly complete axiomatic basis for PC which contains TR1 and TR2 will be strongly complete.

could not be made strongly complete without unfitting them for the work we wish them to do.

The systems we are about to discuss contain formulae which are not purely PC formulae. But of course the general principles of PC hold good for these formulae as well, and we want to be able to help ourselves to such principles whenever it is convenient to do so. One way of ensuring that we can do this is to incorporate the basis of some satisfactory axiomatic PC system into the axiomatic bases of the new systems. This is how we shall normally proceed, and we shall use the PM basis for PC for the purpose[12]. We shall then say that the new system is *built on*, or is *an extension of* PM, or that the new axioms, rules etc., of the system are *subjoined to* those of PM.

[12] But any other consistent and complete basis for PC would do equally well. A considerable number of these are known.

CHAPTER TWO

The System T

For the rest of Part I we shall be concerned with a number of systems of modal propositional logic. The present chapter will deal mainly with the first of these (the System T); but first of all we shall give some general account of the modal notions which the systems are intended to express.

The basic modal notions

Among true propositions we can distinguish between those which merely *happen* to be true and those which are *bound* to be true (or which could not be false). Similarly, among false propositions we can distinguish between those which merely happen to be false and those which are bound to be false (or which could not be true). A proposition which is bound to be true we call a *necessarily true* proposition, or a *necessary truth*, or simply a *necessary* proposition; one which is bound to be false we call an *impossible* proposition; and one which is neither necessary nor impossible we call a *contingent* proposition. Some contingent propositions will of course be true, and others false. If a proposition is not impossible, we say it is a *possible* proposition. 'Possible' here does not mean 'merely possible', if by saying that p is merely possible we mean that it is in fact false though it might have been true. Possible propositions in our sense include all true propositions (and *a fortiori* include all necessary propositions). In fact they include all propositions *except* impossible ones.

By 'necessity', in the present context, we mean what is often called *logical* necessity. We shall not try here to give an adequate account of the nature of logical necessity – this is a topic which bristles with philosophical difficulties in any case. The sense in which we use the term 'necessary' can perhaps be sufficiently indicated by saying that when we say that a certain proposition is necessary, we do not mean that, things being as they are, or the world being as it is, it cannot fail to be true; but rather that it could not fail to be true *no matter how* things were, or no matter what the world turned out to be like. For example, even if the proposition that no body travels faster than light is supported by such weighty scientific evidence that we are inclined to say that in some important sense it is *impossible* for a body to travel faster than light, still this proposition will not count as one which is necessary in our sense; for the reasons which support it consist of facts about the physical universe as it is, and the physical universe might presumably have been other than in fact it is. The propositions that all bachelors are unmarried, that there are no round squares, and that either it is Thursday or it is not Thursday, would, however, count as necessary truths in our sense.

Similarly, by 'impossibility' we mean *logical* impossibility; by 'contingency', *logical* contingency; and by 'possibility', *logical* possibility. The sense of these expressions should be sufficiently clear from what we have said in the case of necessity.

These four notions, necessity, impossibility, contingency and possibility, are said to be *modal*[13] notions. They are clearly closely related to each other; in fact we could explain any three of them in terms of the fourth. Of particular importance is the following connection between necessity and possibility: to say that a proposition, p, is necessarily true is equivalent to saying that it is not possible that p is false; and to say that p is possible (or possibly true) is equivalent to saying that it is not a necessary truth that p is false.

Another important modal notion is that of *entailment*. By this we understand the converse of the relation of *following logically from*: i.e. to say that a proposition, p, entails a proposition, q, is simply an alternative way of saying that q follows logically from p, or that the inference from p to q is logically valid.

Given any proposition, p, we can of course form the proposition that p is necessary, i.e. the proposition we express as 'It is necessary that p'. This proposition will be *true* when p itself is *necessary*, and *false* when p is *not* necessary. 'It is necessary that' is thus a (monadic) proposition-forming operator on propositions.

[13] In mediaeval logic necessity etc. were thought of as the *modes* in which a proposition could be true or false.

It is not, however, a truth-functional operator; for although from the falsity of p it follows that p is not necessary (that is, that 'It is necessary that p' is false), yet given merely that p is true we cannot tell whether or not p is necessary – that is, from the truth of p we can deduce neither the truth nor the falsity of 'It is necessary that p'.

Similarly, 'It is possible that' is also a monadic proposition-forming operator on propositions, and 'It is possible that p' will be true when p is possible, and false when p is impossible. And it too is not truth-functional. 'Entails' and follows 'logically from' are dyadic proposition-forming operators which are also not truth-functional; in fact we drew attention to the non-truth-functionality of 'follows logically from' (and therefore of 'entails') on p. 6.

We call these operators modal operators and the systems of logic which we are about to consider, modal systems or modal logics. The systems will all be based on the propositional calculus as set out in Chapter 1; that is to say, they will include all the wffs of PC, with the same interpretation as before (so that valid wffs of PC remain valid in the modal systems), and the primitive transformation rules of PC, viz. uniform substitution and detachment. But we have observed that the modal operators are not truth-functional, and this means that they cannot be represented by the operators of PC ($\sim$, $\vee$ and complexes of these) since these *are* all truth-functional. So to get a modal logic (i.e., a logic in which we can express such propositions as 'It is necessary that p') we shall have to add further operators to PC and extend our class of formulae. We shall introduce the symbols L and M (as monadic operators) and $\prec$ (as a dyadic operator) with the intention of interpreting them as 'it is necessary that', 'it is possible that' and 'entails' respectively, and allow them to take any formulae as arguments. Thus we can have such wffs as $Lp \vee q$ (meaning 'either p is necessary or q (is true)'), $M(p \,.\, q) \prec Mq$ (meaning 'the proposition that q is possible follows logically from the proposition that p-and-q is possible'), and so on. In view of the intended interpretation we shall call L the *necessity operator* and M the *possibility operator*[14].

[14] Many authors use $\Box$ and $\Diamond$ as we use L and M respectively. For an account of alternative notations in modal logic, *vide* Appendix 4, pp. 347–349.

Systems of modal logic

Which modal formulae are we to count as valid? It is easy to give a general, intuitive account of validity exactly as we initially did for PC, by saying that a formula is valid iff it 'comes out true' for all values of its variables. In PC, because of the truth-functional nature of the operators, this initial account led directly to a quite simple formal definition of validity. We were then able to set up an axiomatic system and enquire whether it measured up to the criterion that the class of theses should exactly coincide with the already defined class of valid formulae. Because of the non-truth-functionality of modal operators, however, the initial account does not lead to any obvious formal definition of validity for modal formulae which will always give us unambiguous results. Nevertheless, there are certain conditions which it seems intuitively reasonable to demand that a system should fulfil if it is to be capable of interpretation as a modal system. These conditions, which we shall list in a moment, will require that certain formulae should count as valid (or as theses, if the system is set out axiomatically) and that certain others should not; but for some formulae they will leave the question of their validity or invalidity undecided. We shall then (in this chapter and the next) construct a number of axiomatic modal systems, each of which satisfies all of these requirements but which differ from one another in the presence or absence as theses of some of the less obviously valid formulae. Only after that shall we turn to the problem of defining validity in a precise way. Thus instead of measuring an axiomatic system up against a definition of validity, as we did in the case of PC, we shall instead measure various definitions of validity up against already constructed axiomatic systems.

The intuitive requirements we have referred to are as follows[15] (some of them have been foreshadowed in our earlier remarks):

1. We have already mentioned the connection between necessity and possibility. This leads us to require that if L and M

[15] What seems intuitively reasonable is notoriously apt to vary from person to person. The reader who does not find all the requirements in our list acceptable will find in Part III that there are several modal systems which do not meet them all. From a formal point of view the requirements simply mark out a class of systems with the purpose of reducing the subject to manageable proportions.

are to be interpreted as necessity and possibility operators, the following equivalences should be valid:

$$Lp \equiv {\sim}M{\sim}p$$

$$Mp \equiv {\sim}L{\sim}p$$

Systems which contain these equivalences need not have both L and M as primitive: such systems could take L as primitive and introduce M by the definition

$$M\alpha =_{\mathrm{Df}} {\sim}L{\sim}\alpha$$

or take M as primitive and define L by

$$L\alpha =_{\mathrm{Df}} {\sim}M{\sim}\alpha$$

A system which does the former we call an *L-based* system; one which does the latter, an *M-based* system. The systems we shall develop in the next few chapters will in fact be L-based, but in each case we could give an exactly parallel exposition by taking M as primitive instead of L.

2. We have also remarked that we wish to be able to interpret $\prec$ as 'entails' or 'necessarily implies'. There has been a good deal of philosophical controversy about the correct analysis of entailment[16], but one thing which is not disputed is that whenever p entails q it is impossible that p should be true without q's being true too. This will lead us to require as valid:

$$(p \prec q) \supset {\sim}M(p \,.\, {\sim}q)$$

What has been disputed is whether or not the converse also holds, i.e. whether in all cases when it is impossible for p to be true without q's being true, we should say that p entails q. Now it certainly makes for a simpler modal logic if we assume that this does hold – or (less controversially) if we interpret $\prec$ as representing that relation which holds between p and q *when and only when* it is impossible for p to be true without q's being true; and this is what we shall do. We shall then have as valid:

$$(p \prec q) \equiv {\sim}M(p \,.\, {\sim}q)$$

and in view of this equivalence we shall not have to take $\prec$ as primitive but can define $(\alpha \prec \beta)$ as ${\sim}M(\alpha \,.\, {\sim}\beta)$. (An alternative,

[16] For a brief account of this, *vide infra*, pp. 335–339.

equivalent, definition of $(\alpha \prec \beta)$ would be $L(\alpha \supset \beta)$, since $\sim M(p \,.\, \sim q)$ easily transforms to $L(p \supset q)$ by $Mp \equiv \sim L \sim p$ and standard PC equivalences[17].) When $\prec$ is interpreted in this way it is usually called the *strict implication* sign and read as 'strictly implies'. To avoid the appearance of begging any philosophical questions, we shall refer to the relation expressed by $\prec$ as 'strict implication' and drop the word 'entailment' from now on. At the very least, strict implication is a concept closely related to that of entailment; and if it is not identical with it (though it may well be), it is both the clearer of the two and the one which is easier to express in a logical system.

When two propositions strictly imply each other we say that

[17] It is important not to confuse $L(p \supset q)$, which means that the whole hypothetical 'if p then q' is a necessary truth, or that q follows logically from p, with $p \supset Lq$, which means that if p is true then q is a necessary truth. Unhappily, these are often confused in ordinary discourse, sometimes with disastrous results; and neglect of the distinction is made all the easier by the ambiguity of such common idioms as 'If . . . then it must be (*or* is bound to be) the case that—'. To make things worse, the structure of such sentences is more closely analogous to that of $p \supset Lq$, but one suspects that most frequently what the speaker intends to assert (or at least all he is entitled to assert) is something of the form $L(p \supset q)$. Thus someone who says, 'If it rains throughout December it is bound to rain on Christmas Day' probably means to assert that 'it will rain on Christmas Day' follows from 'it will rain throughout December' (which is true, since Christmas Day is in December); but he could be taken to be asserting that if it rains throughout December then it is a necessary truth that it will rain on Christmas Day (which, at least if it does rain throughout December, is false because, come what may about the weather, 'it will rain on Christmas Day' is still a contingent proposition, not a necessary one).

Perhaps no one, except in his dullest moments, would be taken in by this example. But people have, it appears, confused the necessary truth of 'If a thing is going to happen it is going to happen' with the view that whatever happens happens by logical necessity, or even argued for Fatalism by inferring illicitly from the former to the latter. And in epistemological discussions the fact (if it is a fact) that, of necessity, if someone knows that p then p is true has sometimes been held to show something which does not follow from it at all, viz. that only necessary truths can ever be known. This transition is facilitated if we express the premiss of the argument by the ambiguous but more colloquial 'If you know something, it must be true (can't be false)'. Even a little study of modal logic can protect us from pitfalls in philosophy and elsewhere.

each is *strictly equivalent* to the other. We use $=$ as the strict equivalence sign and introduce it by the definition:

$$(\alpha = \beta) =_{\mathrm{Df}} ((\alpha \prec \beta) \,.\, (\beta \prec \alpha))^{18}$$

An alternative definition, which in the systems we are about to consider would yield exactly the same results, is:

$$(\alpha = \beta) =_{\mathrm{Df}} L(\alpha \equiv \beta)$$

3. We have also pointed out that modal operators are not truth-functional. This means that in any intuitively plausible modal system Lp must not be equivalent to any truth-function of p. (Since all the other modal operators are definable in terms of L, it is sufficient to state the requirement as it applies to L.) Now there are only four distinct truth-functions of p: one is the familiar negation of p; a second is p itself (true when and only when p is true); a third is the truth-function which is true both when p is true and when p is false; the fourth is the truth-function which is false both when p is true and when p is false. Hence we require that none of the following should count as valid (or be theses):

$$Lp \equiv {\sim}p$$
$$Lp \equiv p$$
$$Lp \equiv (p \vee {\sim}p)$$
$$Lp \equiv (p \,.\, {\sim}p)$$

4. Although $Lp \equiv p$ is not valid, it is clear that one of its 'implicational halves', viz. $Lp \supset p$, is; for this simply expresses the principle that whatever is necessarily true is true. This formula is often called the *axiom of necessity*. An analogous principle is that whatever is true is possible; this is expressed by the formula $p \supset Mp$ (the *axiom of possibility*), which we shall likewise regard as valid. $Lp \supset p$ and $p \supset Mp$ can easily be derived from each other by PC equivalences and the equivalences connecting L and M.

5. Another principle which seems to be intuitively acceptable is that any proposition which has the form of a valid formula is

[18] $=$ and $=_{\mathrm{Df}}$ should not be confused. The former is an operator which occurs in wffs of a modal system; the latter is a meta-logical symbol which never occurs in wffs but is used only in discoursing *about* a system.

not merely true but *necessarily* true. This means that if α is a valid formula, then not merely is every proposition which has the form α true, but so is every proposition which has the form $L\alpha$; and in that case $L\alpha$ will itself be valid. So we shall expect to find in a modal logic the principle that if α is valid, $L\alpha$ is also valid; and in an axiomatic modal system we shall expect to have the (primitive or derived) transformation rule that if α is a thesis, so is $L\alpha$.

6. A final intuitively sound principle is that whatever follows logically from a necessary truth is itself necessarily true. If we were to deny this – if, that is, we were to admit that a contingent proposition (let alone an impossible one) might follow from a necessary proposition – we should be violating a principle which has sometimes been expressed by saying that in a valid inference the conclusion runs no greater risk of falsification than the premisses do. We shall therefore require that whenever p is necessary and p strictly implies q, q shall also be necessary, i.e. that

$$(Lp \,.\, (p \prec q)) \supset Lq$$

shall be valid. An easily obtained variant of this formula, which is often more convenient to use, is:

$$L(p \supset q) \supset (Lp \supset Lq)$$

An example of a formula whose validity is left quite undetermined by the conditions we have mentioned is $Lp \supset LLp$. This formula means that if any given proposition is a necessary truth then the proposition that it is a necessary truth is itself a necessary truth, or more simply, that whatever is necessary is necessarily necessary; and indeed it is difficult to be sure on purely intuitive grounds whether this is valid or not. The first system we shall construct will not have this formula as a thesis, but others will.

It is convenient at this point to explain some of the terminology we shall use in discussing logical systems. When a formula is a thesis of a given system we shall say that it *belongs to*, or is *contained in*, or simply is *in*, that system. If two systems, A and B, have different bases but contain exactly the same theses, we shall say that A and B are *deductively equivalent* or sometimes simply that they are *equivalent*[19]. If every thesis of system A is a thesis of system B but some theses of B are not theses of A, we shall say that A is the *weaker* and B the *stronger* of the two systems. If every thesis of system A is a thesis of system B (whether or not B contains other theses as well), we shall say that B *contains* A. Thus two systems with distinct bases are deductively equivalent iff each contains the other.

The System T

The weakest system complying with all the conditions we have laid down, and the first we shall discuss, is the *System T*, which was first propounded by Robert Feys in 1937[20].

The basis of T is as follows:

Primitive symbols

$p, q, r, \ldots$	[propositional variables]
$\sim, L$	[monadic operators]
$\vee$	[dyadic operator]
(,)	[brackets]

Formation rules

FR1 A variable standing alone is a wff.

FR2 If α is a wff, so are $\sim\alpha$ and $L\alpha$.

FR3 If α and β are wffs, so is $(\alpha \vee \beta)$.

Definitions

Def ., Def $\supset$, Def $\equiv$, as in PC (p. 8), plus:

$$[\text{Def } M] \quad M\alpha =_{\text{Df}} \sim L\sim\alpha$$

$$[\text{Def } \prec] \quad (\alpha \prec \beta) =_{\text{Df}} L(\alpha \supset \beta)$$

$$[\text{Def} =] \quad (\alpha = \beta) =_{\text{Df}} ((\alpha \prec \beta) \cdot (\beta \prec \alpha))$$

Clearly every wff of PC is a wff of T.

[19] For some purposes it is convenient to regard deductively equivalent systems as the same system, and in labelling modal systems this is usually done. For an amplification of this point, *vide infra*, p. 123.

[20] Feys [1937] (*vide* esp. pp. 533–535). Feys' own name for the system is 't' (it was first called 'T' by Sobociński [1953]). Feys derived the system by dropping one of the axioms in a system devised by Gödel [1933] (p. 39), with whom the idea of axiomatizing modal logic by adding to PC originates. Sobociński (op. cit.) showed that T is equivalent to the system M of Von Wright [1951]; for this reason 'M' is often used as an alternative name for T. Cf. *infra*, p. 125.

Axioms
A1–A4 for PM (p. 17), plus:

A5 $Lp \supset p$ [The axiom of Necessity]
A6 $L(p \supset q) \supset (Lp \supset Lq)$

Transformation rules
As for PM – i.e. Uniform Substitution and Modus Ponens (pp. 17f) – plus:

TR3 *The Rule of Necessitation* (N): If α is a thesis, $L\alpha$ is a thesis.

[In future we shall abbreviate 'α is a thesis' to $\vdash\alpha$, and express the derivability of one thesis from another (or from others) by the symbol $\rightarrow$. Hence we could write TR3 succinctly as:

$$\vdash\alpha \rightarrow \vdash L\alpha$$

Similarly we could write the Modus Ponens rule as:

$$\vdash\alpha, \vdash(\alpha \supset \beta) \rightarrow \vdash\beta$$]

The rule of Necessitation must not be confused with the invalid wff

$$(1)\ p \supset Lp$$

Certainly, if we had (1) as a thesis, we could easily derive the rule of Necessitation from it; for by substituting any thesis, α, for p in (1) and applying Modus Ponens we should derive $\vdash L\alpha$. But of course (1) is not a thesis, and we do not want it to be.

It should be noted that whenever we have a thesis of the form $(\alpha \supset \beta)$ we can always use TR3 to obtain $\vdash L(\alpha \supset \beta)$, and hence, by Def $\prec$, $\vdash(\alpha \prec \beta)$. Moreover, whenever we have $\vdash(\alpha \prec \beta)$ we can, by Def $\prec$, substitution of $(\alpha \supset \beta)$ for p in A5, and Modus Ponens, obtain $\vdash(\alpha \supset \beta)$. I.e. whenever $(\alpha \supset \beta)$ is a thesis, so is $(\alpha \prec \beta)$, and vice versa. It is therefore immaterial whether we derive implicative theses in the form $(\alpha \supset \beta)$ or in the form $(\alpha \prec \beta)$. We shall usually derive them in the form $(\alpha \supset \beta)$, since PC theses will then be more easily applicable to them.

Method of setting out proofs
We shall set out proofs of theorems in the following way. At the outset we state the theorem to be proved and give it a reference

number. Each line of the proof itself contains three items: (a) a wff; (b) a justification for writing that wff, written on the left; (c) a reference number for that wff, written immediately before it. Every wff we write down in a proof must be either (i) an axiom or (ii) a previously proved theorem or (iii) a wff derived from one or more axioms, previously proved theorems, or wffs occurring earlier in the proof, by one or more of the transformation rules or by applying a definition. In cases (i) and (ii) the justification entry consists simply of the reference number or name of the axiom or theorem in question. In case (iii) the justification entry refers to the axioms etc. which are being used, records any substitutions being made therein, and indicates which transformation rules or definitions are being applied.

In justification entries we shall use the following abbreviations. The application of the rule of substitution will be indicated by a stroke, to the right of which is written the variable for which substitution is to be made, and to the left the wff to be substituted for it, the whole being enclosed in square brackets. The employment of TR2 will be indicated by '× MP', that of TR3 by '× N'.

Since the basis of PM is contained in the basis of T and PM is a complete basis for PC, every valid wff of PC is a thesis of T. We shall therefore regard any valid wff of PC we need as already proved, and simply enter 'PC' with a number or name referring to the lists in Chapter 1 (pp.14f, 18) as a justification when we want to use one of them [21]. It is worth noting that any implicative thesis of PC can be used to generate a new thesis from one or more old ones. Thus, consider PC21 (Syll), viz.

$$(p \supset q) \supset ((q \supset r) \supset (p \supset r))$$

Suppose we have already derived two theses of T: (1) $(\alpha \supset \beta)$ and (2) $(\beta \supset \gamma)$. By substituting $(\alpha/p,\ \beta/q,\ \gamma/r)$ in Syll we obtain

$$(\alpha \supset \beta) \supset ((\beta \supset \gamma) \supset (\alpha \supset \gamma))$$

Since we have $\vdash(\alpha \supset \beta)$ and $\vdash(\beta \supset \gamma)$, two applications of MP will give us $\vdash(\alpha \supset \gamma)$. To save space we shall indicate this whole

[21] When an equivalence occurs in the lists, we use the same number or name to refer to the corresponding implication. Clearly whenever we have $\vdash(\alpha \equiv \beta)$ we also have $\vdash(\alpha \supset \beta)$.

sequence of derivations simply by '(1), (2) × Syll'. Using Syll in this way can be looked at as using a derived transformation rule which could be formulated as

$$\vdash(\alpha \supset \beta), \vdash(\beta \supset \gamma) \rightarrow \vdash(\alpha \supset \gamma)$$

In a similar way, PC23 (Imp) can be used to transform any thesis of the form $(\alpha \supset (\beta \supset \gamma))$ into the corresponding thesis of the form $((\alpha \,.\, \beta) \supset \gamma)$, and such a transformation will be signalled by '× Imp'.

The reader should now be able to follow the proofs without difficulty.

Proofs of theorems

As a first step we shall establish an extremely useful derived transformation rule. Suppose that $(\alpha \supset \beta)$ is a thesis. Then TR3 will give us $\vdash L(\alpha \supset \beta)$. Substitution in A6 gives $\vdash(L(\alpha \supset \beta) \supset (L\alpha \supset L\beta))$. Hence by Modus Ponens we obtain $\vdash(L\alpha \supset L\beta)$. We therefore have the following rule:

DR1 $\vdash(\alpha \supset \beta) \rightarrow \vdash(L\alpha \supset L\beta)$

We shall establish several other derived rules as we proceed.

T1 $p \supset Mp$

PROOF

A5 $[\sim p/p]$: (1) $L{\sim}p \supset {\sim}p$

(1) × Transp: (2) ${\sim\sim}p \supset {\sim}L{\sim}p$

PC3 (DN): (3) $p \supset {\sim\sim}p$

(3), (2) × Syll, Def M: (4) $p \supset Mp$ **Q.E.D.**

T2 $(p = q) \supset (Lp \equiv Lq)$

PROOF

A6, Def $\prec$: (1) $(p \prec q) \supset (Lp \supset Lq)$

(1) $[q/p, p/q]$: (2) $(q \prec p) \supset (Lq \supset Lp)$

PC26: (3) $(p \supset q) \supset ((r \supset s) \supset ((p \,.\, r) \supset (q \,.\, s)))$

(1), (2) × (3): (4) $((p \prec q) \,.\, (q \prec p)) \supset ((Lp \supset Lq) \,.\, (Lq \supset Lp))$

(4), Def =, Def ≡: (5) $(p = q) \supset (Lp \equiv Lq)$ **Q.E.D.**

T3 $L(p \,.\, q) \equiv (Lp \,.\, Lq)$

(In this proof we first derive $L(p \,.\, q) \supset (Lp \,.\, Lq)$ at line (3), then the converse at line (7), and finally assemble these two results into an equivalence.)

PROOF

PC15 × DR1: (1) $L(p \,.\, q) \supset Lp$

PC16 × DR1: (2) $L(p \,.\, q) \supset Lq$

(1), (2) × Comp: (3) $L(p \,.\, q) \supset (Lp \,.\, Lq)$

PC19 (Adj) × DR1: (4) $Lp \supset L(q \supset (p \,.\, q))$

A6 $[q/p, p \,.\, q/q]$: (5) $L(q \supset (p \,.\, q)) \supset (Lq \supset L(p \,.\, q))$

(4), (5) × Syll: (6) $Lp \supset (Lq \supset L(p \,.\, q))$

(6) × Imp: (7) $(Lp \,.\, Lq) \supset L(p \,.\, q)$

(3), (7) × Adj, Def $\equiv$: (8) $L(p \,.\, q) \equiv (Lp \,.\, Lq)$ **Q.E.D.**

T3 expresses an important property of logical necessity, viz. that a conjunction is necessary iff each conjunct is itself necessary. $Lp \,.\, Lq$ is said to be the *distributed* form of $L(p \,.\, q)$, and T3 may be called the *Law of L-distribution*. It should be compared with T7 below.

T4 $L(p \equiv q) \equiv (p = q)$

PROOF

T3 $[p \supset q/p, q \supset p/q]$: (1) $L((p \supset q) \,.\, (q \supset p)) \equiv (L(p \supset q) \,.\, L(q \supset p))$

(1), Def $\equiv$, Def ⥽, Def $=$: (2) $L(p \equiv q) \equiv (p = q)$ **Q.E.D.**

T4 shows that, as we remarked on p. 28, an alternative definition of $(\alpha = \beta)$ would have been $L(\alpha \equiv \beta)$.

We now establish a further derived transformation rule:

DR2 $\vdash(\alpha \equiv \beta) \rightarrow \vdash(L\alpha \equiv L\beta)$

DERIVATION

Given: (1) $\alpha \equiv \beta$

(1) × N: (2) $L(\alpha \equiv \beta)$

(2), T4$(\alpha/p, \beta/q)$ × MP: (3) $\alpha = \beta$

(3), T2$(\alpha/p, \beta/q)$ × MP: (4) $L\alpha \equiv L\beta$ **Q.E.D.**

Note that at step (3) we have used T4 in its implicational form, viz. $L(p \equiv q) \supset (p = q)$. Clearly whenever we can prove $(\alpha \equiv \beta)$ we can easily prove $(\alpha \supset \beta)$.

We can now show that the rule of Substitution of Equivalents (Eq) holds in T. This rule states (cf. p. 20) that if α is a thesis and β differs from α only in having some wff, δ, at one or more places where α has a wff, γ, then if $(\gamma \equiv \delta)$ is a thesis, β is a thesis. The standard way of establishing Eq in PM proceeds by showing that if $(\gamma \equiv \delta)$ is a thesis, the following are also theses:

$$\sim\gamma \equiv \sim\delta$$
$$(\gamma \vee \zeta) \equiv (\delta \vee \zeta)$$
$$(\zeta \vee \gamma) \equiv (\zeta \vee \delta)$$

DR2 enables us to add to this list:

$$L\gamma \equiv L\delta$$

From this it follows that if α is any wff built up from γ using $\sim$ and L as the only monadic operators and $\vee$ as the only dyadic one, and β is built up from δ in exactly the same way as α is from γ, then if $\vdash(\gamma \equiv \delta)$, $\vdash(\alpha \equiv \beta)$[22]; whence it is easy to prove that if $\vdash\alpha$, then $\vdash\beta$.

Since every wff of T can be written with $\sim$, L and $\vee$ as the only operators, we can apply Eq unrestrictedly in T; i.e., whenever we have proved a theorem of the form $(\gamma \equiv \delta)$, we can replace γ by δ in any already proved thesis, α, no matter where γ occurs in α.

T5 $Lp \equiv \sim M\sim p$

PROOF

PC3 (DN):	(1) $p \equiv \sim\sim p$	
(1) $[Lp/p]$:	(2) $Lp \equiv \sim\sim Lp$	
(2), (1) × Eq:	(3) $Lp \equiv \sim\sim L\sim\sim p$	
(3), Def M:	(4) $Lp \equiv \sim M\sim p$	**Q.E.D.**

Easily proved corollaries of T5 are:

T5a $L\sim p \equiv \sim Mp$

T5b $\sim Lp \equiv M\sim p$

[22] This is of course a proof by induction on the construction of wffs of T (*vide* footnote 10).

T5 and Eq entitle us to replace L by $\sim M \sim$ (or vice versa) in any thesis. They also enable us to form new equivalential theses by taking any wff containing L as one argument of $\equiv$, and the same wff but with $\sim M \sim$ replacing one or more occurrences of L as the other. To obtain such theses we simply substitute the wff in question for p in $p \equiv p$ and apply T5 to one side of the equivalence. T5a and T5b can be used in analogous ways.

Further corollaries of T5 are:

T5c $LLp \equiv \sim MM \sim p$

PROOF

T5 $[Lp/p]$:	(1) $LLp \equiv \sim M \sim Lp$	
(1), T5b $\times$ Eq:	(2) $LLp \equiv \sim MM \sim p$	**Q.E.D.**

and, with similar proofs:

T5d $LL \sim p \equiv \sim MMp$
T5e $MM \sim p \equiv \sim LLp$
T5f $LM \sim p \equiv \sim MLp$
T5g $ML \sim p \equiv \sim LMp$ etc.

From these and their obvious extensions we have the rule: In any sequence of adjacent L's and M's, L may be replaced by M and M by L throughout, provided that a $\sim$ is either inserted or deleted both immediately before and immediately after the sequence. We call this the *Rule of L–M Interchange* ('LMI' for short). When convenient we give the same name to the application of T5, T5a, T5b or Def M, since these are simply the special cases where the sequence contains only a single L or M.

T6 $\sim M(p \vee q) \equiv (\sim Mp \,.\, \sim Mq)$

PROOF

T3 $[\sim p/p, \sim q/q]$:	(1) $L(\sim p \,.\, \sim q) \equiv (L \sim p \,.\, L \sim q)$	
(1) $\times$ LMI:	(2) $\sim M \sim (\sim p \,.\, \sim q) \equiv (\sim Mp \,.\, \sim Mq)$	
(2), de M $\times$ Eq:	(3) $\sim M(p \vee q) \equiv (\sim Mp \,.\, \sim Mq)$	**Q.E.D.**

T6 means that if it is impossible that either-p-or-q, then p and q are both impossible, and conversely.

T7 $M(p \vee q) \equiv (Mp \vee Mq)$

PROOF

PC28: (1) $(\sim p \equiv q) \supset (p \equiv \sim q)$

T6 × (1): (2) $M(p \vee q) \equiv \sim(\sim Mp \,.\, \sim Mq)$

(2), de M × Eq: (3) $M(p \vee q) \equiv (Mp \vee Mq)$ **Q.E.D.**

T7 may be called the *Law of M-distribution.*

T8 $(p \prec q) \supset (Mp \supset Mq)$

PROOF

A6 $[\sim q/p, \sim p/q]$: (1) $L(\sim q \supset \sim p) \supset (L\sim q \supset L\sim p)$

(1), Transp × Eq: (2) $L(p \supset q) \supset (\sim L\sim p \supset \sim L\sim q)$

(2), Def $\prec$, Def M: (3) $(p \prec q) \supset (Mp \supset Mq)$ **Q.E.D.**

By TR3, if $\vdash(\alpha \supset \beta)$, then $\vdash L(\alpha \supset \beta)$, i.e. $\vdash(\alpha \prec \beta)$. Whence by substitution in T8 and MP, $\vdash(M\alpha \supset M\beta)$. So we have the derived rule:

DR3 $\vdash(\alpha \supset \beta) \rightarrow \vdash(M\alpha \supset M\beta)$

T9 $(Lp \vee Lq) \supset L(p \vee q)$

PROOF

PC14: (1) $p \supset (p \vee q)$

(1) × DR1: (2) $Lp \supset L(p \vee q)$

A2: (3) $q \supset (p \vee q)$

(3) × DR1: (4) $Lq \supset L(p \vee q)$

PC27: (5) $(p \supset r) \supset ((q \supset r) \supset ((p \vee q) \supset r))$

(2), (4) × (5): (6) $(Lp \vee Lq) \supset L(p \vee q)$ **Q.E.D.**

T10 $M(p \,.\, q) \supset (Mp \,.\, Mq)$

PROOF

T9 $[\sim p/p, \sim q/q]$ (1) $(L\sim p \vee L\sim q) \supset L(\sim p \vee \sim q)$

(1) × Transp: (2) $\sim L(\sim p \vee \sim q) \supset \sim(L\sim p \vee L\sim q)$

(2) × LMI: (3) $M\sim(\sim p \vee \sim q) \supset \sim(\sim Mp \vee \sim Mq)$

(3), de M × Eq: (4) $M(p \,.\, q) \supset (Mp \,.\, Mq)$ **Q.E.D.**

Note that whereas T3 and T7 are equivalences, T9 and T10 are only implications. The converses of T9 and T10 are not theorems, and in fact can easily be seen to be invalid. For if we replace p by some contingent proposition and q by the negation of that proposition, the consequent of T9 will be true, but its antecedent will be false since neither Lp nor Lq will be true. The same replacements will also make the consequent of T10 true but its antecedent false.

Some important differences between strict and material implication can be brought out by comparing certain pairs of formulae. Sometimes a formula containing occurrences of $\supset$ is a thesis, but when $\supset$ is replaced by $\prec$ the formula ceases to be a thesis. (Of course this will never be the case when the only occurrence of $\supset$ so replaced is the main operator, for then either both formulae are theses or neither is.) For example, in each of the following pairs the first formula is a thesis but the second is not:

$$\text{(1a)}\quad (p \supset q) \vee (q \supset p)$$

$$\text{(1b)}\quad (p \prec q) \vee (q \prec p)$$

$$\text{(2a)}\quad (p \,.\, q) \supset (p \supset q)$$

$$\text{(2b)}\quad (p \,.\, q) \supset (p \prec q)$$

Moreover, sometimes we have an equivalence which is a thesis, but when $\supset$ is replaced by $\prec$ the resulting formula is provable as an implication only. For example,

$$\text{(3a)}\quad ((p \supset r) \vee (q \supset r)) \equiv ((p \,.\, q) \supset r)$$

is a thesis, but while

$$\text{(3b)}\quad ((p \prec r) \vee (q \prec r)) \supset ((p \,.\, q) \prec r)$$

is also a thesis, its converse is not[23].

T11 $(\sim p \prec p) \equiv Lp$

PROOF

PC30:	(1) $(\sim p \supset p) \equiv p$	
(1) × DR2:	(2) $L(\sim p \supset p) \equiv Lp$	
(2), Def $\prec$:	(3) $(\sim p \prec p) \equiv Lp$	**Q.E.D.**

[23] For a more detailed discussion of this theme *vide* Lewis and Langford [1932], Chapter VI.

PC30 (or its implicational form, $(\sim p \supset p) \supset p$) is often called the *Consequentia Mirabilis*. We can give the same name to T11.

Just as whenever we have $\vdash(\alpha \supset \beta)$ we also have $\vdash(\alpha \prec \beta)$, so whenever we have $\vdash(\alpha \equiv \beta)$ we also have $\vdash(\alpha = \beta)$. (This is easily proved by TR3 and T4.) I.e., T11 and all other equivalential theorems are also provable as strict equivalences.

T12 $(p \prec \sim p) \equiv L\sim p$

The proof is similar to that for T11.

T13 $((q \prec p) \cdot (\sim q \prec p)) \equiv Lp$

PROOF

PC31: (1) $((q \supset p) \cdot (\sim q \supset p)) \equiv p$

(1) × DR2: (2) $L((q \supset p) \cdot (\sim q \supset p)) \equiv Lp$

(2), T3 $[q \supset p/p, \sim q \supset p/q]$ × Eq: (3) $(L(q \supset p) \cdot L(\sim q \supset p)) \equiv Lp$

(3), Def $\prec$: (4) $((q \prec p) \cdot (\sim q \prec p)) \equiv Lp$ **Q.E.D.**

T14 $((p \prec q) \cdot (p \prec \sim q)) \equiv L\sim p$

Proof as for T13, but using PC32 instead of PC31.

T11–T14 express important facts about non-contingent propositions (i.e. propositions which are either necessary or impossible). T11 says that a necessary proposition is one which is strictly implied by its own negation. T12 says that an impossible proposition is one which strictly implies its own negation. T13 says that a necessary proposition is one which is strictly implied both by another proposition and by the negation of that other proposition. T14 says that an impossible proposition is one which strictly implies both another proposition and the negation of that other proposition.

T15 $Lp \supset (q \prec p)$

PROOF

PC17: (1) $p \supset (q \supset p)$

(1) × DR1: (2) $Lp \supset L(q \supset p)$

(2), Def $\prec$: (3) $Lp \supset (q \prec p)$ **Q.E.D.**

T16 $L\sim p \supset (p \prec q)$

Proof as for T15, but using PC18 $(\sim p \supset (p \supset q))$ instead of PC17.

T15 and T16 are often called the *paradoxes of strict implication*, just as PC17 ($p \supset (q \supset p)$) and PC18 ($\sim p \supset (p \supset q)$) are known as the *paradoxes of material implication*. The names are unfortunate ones in each case, since the word 'paradox' suggests that the principles in question are false, or at least difficult to believe; and this could only be supposed by someone who was confusing strict or material implication with some other relation. The meaning of PC17 can be expressed by saying that if a proposition is true it is materially implied by any proposition, while PC18 says that if a proposition is false it materially implies any proposition; and when $\supset$ is given its standard truth-functional interpretation it is clear that PC17 and PC18 are simple and uncontroversial truths. Now the analogues of these theses for strict implication, viz. $p \supset (q \prec p)$ and $\sim p \supset (p \prec q)$, are not valid. But T15 says, not that if a proposition is *true* it is strictly implied by any proposition, but that if a proposition is *necessarily true* it is strictly implied by any proposition; and T16 says, not that if a proposition is *false*, but that if it is *impossible*, it strictly implies any proposition. That these are equally uncontroversial truths can be seen when we reflect that, as was pointed out on p. 26, $q \prec p$ is interpreted as meaning no more and no less than that q-and-not-p is impossible. For if p is necessary, not-p is impossible, and therefore so is q-and-not-p; and this is just what T15 says. Similarly, if p is impossible, so is p-and-not-q; and this is what T16 says[24].

T17 $Lp \supset (Mq \supset M(p \, . \, q))$

PROOF

PC19 (Adj):	(1) $p \supset (q \supset (p \, . \, q))$
(1) × DR1:	(2) $Lp \supset L(q \supset (p \, . \, q))$
T8 [q/p, $p \, . \, q/q$], Def $\prec$:	(3) $L(q \supset (p \, . \, q)) \supset (Mq \supset M(p \, . \, q))$
(2), (3) × Syll:	(4) $Lp \supset (Mq \supset M(p \, . \, q))$ **Q.E.D.**

DR4 $\vdash \alpha \rightarrow \vdash (M\beta \supset M(\alpha \, . \, \beta))$

DERIVATION

Given:	(1) α
(1) × N:	(2) $L\alpha$
T17 [α/p, β/q]:	(3) $(L\alpha \supset (M\beta \supset M(\alpha \, . \, \beta)))$
(2), (3) × MP:	(4) $(M\beta \supset M(\alpha \, . \, \beta))$ **Q.E.D.**

[24] The paradoxes of strict implication are discussed further on pp. 335–339.

We shall not go on deriving more theorems or transformation rules in T. Those we have given will serve as an indication of what the system contains.

Consistency of T

We shall prove that T is consistent with respect to $\sim$; i.e. that if α is any thesis of T, $\sim\alpha$ is *not* a thesis of T.

For every wff of T we can construct what we shall call its *PC-transform*. The PC-transform of a wff, α, is formed by re-writing α (if necessary) in primitive notation, and then deleting every occurrence of L. Clearly (i) the PC-transform of any wff of T will be a wff of PC; (ii) every wff of T will have one and only one PC-transform (though two or more wffs of T may have the same PC-transform); and (iii) if the PC-transform of α is α', then the PC-transform of $\sim\alpha$ will be $\sim\alpha'$.

We now show that the PC-transform of every *thesis* of T is a *valid* wff of PC.

This obviously holds for the PM axioms, since they are themselves valid wffs of PC.

It holds for A5 and A6, since their PC-transforms are, respectively, $\sim p \vee p$ and $\sim(\sim p \vee q) \vee (\sim p \vee q)$, which are both valid.

Every thesis of T is either an axiom or a wff obtained from one or more axioms by the rules of Substitution, Modus Ponens and Necessitation.

Let $\alpha', \beta', \ldots$ be respectively the PC-transforms of $\alpha, \beta, \ldots$

If β is obtained from α by uniform substitution of γ for some variable in α, then β' may be obtained from α' by substituting γ' for that same variable in α'. But uniform substitution preserves validity in PC; hence if α' is valid, so is β'.

Suppose β is obtained by Modus Ponens from α and $(\alpha \supset \beta)$. The PC-transforms of α and $(\alpha \supset \beta)$ are respectively α' and $(\alpha \supset \beta)'$. But $(\alpha \supset \beta)'$ is the same wff as $(\alpha' \supset \beta')$; hence β' may be obtained from α' and $(\alpha \supset \beta)'$ by Modus Ponens in PC. But Modus Ponens also preserves validity in PC.

Finally, the PC-transform of α is identical with that of $L\alpha$. Hence if β is obtained from α by Necessitation, and α' is valid, so is β'.

The PC-transform of every thesis of T is therefore a valid wff of PC. It follows that for every wff, α, of T, α and $\sim\alpha$ are not both theses; for if they were, α' and $\sim\alpha'$ would both be valid wffs of PC, which we already know to be impossible.

Hence T is consistent with respect to $\sim$.

We mentioned a little while back that a number of distinct wffs of T can have the same PC-transform. In fact a wff which is not a thesis of T sometimes has the same PC-transform as one which is. This happens, e.g., in the case of $p \supset Lp$ and $Lp \supset p$: the former is not a thesis (we shall show this later) while the latter is, but they both have the PC-transform $\sim p \vee p$. This does not in any way affect the soundness of our consistency proof, but it does give us a further important result, viz. that T is not strongly complete; for the fact that $p \supset Lp$ has a valid PC-transform shows that it could be added to T without the system being thereby made inconsistent.

Exercises – 2

2.1. Prove in T:

(a) $((p \prec q) . (q \prec r)) \supset (p \prec r)$

(b) $((p \prec q) . M(p . r)) \supset M(q . r)$

(c) $(Lp . Lq) \supset (p = q)$

(d) $M(p \supset q) \equiv (Lp \supset Mq)$

(e) $M{\sim}p \vee M{\sim}q \vee M(p \vee q)$

(f) $M(p \supset (q . r)) \supset ((Lp \supset Mq) . (Lp \supset Mr))$

2.2. An *affirmative modality*[25] is a sequence consisting of L's, M's and an even number of $\sim$'s. Where A is any affirmative modality, prove in T:

$$\vdash(\alpha \supset \beta) \rightarrow \vdash(A\alpha \supset A\beta)$$

2.3. Given that $p \supset Lp$ is not a theorem of T, prove that $\vdash M\alpha \rightarrow \vdash \alpha$ is not a derived rule of T (i.e. that it does not hold for every wff α).

2.4. T′ results from T by the addition, as an axiom, of the formula $Lp \vee L{\sim}p$.

(a) Show that T′ is consistent.

(b) Given that $p \supset Lp$ is not a theorem of T, show that T does not contain T′.

[25] Becker [1930].

CHAPTER THREE

The Systems S4 and S5

Introduction

The system T satisfies all the intuitive requirements for a modal logic which we listed in the previous chapter, but it is the weakest system which does. The only theses it contains are ones which someone who accepted those requirements and no others would regard as non-controversial; more perplexing formulae, such as the one we instanced, viz. $Lp \supset LLp$, are not among its theses.

One feature of $Lp \supset LLp$ and many other formulae which makes them hard to pronounce on from an intuitive point of view is that they contain sequences of modal operators one immediately after another; $Lp \supset LLp$, for example, contains the sequence LL. Such sequences are known as *iterated modalities*[26]. Now not all formulae with iterated modalities in them raise difficulties: we need have no qualms, for example, about the validity of $LLp \supset Lp$, since it is a simple substitution-instance of $Lp \supset p$ (A5). But when we ask, informally, whether $Lp \supset LLp$ is valid, the issue we are raising is this: is whatever is necessary, necessarily necessary? I.e., when p is a necessary truth, is the fact that p is a necessary truth always itself a necessary truth? Now this is both a disputed question and one of some obscurity, for it is not altogether clear under what conditions we should say that a proposition is necessarily necessary; but it is at least a reputable and plausible view that it should be answered in the affirmative: it is plausible, that is, to maintain that whenever a proposition is true by logical necessity, this is never a matter of accident but is always something which is logically bound to be the case. We do not, however, have to settle the issue definitely here: the fact that many people would contend for the validity of $Lp \supset LLp$ is

[26] A more precise explanation of 'modality' and 'iterated modality' is given on p. 47.

enough to give us a motive for constructing a system stronger than T, in which this formula would be a thesis, and for seeing what such a system would be like.

We have already noted that $LLp \supset Lp$ is a substitution-instance of A5, and therefore is a theorem of T already; so that the extended system would have as a thesis $Lp \equiv LLp$. An equivalence such as this, which enables us always to replace some sequence of modal operators by a shorter sequence, we shall call a *reduction law* of any system of which it is a thesis. Taking this particular reduction law as valid would be one way of resolving the perplexity about 'necessarily necessary'; for we should then say that p is necessarily necessary precisely when p is necessary, and not otherwise. An extension of T such as we are contemplating would reflect, among other things, the decision to say just this.

Of the various equivalences which could act as reduction laws and which have a certain plausibility under the intended interpretation of L and M, the most important are the following:

R1 $Mp \equiv LMp$

R2 $Lp \equiv MLp$

R3 $Mp \equiv MMp$

R4 $Lp \equiv LLp$

None of these is a thesis of T; in fact one important feature of T is that it contains no reduction laws whatever. If we want to have an extension of T in which R1–R4 are theses, we do not however need to go as far as adding them all as new axioms, for three reasons:

1. As we have already mentioned, $LLp \supset Lp$ is a thesis of T as it stands; and obvious substitutions in A5 or T1 will give $LMp \supset Mp$, $Lp \supset MLp$ and $Mp \supset MMp$. So one half of each equivalence is in T already, and it would therefore be sufficient to add the converses, viz.

R1a $Mp \supset LMp$

R2a $MLp \supset Lp$

R3a $MMp \supset Mp$

R4a $Lp \supset LLp$

2. Secondly, from R4a we could derive R3a and vice versa; and from R1a we could derive R2a and vice versa. (The derivations are given below on pp. 46 and 49.) So it would be sufficient to add as axioms one from each pair, say R1a and R4a.

3. Thirdly, R4a is derivable from R1a, though R1a is not derivable from R4a. (This derivation is also given below.) So we could obtain all four reduction laws by adding R1a to T; while by simply adding R4a we could obtain two of the reduction laws (R3 and R4) but not the other two.

All this suggests the construction of two axiomatic systems, each stronger than T and one of them stronger than the other. The first of these, obtained by adding $Lp \supset LLp$ (R4a) as a new axiom to T, is known as the *System S4*. The second, obtained by adding $Mp \supset LMp$ (R1a) to T, is known as the *System S5* [27].

There is a further reason for being interested in S4 and S5. Some philosophers have maintained not merely that all necessary propositions are necessarily necessary, but that if a proposition possesses *any* modal characteristic (necessity, possibility, impossibility, contingency) it possesses that characteristic by necessity; others, however, have questioned or denied this. The reduction laws R1–R4, together with some simple deductions from them, can be taken as expressing this doctrine. Merely constructing S4 and S5 will not of course enable us to settle the dispute; but these systems provide us with a method of investigating the consequences of the doctrine. Moreover, the fact that S4 and S5 are distinct systems shows something which is certainly not obvious at first sight, viz. that the doctrine could be held in either a stronger or a weaker form; for it would be possible to maintain that all necessary propositions are necessarily necessary (as in S4) without maintaining that all possible propositions are necessarily possible (as in S5).

In the theorems which follow we shall attach to the theorem number one of the symbols '(T)', '(S4)', '(S5)', which will indicate respectively that the theorem belongs to T, that it belongs to

[27] The names 'S4' and 'S5' derive from Lewis and Langford [1932] (p. 501), where systems deductively equivalent to these are the fourth and fifth in a series of modal systems. For an account of the systems as they appear in Lewis and Langford *vide infra*, Chapter 12. Sobociński [1953] has shown that S4 and S5 are respectively equivalent to the systems M′ and M″ in Von Wright [1951]. (Cf. *infra*, pp. 126–129.)

S4 but not to T, and that it belongs to S5 but not to S4. Every thesis of T is of course also a thesis of S4, and every thesis of S4 a thesis of S5.

The system S4

The basis of S4 is that of T plus:

A7 $Lp \supset LLp$

T18 (S4) $MMp \supset Mp$

PROOF

A7 [$\sim p/p$] (1) $L\sim p \supset LL\sim p$

(1) × LMI: (2) $\sim Mp \supset \sim MMp$

(2) × Transp: (3) $MMp \supset Mp$ **Q.E.D.**

T19 (S4) $Lp \equiv LLp$ [R4]

PROOF

A5 [Lp/p]: (1) $LLp \supset Lp$

A7, (1) × Adj, Def $\equiv$: (2) $Lp \equiv LLp$ **Q.E.D.**

T20 (S4) $Mp \equiv MMp$ [R3]

Proof from T1 [Mp/p] and T18.

T21 (S4) $MLMp \supset Mp$

PROOF

A5 [Mp/p]: (1) $LMp \supset Mp$

(1) × DR3: (2) $MLMp \supset MMp$

(2), T18 × Syll: (3) $MLMp \supset Mp$ **Q.E.D.**

T22 (S4) $LMp \supset LMLMp$

PROOF

T1 [LMp/p]: (1) $LMp \supset MLMp$

(1) × DR1: (2) $LLMp \supset LMLMp$

(2), T19 × Eq: (3) $LMp \supset LMLMp$ **Q.E.D.**

T23 (S4) $LMp \equiv LMLMp$

PROOF

T21 × DR1: (1) $LMLMp \supset LMp$

T22, (1) × Adj, Def ≡: (2) $LMp \equiv LMLMp$ **Q.E.D.**

T24 (S4) $MLp \equiv MLMLp$

PROOF

T23 [$\sim p/p$]: (1) $LM\sim p \equiv LMLM\sim p$

(1) × LMI: (2) $\sim MLp \equiv \sim MLMLp$

PC29: (3) $(p \equiv q) \equiv (\sim p \equiv \sim q)$

(2), (3) × Eq: (4) $MLp \equiv MLMLp$ **Q.E.D.**

Modalities in S4

We define a *modality* as any unbroken sequence of zero or more monadic operators ($\sim$, L, M)[28]. We express the zero case by writing '–'. Examples of modalities are: –; $\sim$; L; $M\sim$; LL; $\sim ML\sim M$. It is clear, however, that in any system containing LMI every modality can be expressed either without any negation signs at all or else with only one, and that at the beginning; we shall say that a modality expressed in this way is in *standard form*, and from now on we shall assume that all modalities are expressed in standard form. A modality is said to be an *iterated* modality iff it contains two or more *modal* operators; thus LL and $\sim MLM$ are iterated modalities, but $\sim$ and $\sim L$ are not.

We say that two modalities, A and B, are *equivalent* in a given system iff the result of replacing A by B (or B by A) in any formula is always equivalent in that system to the original formula (otherwise we say that they are *non-equivalent*, or *distinct*, in that system). In a system containing the rules of Uniform Substitution and Substitution of Equivalents, this is so iff $(Ap \equiv Bp)$ is a thesis of that system (where Ap and Bp are the formulae obtained by prefixing A and B respectively to p). If A and B are equivalent in a certain system, and A contains fewer modal operators than B, then B is said to be *reducible* to A in that system. Clearly the formulae we have called *reduction laws* express the reducibility of certain modalities to others in systems of which they are theses.

[28] Cf. Feys [1950].

We are now in a position to prove an important result about S4, viz. that every modality is equivalent to one or other of the following or their negations:

(i) –; (ii) L; (iii) M; (iv) LM; (v) ML; (vi) LML; (vii) MLM

The proof is straightforward. Clearly (ii) and (iii) are the only one-operator modalities (we neglect the negative cases in the meantime). Now T19 and T20 entitle us to replace LL by L and MM by M; hence if we add a modal operator to (ii) or (iii) we shall obtain either a modality equivalent to the original or else (iv) or (v), which are therefore the only irreducible two-operator modalities. In just the same way, if we add a modal operator to (iv) or (v), the only irreducible three-operator modalities we can obtain are (vi) and (vii). If, however, we add a modal operator to (vi) or (vii), the result is always equivalent either to the original as before, or else to (iv) or (v) by T23 or T24; hence there cannot be any irreducible modalities with four or more operators.

Clearly the negative cases can be dealt with in the same way; so what we have shown is that there are at most fourteen distinct modalities in S4. In fact all fourteen are distinct from one another, though we are not yet in a position to prove this.

If we prefix a modality to a wff, α, the result is of course itself a wff. The implication relations which hold (in S4) among the formulae thus obtained from (i)–(vii) are set out in the following diagram [29]. (Implication is symbolized by an arrow for typographical convenience.)

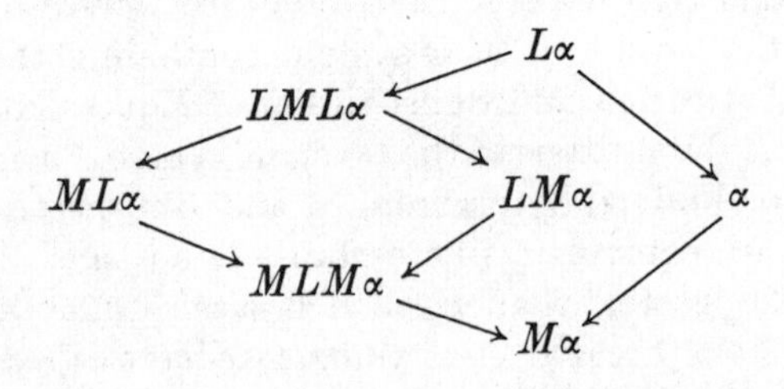

[29] For this diagram *vide* Prior [1957], p. 124. The results were originally obtained by Becker [1930] and Parry [1939].

We can obtain an analogous diagram for the negative cases by negating all the formulae and reversing the direction of all the arrows.

The situation is strikingly different in T. The absence of any reduction laws in that system means that no matter how many modal operators a given modality may contain, we can always construct a longer one which will not be equivalent to it. T therefore contains an infinite number of distinct modalities [30].

The system S5

The basis of S5 is that of T, plus:

A8 $Mp \supset LMp$

The first three theorems of S5 are proved in the same way as T18–T20, but using A8 instead of A7.

T25 (S5) $MLp \supset Lp$

T26 (S5) $Mp \equiv LMp$

T27 (S5) $Lp \equiv MLp$

$Lp \supset LLp$, the characteristic axiom of S4, is not an axiom of S5, but we now prove it as a theorem of S5. Since the two systems have the rest of their bases in common, this constitutes a proof that S5 contains S4.

A7 (S4) $Lp \supset LLp$

PROOF IN S5

T1 $[Lp/p]$:	(1) $Lp \supset MLp$	
(1), T26 $[Lp/p] \times$ Eq:	(2) $Lp \supset LMLp$	
(2), T27 $\times$ Eq:	(3) $Lp \supset LLp$	**Q.E.D.**

Modalities in S5

We have now shown that all the four reduction laws mentioned earlier are theorems of S5. We repeat them here for convenience:

R1 $Mp \equiv LMp$ [T26]

R2 $Lp \equiv MLp$ [T27]

R3 $Mp \equiv MMp$ [T20]

R4 $Lp \equiv LLp$ [T19]

[30] The simplest proof that there are infinitely many distinct modalities in T proceeds by showing that, where L_n is an unbroken sequence of n L's, $(L_n p \supset L_{n+m} p)$ is never a thesis of T, where $m \geqslant 1$. *Vide infra*, p. 65.

A simple way of summarizing these laws is this: in any pair of adjacent monadic modal operators we may delete the first. Since this process may be repeated indefinitely, we have the more comprehensive rule: *in any sequence of monadic modal operators we may delete all but the last*[31].

It is a straightforward consequence of this that S5 contains at most six distinct modalities, viz.

$$\text{(i)} -; \quad \text{(ii)}\ L; \quad \text{(iii)}\ M$$

and their negations. In fact these six modalities are all distinct from one another.

Modal functions and modal degree

Any wff which contains a modal operator is said to be a *modal function* of its variables (just as any wff of PC is a truth-function of its variables). If a wff, α, contains one or more modal operators, but none of these is within the scope of any other modal operator, then α is said to be a modal formula of *first degree* (or a first-degree formula, or a first-degree modal function of its variables). In general a formula of degree n is one in which at least one modal operator has an argument of degree $n-1$ but no modal operator has an argument of any higher degree than $n-1$[32].

The notion of a modal function of degree n is wider than that of a formula containing a modality with n modal operators, and should not be confused with it. Certainly LLp and $MLp \supset Mq$

[31] The difference between S5 and S4 in the matter of iterated modalities can be brought out by saying that in S4 all we are entitled to do is (a) count each unbroken stretch of L's or M's as a single L or M, and then (b) if the sequence is more than three operators long, delete all but the last two or three, according as their number is even or odd.

[32] It is convenient to regard wffs which do not contain any modal operators as modal formulae of degree 0 in much the same way as we have counted – and ~ as modalities. A precise definition of the *modal degree of a formula* can be given as follows (it is assumed that formulae are written in primitive notation):

1. A propositional variable is of degree 0.
2. If α is of degree n, then $\sim\alpha$ is of degree n.
3. If α is of degree n and β is of degree m, then if $n \geqslant m$, $(\alpha \vee \beta)$ is of degree n; otherwise it is of degree m.
4. If α is of degree n, then $L\alpha$ is of degree $n+1$.

This definition is given in Parry [1939], p. 144.

are second degree formulae, and are made so by the presence in them of the modalities LL and ML; but $M(p \supset Lq)$ is also a second-degree formula, though it contains no iterated modalities at all. Any formula containing a modality with n modal operators will be of at least degree n; but a formula can be of a degree n (however great n may be) without containing any modalities with n modal operators, or even any iterated modalities at all.

If a formula of degree n is provably equivalent in a given system to some formula of lower degree than n, we say that it is *reducible* (in that system) to that formula. We have already seen that in S5 any formula which is of higher than first degree *solely* because of the presence in it of iterated modalities can be reduced to a first-degree formula by the reduction laws. It is possible, however, to prove the following much stronger result:

S5 reduction theorem

Every formula of higher than first degree is reducible in S5 to a first-degree formula.

As a preliminary to proving this we derive a few more theses. (In their proofs we refer to the reduction laws as R1–R4 rather than by their theorem numbers, since they are probably most easily remembered in that way.)

T28 (T) $L(p \vee q) \supset (Lp \vee Mq)$

PROOF

A6 [$\sim q/p, p/q$]: (1) $L(\sim q \supset p) \supset (L\sim q \supset Lp)$

(1), Def $\supset$, × DN: (2) $L(q \vee p) \supset (\sim L\sim q \vee Lp)$

(3) × LMI, × Comm: (3) $L(p \vee q) \supset (Lp \vee Mq)$ **Q.E.D.**

T29 (S5) $L(p \vee Lq) \equiv (Lp \vee Lq)$

PROOF

T28 [Lq/q], R2 × Eq: (1) $L(p \vee Lq) \supset (Lp \vee Lq)$

T9 [Lq/q]: (2) $(Lp \vee LLq) \supset L(p \vee Lq)$

(2), R4 × Eq: (3) $(Lp \vee Lq) \supset L(p \vee Lq)$

(1), (3) × Adj, Def $\equiv$: (4) $L(p \vee Lq) \equiv (Lp \vee Lq)$ **Q.E.D.**

T30 (S5) $L(p \vee Mq) \equiv (Lp \vee Mq)$

PROOF

T29 [Mq/q]: (1) $L(p \vee LMq) \equiv (Lp \vee LMq)$

(1), R1 × Eq: (2) $L(p \vee Mq) \equiv (Lp \vee Mq)$ **Q.E.D.**

T31 (S5) $M(p \,.\, Mq) \equiv (Mp \,.\, Mq)$

PROOF

T29 [$\sim p/p, \sim q/q$]:	(1) $L(\sim p \vee L\sim q) \equiv (L\sim p \vee L\sim q)$	
(1) × PC29:	(2) $\sim L(\sim p \vee L\sim q) \equiv \sim(L\sim p \vee L\sim q)$	
(2) × LMI:	(3) $M\sim(\sim p \vee \sim Mq) \equiv \sim(\sim Mp \vee \sim Mq)$	
(3), Def.:	(4) $M(p \,.\, Mq) \equiv (Mp \,.\, Mq)$	**Q.E.D.**

T32 (S5) $M(p \,.\, Lq) \equiv (Mp \,.\, Lq)$

PROOF

T31 [Lq/q]:	(1) $M(p \,.\, MLq) \equiv (Mp \,.\, MLq)$	
(1), R2 × Eq:	(2) $M(p \,.\, Lq) \equiv (Mp \,.\, Lq)$	**Q.E.D.**

We can now prove the S5 reduction theorem by describing an effective procedure for reducing any formula of higher than first degree to one of first degree by equivalence transformations. The only equivalences required (apart from PC equivalences) are those given by LMI, the laws of L- and M-distribution (T3 and T7), the reduction laws R1–R4, and T29–T32. All are of course in S5.

The law of L-distribution ($L(p \,.\, q) \equiv (Lp \,.\, Lq)$) entitles us to distribute L over any conjunction whatsoever. If either conjunct already begins with a modal operator, the appropriate reduction law will enable us to delete the L when it meets this operator. Thus $L(p \,.\, Mq)$ becomes not merely $(Lp \,.\, LMq)$ by L-distribution, but $(Lp \,.\, Mq)$ by R1. In such a case we shall say that the L has been *absorbed* by the M. T29 and T30 entitle us to practise the same kind of distribution and absorption when L precedes a disjunction, *provided that* at least one of the original disjuncts begins with a modal operator[33]. The law of M-distribution ($M(p \vee q) \equiv (Mp \vee Mq)$) and T31 and T32 similarly allow us to practise distribution and absorption of M unrestrictedly over any disjunction, and, subject to the same proviso as before, over a conjunction. These manoeuvres are key steps in the process of reduction to first degree.

[33] T29 and T30 are stated for two-membered disjunctions only. If we want to practise L-distribution over an n-membered disjunction we must gather together all the *unmodalized* members of the disjunction and treat them as a single disjunct. E.g., if we have $L(p \vee Mq \vee r)$, we form $L((p \vee r) \vee Mq)$ and then distribute to get $L(p \vee r) \vee Mq$. We do *not* go to $Lp \vee Mq \vee Lr$.

It will be sufficient if we show how any second-degree formula can be reduced to first degree, since repetition of the procedure will then enable us to deal with a formula of any higher degree.

There are four steps in the procedure, though of course not all will be needed in every case. The first three are straightforward and by now familiar.

1. We first eliminate all operators except $\sim$, L, M, $\vee$ and $.$ by using the appropriate definitions.

2. We then eliminate every occurrence of $\sim$ immediately before a bracket or a modal operator by the de Morgan laws and LMI. (As a result $\sim$ will be prefixed only to variables.)

3. We next reduce all iterated modalities to single modal operators by the reduction laws.

4. If the formula we have as a result of steps 1–3 is still of second degree, this can only be because it, or some part of it, is of the form $L\alpha$ or $M\alpha$, where α is of first degree and is either a conjunction or a disjunction.

We consider the case of $L\alpha$. There are three possibilities: (a) α is a conjunction; in that case we distribute L over the conjuncts, letting it be absorbed by any modal operator it meets in the process. (b) α is a disjunction at least one of whose disjuncts begins with a modal operator; in that case we again distribute L and let it be absorbed. (c) α is a disjunction none of whose disjuncts begins with a modal operator. Since α is of first degree, this can only be because some disjunct in α is a conjunction with a modal operator inside it. To handle this case we transform α into a conjunction by the PC distributive law, $(p \vee (q \,.\, r)) \equiv ((p \vee q) \,.\, (p \vee r))$, and distribute L over the conjunction so obtained. (Remember that L distributes unrestrictedly over conjunctions.) Thus if $L\alpha$ is $L(p \vee (q \,.\, Mr))$ we transform this by Distrib to:

$$L((p \vee q) \,.\, (p \vee Mr))$$

and then by L-distribution to:

$$L(p \vee q) \,.\, L(p \vee Mr)$$

We can then either (as in this case) proceed as in (b), to obtain

$$L(p \vee q) \,.\, (Lp \vee Mr)$$

or, if this is impossible, apply Distrib and L-distribution once more. Repetition of these moves will always allow the L to meet each modal operator, no matter how deeply it is embedded in α, and be absorbed by it.

The case of $M\alpha$ can be dealt with analogously, except that this time it is when α is a conjunction none of whose conjuncts begins with a modal operator that we cannot proceed directly, and that the PC distributive law we then need is

$$(p \cdot (q \vee r)) \equiv ((p \cdot q) \vee (p \cdot r))$$

(To make all this clearer we shall give one or two examples of reduction to first degree on p. 56.)

Every wff, then, is equivalent in S5 to some first-degree modal function of its variables[34]. Now it is not difficult to see that there can be only a finite number of distinct first-degree modal functions of any finite set of variables. For every first-degree formula (written in primitive notation) is a truth-function of (i) propositional variables and (ii) wffs consisting of L followed by a truth-function of propositional variables; and there is only a finite number of non-equivalent truth-functions of any finite number of formulae. Hence the S5 reduction theorem shows that in S5 there are only a finite number of non-equivalent modal functions of any finite number of variables.

(It is worth noting that in showing that there are only a finite number of distinct *first-degree* modal functions of a finite number of variables we do not make use of any principles belonging specifically to S5; this result holds equally for T and S4. Moreover it can easily be generalized to show that there are only a finite number of distinct modal functions (of a finite number of variables) of *any* given finite degree. Hence if we had a system in which, although we could not reduce every wff (as in S5) to first degree, yet we could reduce them all to some specified finite degree (say, fourth), that would be enough to show that in that system there were only a finite number of distinct modal functions (of any finite number of variables).)

Modal conjunctive normal form

As we explained on p. 16, we say that a wff is in conjunctive normal form (CNF) iff it is a conjunction (possibly degenerate) of disjunctions (again possibly degenerate), each disjunct of

[34] This is true even of a wff containing no modal operators; for any wff, α, is equivalent to $\alpha \cdot (Lp \vee {\sim}Lp)$, where p is some variable in α.

which is of a certain specified form. We shall say that a wff is in *modal conjunctive normal form* (MCNF) iff it is in CNF and each disjunct is *either* (a) a wff of PC, *or* (b) a wff of PC preceded by a single L or M[35]. Thus the following wffs are in MCNF:

$$(p \vee Lp) \,.\, q$$
$$[M((p \vee q) \supset r) \vee Lp \vee (r \,.\, s)] \,.\, [M(p \vee q) \vee Lr]$$

but the following are not:

$$(M(p \vee q) \,.\, r) \vee s$$
$$L(M(p \vee q) \vee r) \,.\, (Lp \vee Mq)$$

MCNF theorem

Any wff can be reduced in S5 to MCNF.

I.e., there is an effective procedure whereby for any wff, α, we can find a wff, α', such that α' is in MCNF and $(\alpha \equiv \alpha')$ is a thesis of S5.

PROOF

(i) If α is a wff of PC, it is in MCNF already.

(ii) If α is a first-degree formula, we first eliminate all modal operators except L and M by the appropriate definitions. The resulting formula will be a truth-function of wffs each of which is either a wff of PC or a wff of the form $L\beta$ or $M\beta$, where β is a wff of PC. Taking each such wff as an indivisible unit, we reduce the whole formula to CNF by PC methods. Finally we replace $\sim L$ everywhere by $M\sim$ and $\sim M$ everywhere by $L\sim$. The resulting formula, α', is in MCNF.

(iii) If α is of higher than first degree, we begin by reducing it to first degree by the method explained in the previous section, and then obtain α' by proceeding as in (ii). (In fact the only further step required in this case will be the application of the PC distributive law.)

Since the only transformations involved are licensed by equivalences which are in S5, $(\alpha \equiv \alpha')$ is a thesis of S5 in every case.

[35] The name 'modal conjunctive normal form' is ours, but the idea derives from Carnap [1946]. Carnap calls the formula in MCNF to which a wff, α, can be reduced the *MP-reductum* of α. In Wajsberg [1933], p. 122, a slightly more complicated normal form is described, in which each disjunct consists of L or $\sim L$ followed by a disjunction of variables (negated or un-negated).

We give here some examples of reduction to MCNF. These will also illustrate reduction to first degree.

EXAMPLE 1

$$(MMp ⥽ p) \supset (p ⥽ Lp)$$

We first reduce to first degree as follows:

Step 1: $L(MMp \supset p) \supset L(p \supset Lp)$
$\sim L(\sim MMp \vee p) \vee L(\sim p \vee Lp)$

Step 2: $M\sim(\sim MMp \vee p) \vee L(\sim p \vee Lp)$
$M(MMp \,.\, \sim p) \vee L(\sim p \vee Lp)$

Step 3: $M(Mp \,.\, \sim p) \vee L(\sim p \vee Lp)$

Step 4: $(Mp \,.\, M\sim p) \vee (L\sim p \vee Lp)$

We now have a first degree formula. To put it into MCNF we apply Distrib and obtain

$$(Mp \vee L\sim p \vee Lp) \,.\, (M\sim p \vee L\sim p \vee Lp)$$

EXAMPLE 2

$$(p ⥽ (q \,.\, Mr)) ⥽ \sim M(p \,.\, \sim q \,.\, \sim Mr)$$

We again begin by reducing to first degree.

Step 1: $L(L(p \supset (q \,.\, Mr)) \supset \sim M(p \,.\, \sim q \,.\, \sim Mr))$
$L(\sim L(\sim p \vee (q \,.\, Mr)) \vee \sim M(p \,.\, \sim q \,.\, \sim Mr))$

Step 2: $L(M\sim(\sim p \vee (q \,.\, Mr)) \vee L\sim(p \,.\, \sim q \,.\, \sim Mr))$
$L(M(p \,.\, \sim(q \,.\, Mr)) \vee L(\sim p \vee q \vee Mr))$
$L(M(p \,.\, (\sim q \vee \sim Mr)) \vee L(\sim p \vee q \vee Mr))$
$L(M(p \,.\, (\sim q \vee L\sim r)) \vee L(\sim p \vee q \vee Mr))$

Step 4: $M(p \,.\, (\sim q \vee L\sim r)) \vee L(\sim p \vee q \vee Mr)$
$M((p \,.\, \sim q) \vee (p \,.\, L\sim r)) \vee L((\sim p \vee q) \vee Mr)$
$M(p \,.\, \sim q) \vee M(p \,.\, L\sim r) \vee L(\sim p \vee q) \vee Mr$
$M(p \,.\, \sim q) \vee (Mp \,.\, L\sim r) \vee L(\sim p \vee q) \vee Mr$

This is a first-degree formula. Distrib gives us the following formula in MCNF:

$$(Mp \vee M(p \,.\, \sim q) \vee L(\sim p \vee q) \vee Mr) \,.\, (L\sim r \vee M(p \,.\, \sim q) \vee L(\sim p \vee q) \vee Mr)$$

Modal functions in S4

Clearly any system which, like T, contains an infinite number of distinct modalities also contains an infinite number of distinct modal functions even of a single variable. In S5, as we have seen, there are only a finite number of modalities, and also only a finite number of distinct modal functions of any finite number of variables. This might lead us to expect that since S4 is like S5 in having only a finite number of modalities, it will also be like S5 in having only a finite number of modal functions. This, however, is not so: in S4, as in T, there are infinitely many distinct modal functions even of a single variable[36].

We have already remarked (p. 54) that in any of the systems we have considered there are only a finite number of distinct modal functions (of a finite number of variables) *of any given degree.* Hence the fact that in S4 there are infinitely many distinct modal functions of even a single variable shows that no universally applicable reduction of wffs to any specified degree (and *a fortiori* no reduction to a normal form of some specified degree, corresponding to reduction to MCNF in S5) is possible in S4.

The Brouwerian system

A special interest attaches to the following pair of theorems:

T33 (S5) $p \supset LMp$

PROOF

T1, A8 × Syll.

T34 (S5) $MLp \supset p$

PROOF

T33 [$\sim p/p$]: (1) $\sim p \supset LM\sim p$

(1) × LMI: (2) $\sim p \supset \sim MLp$

(2) × Transp: (3) $MLp \supset p$ **Q.E.D.**

[36] Makinson [1966a], p. 406, cites this as a 'well known' result about S4, though we have been unable to find it in the literature. It does, however, follow easily from his own proof of the same result for a system which he calls D* and which contains S4.

Neither theorem is in S4. Indeed, if we were to add either as an extra axiom to S4 we should obtain a system at least as strong as S5 (in fact we should obtain exactly S5). In the case of T33 we only need to substitute Mp for p and then apply R3 to obtain the S5 axiom A8, and the case of T34 is not much more complicated. If, however, we were to add T33 (or T34) to T instead of to S4 we should not obtain S5, but a system which is weaker than S5 and which neither contains nor is contained in S4. This system has been called the *Brouwerian system* and T33 the *Brouwerian axiom*[37]. We shall sometimes refer to each simply as 'B'.

The following is a derived transformation rule of the Brouwerian system (and also of course of S5):

DR5 $\vdash(M\alpha \supset \beta) \rightarrow \vdash(\alpha \supset L\beta)$

DERIVATION

Given:	(1) $M\alpha \supset \beta$
(1) × DR1:	(2) $LM\alpha \supset L\beta$
T33 $[\alpha/p]$:	(3) $\alpha \supset LM\alpha$
(3), (2) × Syll:	(4) $\alpha \supset L\beta$

Q.E.D.

Yet another way of obtaining S5 would be to add DR5 as a primitive transformation rule to S4, without any new axioms; for then, since $MMp \supset Mp$ (T18) is a thesis of S4, DR5 would immediately give us $Mp \supset LMp$ (i.e. A8).

[37] This formula derives from Becker [1930], p. 509. Some authors have called T33 the *Brouwersche* axiom, and the system the *Brouwersche* system, perhaps because in Lewis and Langford [1932], p. 497, Becker's phrase 'Brouwersche Axiom' is quoted untranslated. The name derives from L. E. J. Brouwer, the founder of the intuitionist school of mathematics. In the intuitionist propositional calculus (*vide* p. 305) the law of double negation is not valid as an equivalence. More precisely, $p \supset {\sim}{\sim}p$ is valid but ${\sim}{\sim}p \supset p$ is not. One way of making this sound reasonable has been to suppose that in this calculus $\sim$ means something like 'it is not possible that', i.e. that it means what we mean by $L{\sim}$ (*vide* p. 306). Now if we replace $\sim$ by $L{\sim}$ then the law of double negation does indeed seem not to be generally valid. For ${\sim}{\sim}p \supset p$ becomes $L{\sim}L{\sim}p \supset p$, i.e. $LMp \supset p$. If this is added to S5 we can easily derive (by R1) the clearly invalid $Mp \supset p$, and so any argument for S5 must count against this formula. The converse, however (viz. $p \supset {\sim}{\sim}p$, the formula which *is* valid in the intuitionist calculus) becomes $p \supset LMp$, i.e. T33, which *is* in S5. Thus although the connection with Brouwer is somewhat tenuous, historical usage has continued to associate his name with this formula.

We shall have more to say about the Brouwerian system later on (pp. 74f, 257f).

Consistency

That the systems S4, S5 and B are all consistent with respect to $\sim$ can be shown by merely adding to the consistency proof for T (pp. 41f) the observation that the PC-transforms of A7, A8 and T33 are valid. (The PC-transform in each case is of course simply $\sim p \vee p$.) The systems are not strongly complete, for the same reason as we gave in the case of T.

Collapsing into PC

It is of some interest to see what kind of system we should obtain if we were to add, even to T, and *a fortiori* to S4 or S5, the extra axiom $p \supset Lp$. This formula is of course intuitively invalid, but as we saw on p. 42, adding it would not make the system inconsistent.

In such a system the new axiom, together with A5, would immediately yield $\vdash(Lp \equiv p)$ and then by simple steps, $\vdash(Mp \equiv p)$. By the rules of Uniform Substitution and Substitution of Equivalents we should then have the result that every formula would be equivalent to its PC-transform; so in any formula we could delete or insert L's and M's to our heart's content (provided we preserved well-formedness), and the result would be equivalent to the original. In such a system, therefore, the modal operators would merely 'idle'; in interpreting the system we could draw no significant distinction between necessity, possibility and truth; and for all practical purposes the system could be regarded simply as the Propositional Calculus itself, encrusted with L's and M's as mere typographical embellishments. A system like this, in which every wff is equivalent to its PC-transform, is sometimes said to *collapse into PC*.

Note that we obtain these results from the new axiom and A5 alone (together with the PC part of the basis of T): we should not need A6 as an axiom, since its PC-transform is a thesis, nor should we need the rule of Necessitation, as we explained on p. 31. Moreover, the system would clearly contain even S5, since the PC-transform of A8 is a thesis. We shall prove later on (p. 71) that none of the systems we have considered collapses into PC.

Exercises – 3

3.1. Prove in S4:

(a) $(Lp \vee Lq) \equiv L(Lp \vee Lq)$

(b) $((p = q) \prec r) \supset ((p = q) \prec Lr)$

3.2. Where A is any affirmative modality (*vide* Exercise 2.2) prove that $(p \prec q) \supset (Ap \prec Aq)$ is a theorem of S4.

3.3. Prove that the addition to T, as an axiom, of the formula $(p = q) \supset (Lp = Lq)$ gives a system deductively equivalent to S4.

3.4. Prove in S5:

(a) $(Lp \prec Lq) \vee (Lq \prec Lp)$

(b) $(Mp \prec q) = (p \prec Lq)$

3.5. Reduce the following formulae to Modal Conjunctive Normal Form (MCNF):

(a) $L(p \vee (q \,.\, (r \vee Ls)))$

(b) $M(p \,.\, q) \supset [(Lp \prec Lq) \prec Mq]$

(c) $[p \prec (q \supset (p \prec q))] \supset [\sim(p \prec q) \supset (p \prec \sim q)]$

(d) $L(\sim p \,.\, \sim q) \supset [(L(p \vee q) \prec r) \,.\, (r \supset (p \prec p))]$

(e) $(p \prec q) \supset [M(p \,.\, \sim Lp) \prec M(q \,.\, (p \prec Lp))]$

CHAPTER FOUR

Validity in T, S4 and S5

We have already alluded to the problem of defining validity for modal formulae. Up to now we have used a few intuitive criteria which have indeed been sufficient to settle the validity or otherwise of a number of formulae; but we have lacked anything corresponding to the account we gave for PC formulae when we said (p. 11) that a wff α is PC-valid iff $V(\alpha) = 1$ for every PC-assignment, V, with respect to the variables in α. Later in this chapter we shall give a number of definitions of validity for modal formulae in the same style as this, but to make them more easily comprehensible we shall first of all describe a series of parlour games which, it will turn out, exactly reflect their structure. Indeed validity for modal formulae can be directly defined in terms of these parlour games themselves.

The PC game

As a preliminary, consider how we might devise a simple game based on the definition of PC-validity which we have just mentioned. The game could take this form. We give a player a sheet of paper on which we have previously written a number of letters of the alphabet (preferably taken from the series p, q, r . . . etc.). We shall refer to the player and his sheet as a *setting* of the PC game, or more succinctly a *PC-setting*. PC-settings will differ only in the list of letters on the sheet of paper.

We then call out to the player wffs of PC, to which he is to respond by either raising his hand or keeping it down. But each call must be appropriately prepared for, in that before a wff α is called we must have previously called all the formulae which occur as parts of α, beginning with the variables. E.g., if $(\sim p \vee p)$ is to be called we must first call p, and then $\sim p$ and only then may we call $(\sim p \vee p)$. The player's instructions are as follows (for simplicity we shall first consider only formulae in primitive notation):

1. If a single letter (variable) is called, raise your hand if that letter is on the sheet; keep it down if it is not.

2. If $\sim\alpha$ is called (where α is a wff) raise your hand if you kept it down when α was called; keep it down if you raised it when α was called. (Remember that if $\sim\alpha$ has been appropriately prepared for, α must have already been called.)

3. If $(\alpha \vee \beta)$ is called, raise your hand if you raised it for α or for β; keep it down if you kept it down for both α and β.

(Using the definitions of $\supset$, . and $\equiv$ we can easily derive rules for responding to formulae containing these operators. Alternatively we can transform all formulae into primitive notation before we begin.)

It is not difficult to see that in any PC-setting the rules enable the player to respond unambiguously to any PC formula, provided that it is appropriately prepared for. If the player in a PC-setting raises his hand when a PC wff α is called, we shall say that α is *successful* in that setting. Many formulae will be successful in some settings but not in others (depending of course on which letters appear on the sheet for a given setting). But there will be some formulae which will be successful in every PC-setting (e.g. $p \vee \sim p$). These we call *PC-successful*.

To make explicit what must be becoming an obvious parallel, let us call a setting V and write $V(\alpha) = 1$ to mean 'the player in setting V raises his hand when α is called', and $V(\alpha) = 0$ to mean 'the player in setting V keeps his hand down when α is called'. The rules 1, 2 and 3 for responding to formulae when thus translated become the conditions 1, 2 and 3 of p. 10 under which V is a PC-assignment. A formula will be successful in a PC-setting V iff it is verified by the corresponding assignment. And a formula will be PC-successful iff it is verified by every PC-assignment. I.e., the PC-successful calls are precisely those which are PC-valid.

Since for any wff α containing n variables we need only consider sheets which contain a selection (possibly all or possibly none) of those n variables (for clearly the responses to variables not in α cannot affect the response to α), we can set out all the relevantly different PC-settings on 2^n sheets. So we could check whether α is valid by preparing such a set of sheets and calling α (with the appropriate preparatory calls) for each of them. Each sheet together with the responses to α and to all its well-formed parts, corresponds exactly to a row of the truth-table for α.

The T game

The game which, anticipatorily, we shall call the *T game* requires a caller and any number of players from one upwards. Each player is provided with a sheet of letters as in the PC game. The contents of these sheets may vary from player to player in any way whatever. The players are seated in some way which determines precisely which of the others, if any, each player is to be able to see during the course of the game (screens or some other devices might be used for this purpose). Any 'seeing arrangement' whatever may be made, from one in which no player can see any other player to one in which each can see everyone else. The arrangement does not have to be reciprocal; i.e., if player A can see player B, B may or may not be allowed to see A.

A setting of the T game will thus be a set of players, each with his sheet and each able (or unable) to see certain of the other players. In the T game any wff of T may be called, provided that as with the PC game its well-formed parts, beginning with the variables, are called first. (We can again assume that the wffs are written in primitive notation, with M, $\prec$ and $=$ eliminated, though we shall in fact state the rule for M explicitly.)

The instructions to each player are those numbered 1, 2 and 3 in the PC game, with the following two for calls involving L and M.

4. If $L\alpha$ is called (where α is a wff of T) raise your hand if *every player you can see* (including yourself[38]) raised his hand when α was called; otherwise keep your hand down.
5. If $M\alpha$ is called, raise your hand if *at least one of the players you can see* (yourself or any other) raised his hand when α was called; otherwise keep your hand down.

The difference between calls for L and M and calls for truth-

[38] In other words we assume that every player can see himself.

functions is that for L and M a player has to know not merely what *he* has done for previous calls but also what the other players he can see have done for previous calls. As with the PC game it should be clear that in each T-setting each wff of T (when appropriately prepared for) will get (from each player) a unique response. In a given T-setting a call may of course lead some players but not others to raise their hands. If it leads every player without exception to raise his hand we shall say that it is a *successful* call in that T-setting. It is obvious that a call might be successful in one T-setting but not successful in another. There are some calls, however, which would be successful in any T-setting whatever – i.e., no matter how many players there are, what letters are on their sheets, or what the seeing arrangement is. We shall say that such calls are *T-successful* calls.

We now define a *T-valid* formula as one which would form a T-successful call.

A simple example of a T-successful call is $Lp \supset p$ (in primitive notation, $\sim Lp \vee p$). For consider any player, A. By the restriction on order, the first call must be p. If p is on A's sheet, he will raise his hand for this call, and hence he will also raise it for $\sim Lp \vee p$ (by rule 3). If p is not on his sheet, he will not raise his hand when p is called, and hence he will not raise it for Lp either (by rule 4); in that case he must raise it for $\sim Lp$ (by rule 2) and therefore also for $\sim Lp \vee p$ (by rule 3). So A must raise his hand for $\sim Lp \vee p$, whether p is on his sheet or not; and every other player in any T-setting must do likewise, for the same reason.

On the other hand, consider $Lp \supset LLp$. In some T-settings this would be a successful call, but in others it would not. Here is one in which it would not. There are three players, A, B and C. A's and B's sheets contain p but C's does not. A can see B, B can see C, but A cannot see C (the remaining details are irrelevant). We can represent this by the following diagram:

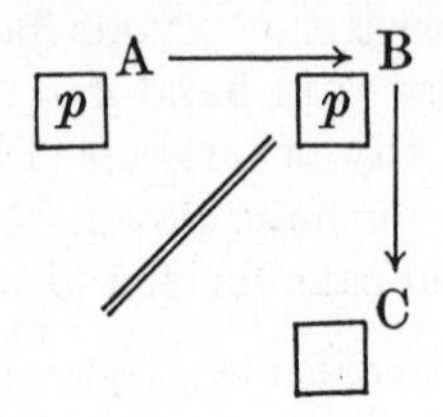

The arrows stand for the seeing-relation, the double line for a screen blocking the view, and the squares for the players' sheets (C's is empty). This is how the play would go:

1st call: p. A and B raise their hands; C does not.

2nd call: Lp. A raises his hand (since all the players he can see, viz. himself and B, raised their hands for p); but B and C do not.

3rd call: LLp. No one raises his hand (not even A, since B, whom he can see, did not raise his hand for Lp).

4th call: $Lp \supset LLp$. B and C raise their hands, but A does not (because he raised his hand for Lp but not for LLp).

We can represent these responses in our diagram by making a list beside each player of all the well-formed parts of $Lp \supset LLp$, and after each of these writing '$= 1$' if that player raises his hand for it and '$= 0$' if he does not. What we get is this:

A: $p = 1$, $Lp = 1$, $LLp = 0$, $(Lp \supset LLp) = 0$

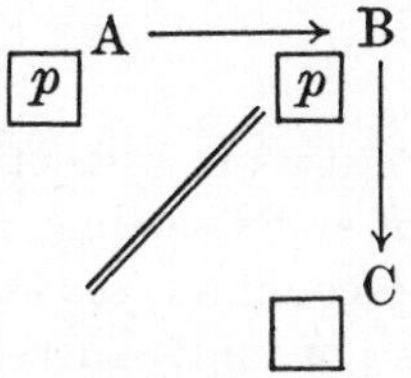

B: $p = 1$, $Lp = 0$, $LLp = 0$, $(Lp \supset LLp) = 1$

C: $p = 0$, $Lp = 0$, $LLp = 0$, $(Lp \supset LLp) = 1$

$Lp \supset LLp$ is not T-successful since there is at least one player (viz. A) in at least one T-setting (viz. the one we have just described) who does not raise his hand for it. Hence $Lp \supset LLp$ is not T-valid.

We can in fact prove a more general result than this. Where L_n represents a sequence of n consecutive L's, we can show that $L_np \supset L_{n+m}p$ is never T-successful where $m \geqslant 1$ ($Lp \supset LLp$ is of course just the special case when $n = 1$ and $m = 1$). Let there be $n + m + 1$ players, arranged in a series so that each sees the next player but no others (and the last sees no one but himself). Let p be on the sheets of all the players except the last. Then a generalization of the method we used for $Lp \supset LLp$ will show that the first player will raise his hand for L_np but not for $L_{n+m}p$.

By our definition of T-validity, therefore, $Lp \supset p$ is T-valid but no wff of the form $L_np \supset L_{n+m}p$ (where $m \geqslant 1$) is T-valid.

The S4 game

The *S4 game* is exactly like the T game except in the following respect. We are not free to make any seeing arrangement we like, but must observe the restriction that the seeing-relation be transitive: i.e., should it happen that of any three players, A, B and C, A can see B and B can see C, then A must be able to see C. (Note that any setting with fewer than three players automatically fulfils this condition.) Any setting in which the seeing arrangement complies with this condition we call an *S4-setting*. (Thus every S4-setting is also a T-setting, but not every T-setting is an S4-setting.) A call which would be successful in every S4-setting is an *S4-successful* call. Clearly every T-successful call is also S4-successful; but a call which would succeed in every S4-setting might fail in some T-setting and so not be T-successful.

We define an *S4-valid* formula as one which would form an S4-successful call.

Now the T-setting we used in order to show that $Lp \supset LLp$ is not T-successful is not an S4-setting, since in it A could see B and B could see C but A could not see C. So if we were to try to show that $Lp \supset LLp$ is not an S4-successful call, we could not use that setting for the purpose. And as a matter of fact this call *is* an S4-successful one, as we can show in the following way. We ask: in what circumstances could a player, A, raise his hand for Lp but not for LLp? He could do so only if everyone he sees raises his hand for p but some of these keep their hands down for Lp. But they in turn can keep their hands down for Lp only if some of the players *they* can see keep theirs down for p. This however is impossible, in any S4-setting, since in such a setting everyone whom they can see A can see too; and everyone A can see *raised* his hand for p.

On the other hand the call $Mp \supset LMp$ is not S4-successful. A very simple S4-setting will show this. Let there be only two players, A and B, of whom A can see B but B cannot see A. A's sheet contains p but B's does not. When p is called, A's hand goes up but not B's. When Mp is called, A's hand goes up again; but B's does not, since the only player he can see is himself and he did not raise his hand for p. Hence when LMp is called, A's hand does not go up, since B, whom he can see, did not raise his hand for Mp. So A raises his hand for Mp but keeps it down for LMp; hence he does not raise it for $Mp \supset LMp$.

The S5 game

The S5 game differs from the T game or the S4 game only in this respect, that in an S5-setting every player must be able to see every other player[39]. Clearly every S5-setting is also an S4-setting (and therefore a T-setting), but some S4-settings are not S5-settings. A call which would be successful in every S5-setting is an *S5-successful* call. Clearly every S4-successful call is also S5-successful, but an S5-successful call might fail in some S4-setting and so not be S4-successful.

We define an *S5-valid* formula as one which would form an S5-successful call.

The last call we discussed ($Mp \supset LMp$), though not S4-successful, is S5-successful. Consider any player, A. We show that, in any S5-setting, if he raises his hand for Mp he will also raise it for LMp. If he raises his hand for Mp, this means that someone whom he can see raised his hand when p was called. But since in every S5-setting every player can see every other player, if A saw someone raise his hand for p, so did everyone else. Hence not merely A, but everyone, raises his hand for Mp; and therefore A raises his hand for LMp.

[39] It is of some interest to note that in order to obtain an S5 game (i.e. a game in which just the same calls will be S5-successful as in the game described in the text) it is not necessary to insist that every player should be able to see every other player. It is sufficient to add to the S4 game the further requirement that the seeing-relation be symmetrical or reciprocal, in the sense that whenever any player, A, can see another player, B, B must be able to see A too. This will allow a certain freedom in the seeing arrangements; but the only alternative to allowing all the players to see each other will consist in splitting them up into two or more groups in such a way that every player can see every other player within his group, but no player can see any player in any other group.

To see that this version of the game makes the same calls S5-successful, we need only reflect that if a set of players is divided into groups in this way, then each of these groups would count as a setting in the S5-game as described in the text. So what we should have could equally well be thought of as a number of simultaneous settings of the original S5 game. An S5-successful call has to be successful in every setting, and clearly it makes no difference whether these settings are arranged at the same or different times.

As an example of a call which is not S5-successful (and therefore, of course, not S4-successful or T-successful either) we take $p \supset Lp$. Suppose there are two players, A and B, in an S5-setting, and that p is on A's sheet but not on B's. Then A will raise his hand for p but will not raise it for Lp, since B, whom he can see, does not raise his for p.

Some results

We now have three distinguishable though related definitions of validity for modal formulae. Later in this chapter we shall show how to express these definitions in a purely formal way, without the adventitious trappings of sheets of letters, raising hands and so forth. But we shall speak in terms of the parlour games in the meantime and prove a number of important results.

THEOREM 1

Every thesis of T is T-valid.

We prove this by showing that every axiom of T is T-valid and that the transformation rules are T-validity-preserving.

The PM axioms present no difficulty. Since they are PC formulae each player works from his own sheet only and is simply playing the PC game. And since they are all *valid* PC formulae, they will lead each player to raise his hand, for the reasons we gave earlier.

We have already shown that A5 is T-valid in our explanation of the T game.

For A6 ($L(p \supset q) \supset (Lp \supset Lq)$) we proceed as follows. If this call is not T-successful, then in some T-setting some player, say A, must fail to raise his hand. Now he can only do this if he raises his hand for $L(p \supset q)$ and Lp but not for Lq. But if he raises his hand for $L(p \supset q)$, then everyone he can see raised his hand for $p \supset q$; i.e., everyone he can see who raised his hand for p raised it for q also. A, however, also raises his hand for Lp, and therefore everyone he can see raised his hand for p. Hence everyone he can see raised his hand for q. Therefore if A raises his hand for $L(p \supset q)$ and for Lp, he must raise it for Lq too. Hence A6 is T-valid.

The rule of Uniform Substitution is T-validity-preserving. In terms of the T game, what we have to prove here is this: that if α is any T-successful call and we uniformly replace any letter (say p) in α by any wff of T, the result is another T-successful call. Now if a call, α, is T-successful this means that every player's hand is raised for α in every T-setting, whether it was raised for p or not; hence it will still be raised for α whether or not it is raised for the formula which replaces p.

Modus Ponens is T-validity-preserving. Any player in any T-setting who raises his hand for α must keep it down for $\sim\alpha$. Hence if he also raises his hand for $(\sim\alpha \vee \beta)$, he must raise it for β. But $(\alpha \supset \beta)$ is defined as $(\sim\alpha \vee \beta)$. Hence if all hands are raised in every T-setting both for α and for $(\alpha \supset \beta)$, all hands must be raised for β.

The rule of Necessitation is also T-validity-preserving. In every T-setting, if *all* hands are raised for α, then every player finds that every player seen by him raises his hand for α. Hence all hands are raised for $L\alpha$.

We have therefore proved that every thesis of T is T-valid.

THEOREM 2

Every thesis of S4 is S4-valid.

PROOF

The basis of S4 is the same as that of T but with the additional axiom A7 $(Lp \supset LLp)$. Since every call that is T-successful is S4-successful, every wff which is T-valid is S4-valid. We showed in our exposition of the S4 game that the call $Lp \supset LLp$ is S4-successful, i.e. that A7 is S4-valid. Our proofs that the transformation rules are T-validity-preserving do not involve any features which distinguish the T game from the S4 game or the S5 game, and therefore show equally that they preserve S4-validity (and S5-validity too for that matter). Hence every thesis of S4 is S4-valid.

THEOREM 3

Every thesis of S5 is S5-valid.

Since we showed in the exposition of the S5 game that the characteristic S5 axiom, $Mp \supset LMp$, is S5-valid, this theorem can be proved in the same way as Theorem 2, with obvious minor changes.

(We shall prove at a later stage that every T-valid formula is a thesis of T, every S4-valid formula a thesis of S4 and every S5-valid formula a thesis of S5. We shall then have shown that the three definitions of validity match the relevant axiomatic systems exactly.)

We can now also prove certain assertions made without proof in Chapter 3.

THEOREM 4

T, S4 and S5 are distinct systems.

PROOF

We showed in the exposition of the T game that A7 ($Lp \supset LLp$) is not T-valid. Hence, by Theorem 1, A7 is not a thesis of T. It is, however, a thesis of S4. Therefore T does not contain S4. Similarly, since A8 is not S4-valid (and therefore not T-valid) neither S4 nor T contains S5.

THEOREM 5

There are infinitely many distinct modalities in T[40].

PROOF

We showed on p. 65 that if L_n is an unbroken sequence of n L's, then for any $m \geqslant 1$, $L_n p \supset L_{n+m} p$ is not T-valid. It follows, by Theorem 1, that it is not a thesis of T; and hence neither is $L_n p \equiv L_{n+m} p$ – i.e. L_n and L_{n+m} are distinct modalities in T. Clearly there are infinitely many modalities each consisting solely of L's, and what we have just shown is that no two of these are equivalent in T.

We can also prove that all the fourteen modalities listed on p. 48 are distinct in S4. Briefly, for each pair of these modalities, A and B, we can find an S4-setting in which ($Ap \equiv Bp$) is not successful. From this it will follow that ($Ap \equiv Bp$) is not S4-valid, and hence, by Theorem 2, not a thesis of S4; so A and B will be distinct in S4. We can show similarly that the six modalities listed on p. 50 are distinct in S5. Details are left to the reader.

40 This result was first proved by Sobociński [1953], using a method adapted from McKinsey [1940].

THEOREM 6

S5 does not collapse into PC (*vide* p. 59).

PROOF

As we showed on p. 68, $p \supset Lp$ is not S5-valid; hence, by Theorem 3, it is not a thesis of S5. Its PC-transform, however, viz. $\sim p \vee p$, is a thesis of PC and therefore of S5. Hence some wff is not equivalent in S5 to its PC-transform.

Since S5 contains both S4 and T, these systems do not collapse into PC either.

Formal definitions of validity

We shall now fulfil the undertaking we gave earlier to re-express the definitions of T-, S4- and S5-validity in a purely formal way.

A setting for the T game consists of three elements: (i) a group of players; (ii) a seeing arrangement; (iii) a set of instructions for responding to calls of wffs. (We do not distinguish the sheets of letters as a separate element, but include them under (iii), since their function is simply to tell each individual player how he is to respond to calls of a single letter.) Our formal definitions of T-validity will express the structure of these elements and their relation to each other.

Instead of saying that we have a group of players we shall simply say that we have a set, W, of objects of some kind. (These objects are often called *worlds*, for a reason we shall come to later.) We shall use the symbols $w_1, \ldots, w_i, \ldots$ to designate members of W. The features of the seeing-relation which are relevant to the T game are these: (a) it is a *dyadic* relation (i.e., one which requires two terms); (b) in any T-setting it is *defined over the players* (i.e., for each pair of players, A and B, it is specified either that A can see B or that A cannot see B); (c) it is a *reflexive* relation (i.e., each player without exception can see himself). Otherwise we impose no restrictions on the nature of the relation. We can therefore represent it formally simply by a dyadic reflexive relation, R, defined over the members of W. Finally, our description of the PC game and its relation to PC value-assignments should make it clear that the instructions for responding to calls have the structure of a value-assignment to wffs, though a more complex kind than we have in PC. As in the case of the PC game, raising a hand for a wff can be taken to represent the assignment of the value 1 to that wff, and keeping it down will represent the assignment of the value 0. However, in the T game we have not simply a single player but a group of players; and for a given call we have not simply a hand-raising or a non-hand-raising but a pattern of hand-raising (some players may raise their hands, others may leave them down). So to describe the response to a call adequately we have to say things like, 'Player A raises his hand, player B does not, . . .' – and this all in the same setting of the T game. If then we are to represent such a situation by a value-assignment, we cannot let $V(\alpha)$ simply be 1 or 0 without qualification, but must let it be 1 or 0 *with respect to a member of W* (for it is the members of W which correspond to the players). That is, where w_i is a member of W we have $V(\alpha) = 1$ (or 0) with respect to (or 'in') w_i, and we write this as $V(\alpha, w_i) = 1$ or $V(\alpha, w_i) = 0$.

We now have to give rules for evaluating any formula whatever. The assignment V must first (as in the case of PC) give a value to each of the variables in the formula, but here it must give it a value in each member of W separately; e.g., if $Lp \supset p$ is to be evaluated in (say) three worlds, w_1, w_2 and w_3, we must require that V assign to p a value (1 or 0) in w_1, in w_2 and in w_3. (This will correspond to the presence or absence of p on each player's sheet.) For complex formulae, just as a given player's response to $\sim\alpha$, $\alpha \vee \beta$, $L\alpha$ or $M\alpha$ depends on the responses of players to α and β, so the assignment to $\sim\alpha$, $\alpha \vee \beta$, $L\alpha$ or $M\alpha$ in a given member of W depends on the assignments to α and β in members of W. For $\sim$ and $\vee$ the rules are simply [V$\sim$] and [V$\vee$] generalized for all members of W, and for L, where w_i is any member of W, we let $V(L\alpha, w_i)$ be 1 iff α has been assigned 1 in every w_j such that $w_i R w_j$.

We shall say that a set of objects, W, a relation, R, and a value-assignment, V, when they satisfy the conditions we have just explained, together constitute a *T-model*[41]. It should be clear that a T-model expresses exactly the structure of a setting

[41] We could speak analogously of a PC value-assignment as a *PC-model*; but it seems pointful to distinguish between a value-assignment and a model only when (as in the case of T) the value-assignment is not the sole element in the model. Formally a value-assignment V in a T-model is simply a dyadic function whose first argument is a wff of T, whose second argument is a member of W, and whose value is 1 or 0.

for the T game. Our earlier definition of T-validity as successfulness in every T-setting can therefore now be re-expressed by saying that a wff, α, is T-valid iff in every T-model, $V(\alpha, w_i) = 1$ for every w_i in W.

We can set all this down as follows[42]:

We define a T-model as an ordered triple $\langle W, R, V \rangle$, where W is a set of objects (worlds), R is a dyadic reflexive relation defined over the members of W, and V is a value assignment satisfying the following conditions:

1. For any propositional variable, p_j, and for any w_i which is a member of W, either $V(p_j, w_i) = 1$ or $V(p_j, w_i) = 0$.

2. [V$\sim$]. For any wff, α, and for any $w_i \in W$[43], $V({\sim}\alpha, w_i) = 1$ if $V(\alpha, w_i) = 0$; otherwise $V({\sim}\alpha, w_i) = 0$.

3. [V$\vee$]. For any wffs, α and β, and for any $w_i \in W$, $V((\alpha \vee \beta), w_i) = 1$ if either $V(\alpha, w_i) = 1$ or $V(\beta, w_i) = 1$; otherwise $V((\alpha \vee \beta), w_i) = 0$.

4. [VL]. For any wff, α, and for any $w_i \in W$, $V(L\alpha, w_i) = 1$ if for every $w_j \in W$ such that $w_i R w_j$, $V(\alpha, w_j) = 1$; otherwise $V(L\alpha, w_i) = 0$[44].

A wff, α, is T-valid iff for every T-model $\langle W, R, V \rangle$ and for every $w_i \in W$, $V(\alpha, w_i) = 1$.

42 The account we give here follows in essence that of Kripke [1963a]. (For Kripke's own terminology and method *vide* Appendix 5.) Early attempts to define validity for modal systems occur in McKinsey [1945] and Carnap [1946]. Some more recent definitions appear in Kanger [1957a], Hintikka [1961] (and elsewhere), and Bayart [1958]. For definitions in terms of Boolean algebra *vide* Chapter 17. The analogy between modal systems and (non-modal) monadic quantification theory with one individual-variable (*vide* Chapter 8) has been noted by several authors: in the case of S5 by Wajsberg [1933] (p. 125), and in the case of other systems by Meredith [1956b], Thomas [1962] and others.

43 'Any $w_i \in W$' is a convenient abbreviation for 'any object, w_i, which is a member of W'. Other set theory terminology we use is $\{x_1, \ldots, x_n\}$ for the set consisting of the objects $x_1, \ldots, x_n$ and $\langle x_1, \ldots, x_n \rangle$ for the set consisting of $x_1, \ldots, x_n$ in that order (i.e. the ordered n-tuple whose members are $x_1, \ldots, x_n$).

44 Since $M\alpha =_{Df} {\sim}L{\sim}\alpha$, we do not need a separate rule for M. It is, however, convenient to state such a rule, and in view of the definition it would run: [VM] For any wff, α, and for any $w_i \in W$, $V(M\alpha, w_i) = 1$ if for at least one $w_j \in W$ such that $w_i R w_j$, $V(\alpha, w_j) = 1$; otherwise $V(M\alpha, w_i) = 0$. The parallel with rule 5 of the T game should be obvious.

We now move on to S4. The only difference between the T game and the S4 game is that in the latter the seeing-relation must also be *transitive*. (A dyadic relation, R, is said to be transitive iff for every x, y and z, if $x\mathrm{R}y$ and $y\mathrm{R}z$, then $x\mathrm{R}z$.) An *S4-model* is therefore an ordered triple $\langle \mathrm{W}, \mathrm{R}, \mathrm{V} \rangle$, where W and V are defined as for a T-model and R is a (dyadic) reflexive and transitive relation defined over the members of W.

A wff, α, is S4-valid iff for every S4-model $\langle \mathrm{W}, \mathrm{R}, \mathrm{V} \rangle$ and for every $w_i \in \mathrm{W}$, $\mathrm{V}(\alpha, w_i) = 1$.

There are two ways of dealing with S5, corresponding to the two versions of the S5 game described respectively in the text and in the footnote on p. 67, though both, as we explained there, yield exactly the same formulae as valid. The footnote version follows more closely the method we have been using up to now, so we shall consider it first. In this version the difference between the S4 game and the S5 game is that in the latter the seeing-relation must also be *symmetrical*. (A dyadic relation, R, is said to be symmetrical iff for every x and y, if $x\mathrm{R}y$ then $y\mathrm{R}x$.) We can therefore say that an S5-model is an ordered triple $\langle \mathrm{W}, \mathrm{R}, \mathrm{V} \rangle$, where W and V are defined as before and R is a (dyadic) reflexive, transitive and symmetrical relation[45] defined over the members of W.

A wff, α, is S5-valid iff for every S5-model $\langle \mathrm{W}, \mathrm{R}, \mathrm{V} \rangle$ and for every $w_i \in \mathrm{W}$, $\mathrm{V}(\alpha, w_i) = 1$.

In the other version of the S5 game it is simply laid down that every player sees every other player (as well as, of course, himself). What this comes to is that we could dispense with reference to the seeing-relation altogether, since in rule 4 the phrase 'every player you can see' could simply read 'every player'. Hence an alternative and simpler definition of an S5-model would be as an ordered couple $\langle \mathrm{W}, \mathrm{V} \rangle$, where W and V are defined as before except that in place of [V*L*] we have:

[V*L*′]. For any wff, α, and for any $w_i \in \mathrm{W}$, $\mathrm{V}(L\alpha, w_i) = 1$ if for every $w_j \in \mathrm{W}$, $\mathrm{V}(\alpha, w_j) = 1$; otherwise $\mathrm{V}(L\alpha, w_i) = 0$.

A wff, α, will then be said to be S5-valid iff for every S5-model $\langle \mathrm{W}, \mathrm{V} \rangle$ and for every $w_i \in \mathrm{W}$, $\mathrm{V}(\alpha, w_i) = 1$.

We can give an analogous account of validity for the Brouwerian system (B-validity). The only difference between

[45] Such a relation is often called an *equivalence* relation.

a T-model and a B-model is that in the latter R must be symmetrical as well as reflexive (it need not be transitive, as in an S4-model). B-validity is then defined as in the other cases. The B game will of course be the same as the T game with a corresponding restriction on the seeing-relation. It is left to the reader to prove that the Brouwerian axiom, and hence every thesis of B, is B-valid, and that B is a distinct system from T, S4 and S5.

The intuitive interpretation of models and validity

It is important to notice that validity as we have defined it is not a structural property of a formula itself. Nor does it consist in the relation of the formula to other formulae, as thesishood in an axiomatic system does (even though the set of theses and the set of valid formulae can be proved to coincide exactly). Our method of defining validity has been based on assigning to elements in formulae certain values which are not themselves parts of the system to which the formulae belong, and evaluating the formulae in terms of those assignments. This procedure is like attaching meanings to formulae: in this chapter, that is, in contrast to the earlier ones, we have been engaging in a kind of semantical study of modal systems[46]. (Hence the phrases 'a semantical definition of validity', 'a semantical model', 'a semantics for a system', which will occur frequently in later chapters.)

What we have just said holds good whether or not we go on to attach a meaning to the values 1 and 0 themselves. Intuitively, however, we think of them as representing truth and falsity respectively; and if we do so we can say, conveniently if a little loosely, that we have defined the validity of a formula as its truth in every world in every model (of the appropriate kind).

The word 'world' has been used by a number of logicians in this connection, and seems to be the most convenient one, but perhaps some such phrase as 'conceivable or envisageable state of affairs' would convey the idea better. What is meant can be explained in the following way. Suppose that in the modal

[46] Modal systems have been investigated in three main ways: axiomatically, semantically and algebraically. The axiomatic and semantical approaches, and the relations between them, dominate this book, but some account of the algebraic approach will be found in Chapter 17.

parlour games the letters occurring in the calls represent certain propositions. We could specify how things are, or what the state of the world is, as far as these propositions are concerned, by giving a list of which of them are true and which are false. By varying this list we could similarly specify various states of affairs which might have obtained in place of the one which actually does. Each of these (actual or merely conceivable) states of affairs is what we mean by a 'world'. A player's sheet of letters can then be thought of as representing that world in which the propositions on his sheet are true and the propositions not on his sheet are false. For the duration of the game he is to suppose that things are as indicated on his sheet and his response to complex formulae will indicate whether the propositions so formed are true or not, granted this assumption. He regards the sheets of the other players whom he can see as indicating various alternative, conceivable, though perhaps not actual, states of affairs. He is further to suppose that these other sheets represent *all* the alternative states of affairs of which he can conceive.

If then a player is asked whether a certain proposition, p, is *true*, he simply enquires whether p is true in his own world (i.e., whether p is on his sheet). If, however, he is asked whether p is *possible* (i.e., whether Mp is true), it will be sufficient for p to be on one of the other players' sheets; for this will indicate that there is a conceivable state of affairs in which p is true (i.e., that a state of affairs in which p is true is a conceivable or possible one), and hence that p is conceivable or possible. And if he is asked whether p is necessary (i.e., whether Lp is true) he must satisfy himself that p is not merely on his own sheet but on the others as well, for if it is absent from any of them this will indicate that p, even if in fact true, *might* have been false.

Looked at in this way, the semantical models reflect in an obvious way a familiar philosophical idea which is often credited to Leibniz, that a necessary proposition is one which is true not merely in the actual world but in every other possible world as well. Of the systems we have considered, it is S5 which seems to express this idea most directly, since in an S5-model the truth of p in every world in W is required for the truth of Lp in any of them. In the cases of T and S4 there is also a relation which in the games we called a seeing-relation and in the semantic models simply a reflexive (or for S4 a reflexive and transitive) relation, R. The effect of this is that in deciding whether p is necessary in a given world we have only to take account of the value of p in a specified set of worlds, which need not include all the worlds in the model. By those who speak of the members of W as worlds, R is often called the *accessibility* relation (a world, w_j, is said to be accessible to a world, w_i, iff $w_i \mathrm{R} w_j$). Adopting this usage, we shall say that a proposition is necessarily true in a given world iff it is true in every world accessible to that world.

This notion of one possible world's being accessible to another has at first sight a certain air of fantasy or science fiction about it, but we might attach quite a sober sense to it in the following way. We can conceive of various worlds which would differ in certain ways from the actual one (a world without telephones, for example). But our ability to do this is at least partly governed by the kind of world we actually live in: the constitution of the human mind and the human body, the languages which exist or do not exist, and many other things, set certain limits to our powers of conceiving. We could then say that a world, w_2, is accessible to a world, w_1, if w_2 is conceivable by someone living in w_1; and this will make accessibility a relation between worlds, as we want it to be. Now if we were to find ourselves in one of the worlds of which we can conceive, then in that world our powers of conceiving might remain just as they were, or they might not – they might become enhanced or restricted in all sorts of ways; we might or might not even be able to conceive of the world in which we formerly lived. So the worlds which are accessible to w_2 need not be the same as those which are accessible to w_1. Suppose now that someone living in w_1 is asked whether a certain proposition, p, is possible (whether p *might* be true). He will regard this as the question whether in some conceivable world (conceivable, that is, from the point of view of his world, w_1), p would be true; but if there are worlds which are not conceivable to anyone living in w_1, then he will not (because he cannot) take these into account. Similarly, if he is asked whether p is necessary (whether p is *bound* to be true), he will construe this as the question whether in all worlds accessible to his own, p would still be true; and once more, if there are worlds not accessible to his he can take no account of them. At any rate, this is one way in which we can interpret the notions of possibility and necessity, and it seems to be reflected in T.

It is not, however, the only way. In a stronger sense of 'conceive', one would not conceive (or at least fully conceive) of a world of a certain kind unless one had a complete insight into what it would be like to live in such a world – and that would include knowing what powers of conceiving someone who lived in that world would have. If we were to define accessibility in terms of conceivability in *this* sense, accessibility would be a transitive relation. For if someone in w_2 can conceive of w_3, then anyone in w_1 who can conceive (in the strong sense) of w_2 will know what it is to conceive of w_3, and to do this he must be able to conceive of w_3 himself. So if w_2 is accessible to w_1, then so is every world accessible to w_2; and with accessibility as a transitive relation we have S4, not T. Someone in w_1 who construes 'possible' in terms of this stronger sense of 'conceivable' will reckon p to be possible if it is true in any of the worlds accessible to w_1, or in any world accessible to any of *those* worlds, for they are also accessible to w_1 itself; he will, that is, equate the possible with the possibly possible. Similarly, he will equate the necessary with the necessarily necessary; for what counts as necessarily necessary in w_1 is whatever is true in all worlds accessible to worlds accessible to w_1; but these are precisely the worlds accessible to w_1 itself – that is, they are precisely the worlds in which p has to be true for p to be necessary in w_1. These principles, of course, find expression in the S4 reduction laws, R3 and R4.

We have been distinguishing between T and S4 in terms of a difference between two senses of 'conceivable' (and hence of 'possible' and 'necessary'). Roughly speaking, the difference is that between knowing what a certain state of affairs would be like, and knowing what it would be like to live in that state of affairs. But this stronger sense of 'conceivable', although it makes accessibility transitive, does not make it symmetrical. For example, we can conceive of a world without telephones. (Don't many of us sometimes wish we lived in such a world?) But if there had been no telephones, it might surely have been the case that in such a world no one would know what a telephone was, and so no one could conceive of a world (such as ours) in which there are telephones; i.e. the telephoneless world would be accessible to ours but ours would not be accessible to it. And this would seem to be so even when 'conceive' is used in the stronger sense which leads to the validation of the S4 axioms.

Thus if the necessity and possibility of propositions is defined in terms of the conceivability of worlds in this sense, then of the systems we have discussed it is S4, rather than T or S5, which reflects validity.

S5, as we have already mentioned, reflects on the other hand an 'absolute' sense of the word 'conceivable' – a sense in which to say that a state of affairs is conceivable is to say something about *it*, without reference to the powers of conceiving which may or may not exist in any other state of affairs.

Thus we seem to be able to distinguish at least three different senses in which a state of affairs might be said to be conceivable, and we could define necessity and possibility in terms of any of them; and it is at least arguable that if we define them in one of these ways it is T which gives us a complete body of sound modal principles, while if we define them in another way it is S4 which does this, and if we define them in still another way it is S5.

All this has a wider importance. We have so far encountered three modal systems – four if we count the Brouwerian system – and in later chapters we shall come across a great many more. This multiplicity of systems is apt to provoke the question, Which system is the correct one? Now the assumption behind this question seems to be that we have in mind some single sense of 'necessity' and 'possibility', and that systems weaker than the correct one will give us less than the whole truth, while stronger systems will contain theses which even if plausible are really false. But perhaps the systems are not rivals in this way. It is at least possible that a number of systems may each give us the truth about necessity and possibility, though each in a somewhat different sense of those terms. Merely constructing semantic models will not by itself give us an adequate characterization of these different senses[47]; for that a great deal of intricate philosophical work would be required, though as we have seen the semantic models can give us valuable help in this task.

Another method of showing various modal systems not to be rivals would consist in correlating the use of L in each of them with the use of 'necessary', or some related expression, in some already established sphere of discourse. Some suggestions along

[47] Cf. Pollock [1967a].

these lines have been made by Lemmon [47a]. One of these is that if L is taken to mean 'it is informally provable in mathematics that', then S4 is the correct modal system; another (stated more tentatively) is that if L means 'it is analytically the case that', then it is S5 that is correct.

Thus in one way or another the multiplicity of modal systems, far from being a symptom of confusion or a source of perplexity, can in fact be a positive help in drawing our attention to distinctions which might otherwise go unnoticed.

Exercises – 4

4.1. (a) Construct a T-setting in which $LLp \supset LLLp$ is successful but $Lp \supset LLp$ is not.

(b) State necessary and sufficient conditions for the seeing-relation in such a setting and show them to be necessary and sufficient.

(c) Show that the 14 modalities referred to on p. 48 are distinct in S4.

4.2. (a) Show that $p \supset LMp$ is successful in every Brouwerian setting.

(b) Construct a Brouwerian setting in which $Mp \supset LMp$ is not successful. Explain why this cannot be an S4-setting.

4.3. Consider an S4-setting in which there are three players: A, B and C. A can see both B and C but neither B nor C can see anyone but himself. B's sheet contains p but not q, C's sheet contains q but not p. (Other details are irrelevant.)

(a) Show that in such a setting $(Lp \prec Lq) \vee (Lq \prec Lp)$ is not successful.

(b) Show that it would be successful if either B can see C or C can see B.

4.4. Imagine a game which is like the T game except that some of the players ('pink players') may have pink sheets of paper (in contrast to the other players who have white sheets and whom we shall call 'normal players'). Suppose that every pink player can be seen by at least one normal player but that a pink player cannot see anyone. A normal player's instructions are as in the T game; a pink player's instructions differ from these in the following respect only: Whenever $L\alpha$ is called he is to keep his

[47a] Lemmon [1959].

hand down, no matter what formula α may be. A call is to be successful in a setting of this game iff every *normal* player raises his hand. (Validity, as usual, is simply success in every setting.)

(a) Show that the rule of Necessitation is not validity-preserving in such a game.

(b) Show that if α is PC-valid then $L\alpha$ is valid in this game.

(c) Show that $\vdash \alpha \prec \beta \rightarrow \vdash L\alpha \prec L\beta$ is validity-preserving.

4.5. Assume a game of the kind described in the previous exercise but with the condition that the seeing relation be transitive.

(a) Show that $Lp \supset LLp$ is not valid in such a game.

(b) Show that $(p \prec q) \prec (Lp \prec Lq)$ *is* valid in such a game.

4.6. Assume a game of the kind described in exercise 4.4 except that there *must* always be at least one pink player and that a call is to be successful iff every *normal player who can see a pink player* raises his hand.

(a) Show that MMp is valid in such a game.

(b) Show that $LMMp$ is valid iff every normal player (in every setting) can see at least one pink player.

CHAPTER FIVE

T: Decision Procedure and Completeness

Testing for T-validity

As we explained in the previous chapter, a formula, α, is T-valid iff it is true in every world in every T-model; i.e., iff in every T-model $\langle W, R, V \rangle$, $V(\alpha, w_i) = 1$ for every $w_i \in W$. In this chapter we shall show how to test formulae for T-validity.

In testing a PC formula for validity by the truth-table method outlined in Chapter 1 we list all the distinct PC-assignments with respect to the variables in the formula, and then check whether the formula is true for each of them. This method can in theory be applied to any PC formula whatsoever; and even for moderately complicated formulae it is a practical method since for a formula containing n variables there are only 2^n distinct value-assignments. The corresponding method for T would be to list all the relevantly different T-models for the formula with which we were concerned, and then check whether the formula was true in every world in each of them. Now this would *in theory* be a sound procedure since, as we shall prove later (p. 324), for any particular formula, α, only models with no more than a certain finite number of members of W (depending on the structure of α) need be considered, and for any finite number of members of W only a finite number of distinct T-models can be constructed. Nevertheless, the number of distinct T-models, though always finite for any formula, is apt to be extremely large, and this method would involve us in millions of calculations in order to test even a quite simple formula.

Fortunately there are shorter methods. The one we shall describe[48] is an extension of the Reductio test for PC-validity

[48] Our procedure is similar in essentials to the method of semantic tableaux found in Kripke [1963a] and elsewhere. For other decision procedures for T, *vide* Von Wright [1951], Anderson [1954] and Ohnishi and Matsumoto [1957].

outlined on pp. 12–14, with which we shall assume that the reader is familiar. Briefly, we attempt to find, for a given wff, α, a falsifying T-model (i.e., a T-model in which, for at least one $w_i \in W$, $V(\alpha, w_i) = 0$). The method will enable us to construct such a T-model if this is possible, or else it will demonstrate the impossibility of there being such a T-model. In the former case, of course, α is invalid; in the latter case, α is valid.

We shall describe the method by working through a number of examples: after that we shall show that the method provides a decision procedure for T. For convenience we shall assume that the only modal operators occurring in formulae are L and M; we can of course always eliminate other modal operators by using their definitions.

First example

$$[1] \quad L(p \supset L(q \supset r)) \supset M(q \supset (Lp \supset Mr))$$

We begin by supposing that in some T-model there is a world (say w_1) such that $V([1], w_1) = 0$. We represent this by writing 0 under the main operator. The PC rule [V⊃] then immediately gives 1 as the value (in w_1) of the antecedent and 0 as the value of the consequent; i.e., we have $V(L(p \supset L(q \supset r)), w_1) = 1$ and $V(M(q \supset (Lp \supset Mr)), w_1) = 0$. This is as far as purely PC principles can take us at this stage; but by [VL], if $L\alpha$ has the value 1 in w_1, α itself has the value 1 in w_1 (as well as in all other worlds accessible to w_1), and by [VM], if $M\alpha$ has the value 0 in w_1, α itself has the value 0 in w_1 (as well as in all other worlds accessible to w_1). Hence we have the following values so far:

$$\begin{array}{cccccccccccccccccccccc} L & (& p & \supset & L & (& q & \supset & r &) &) & \supset & M & (& q & \supset & (& Lp & \supset & Mr &) &) \\ 1 & & & & 1 & & & & & & & 0 & 0 & & & 0 & & & & & & \end{array}$$

It is now a straightforward matter to apply [V⊃], [VL] and [VM] again. Note that [VL] not merely obliges us to give α the value 1 when $L\alpha$ has the value 1, but also obliges us to give $L\alpha$ the value 0 when α has the value 0; similarly by [VM] we must give $M\alpha$ the value 1 if we have given the value 1 to α.

Ultimately we reach the following values (the numbers

written above the formula should indicate sufficiently the order in which we have taken the subsequent steps):

$$\begin{array}{c} \scriptstyle 5\quad 7\ 5\ 6\ 5 \qquad\quad 1 \qquad 2\ 3\ 1\quad 2\ 4 \\ L(p \supset L(q \supset r)) \supset M(q \supset (Lp \supset Mr)) \\ \scriptstyle 1\ \underline{1\ 1\ 0}\ 1\ 0\ 0 \quad 0\ 0\ 1\ 0\quad 1\ 1\ 0\ 0\ 0 \end{array}$$

But here we strike an inconsistency, as the underlining indicates; for we have $V(p \supset L(q \supset r), w_1) = 1$ and $V(p, w_1) = 1$, but $V(L(q \supset r), w_1) = 0$, and this violates $[V\supset]$. What this shows is that there could be no world in any T-model in which $V([1]) = 0$; hence [1] is T-valid.

(We shall usually call attention to an inconsistency in a value-assignment by underlining, as we have done here. Every such inconsistency can, however, be expressed as the assignment both of the value 1 and the value 0 to some well-formed part of a formula; in the present case, e.g., we could have made step 8 the assignment of 0 to p. The most convenient way of defining an inconsistent value-assignment, in fact, is to say that V is an inconsistent value-assignment to a formula (or set of formulae) iff for some well-formed part, α, of that formula (or set of formulae), $V(\alpha) = 1$ and $V(\alpha) = 0$.)

We have been able to show the validity of [1] without considering more than one world. This seldom happens, and our other examples will involve steps of another kind. As our next example we take a formula which differs from [1] only in that the second L is replaced by M.

Second example

$$[2] \quad L(p \supset M(q \supset r)) \supset M(q \supset (Lp \supset Mr))$$

The same steps as with [1] lead to the following value-assignments in w_1:

$$\begin{array}{c} L(p \supset M(q \supset r)) \supset M(q \supset (Lp \supset Mr)) \\ \scriptstyle 1\ 1\ 1\ 1\ 1\ 0\ 0 \quad 0\ 0\ 1\ 0\quad 1\ 1\ 0\ 0\ 0 \\ \scriptstyle * \qquad * \end{array}$$

Here we do not have any immediate inconsistency; but at the place marked by asterisks we have $M(q \supset r)$ as true but $(q \supset r)$

as false; i.e. we have a formula, α, such that $V(M\alpha, w_1) = 1$ but $V(\alpha, w_1) = 0$. Now [V*M*] does not forbid this combination of values absolutely; all it insists on is that if $M\alpha$ is to be true in w_1 then there must be *some* world accessible to w_1 (not necessarily w_1 itself) in which α is true. Clearly in the present case the world in question cannot be w_1, so we must suppose the existence of another world, w_2, in which $(q \supset r)$ is true. w_2 must of course be accessible to w_1 and this means that if we are to satisfy [V*L*] and [V*M*], not only must $(q \supset r)$ be true in w_2 but for every wff, α, if $L\alpha$ is true (is assigned 1) in w_1 then α must be assigned 1 in w_2, and if $M\alpha$ is false (is assigned 0) in w_1 then α must be assigned 0 in w_2. We then work out what values would have to obtain in w_2 for this to be so, exactly as we worked out what values were needed in w_1. We thus find that in w_2 $V(q \supset r) = 1$, $V(p \supset M(q \supset r)) = 1$, $V(q \supset (Lp \supset Mr)) = 0$, $V(p) = 1$, $V(r) = 0$. But it only requires a little calculation to see that this situation cannot obtain in any world; for $V(q \supset (Lp \supset Mr)) = 0$ gives us $V(q) = 1$, and this together with $V(r) = 0$ gives $V(q \supset r) = 0$, which is in conflict with $V(q \supset r) = 1$. So there could be no world fulfilling the conditions for w_2 and therefore [2] could not have the value 0 in any world – i.e. [2] is T-valid.

We can set out the whole calculation diagrammatically as follows:

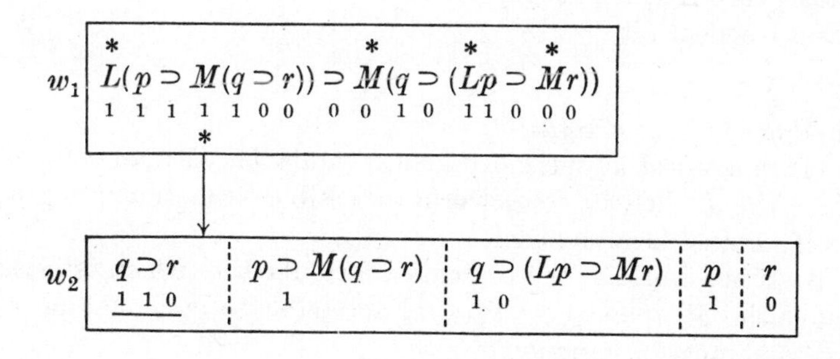

In the diagram each rectangle represents a world and the arrow represents the accessibility relation. In rectangle w_1 is written the formula to be tested, [2], with the values which have to be assigned to its well-formed parts if we are to have $V([2], w_1) = 0$. The asterisk below the M indicates the need for another world, w_2, accessible to w_1, in which the argument of M will be true, if [2] is to be false in w_1. The asterisks *above* the L's and M's serve as a reminder that in any world accessible to w_1 the arguments of these operators must be true and false respectively. In the second rectangle, we write the value-assignments which must be made in w_2; and the underlining marks the inconsistency which arises when we try to make these assignments, and hence indicates the validity of [2].

We shall call a diagram of the kind we have just constructed a *semantic T-diagram*, and the whole method the *method of semantic diagrams*.

In this example we have used [VM] to show that when we have $V(M\alpha, w_1) = 1$ but $V(\alpha, w_1) = 0$, the model must contain a world, w_2, accessible to w_1, such that $V(\alpha, w_2) = 1$. Similarly if in w_1 we had $V(L\alpha, w_1) = 0$ but $V(\alpha, w_1) = 1$, [VL] would require us to assume a world accessible to w_1 in which $V(\alpha) = 0$; for if in *every* world accessible to w_1 we had $V(\alpha) = 1$, we should have $V(L\alpha, w_1) = 1$. We now state these rules explicitly and generally:

I *Rule for putting in asterisks*

An asterisk is put *above* every L which has a 1 beneath it and above every M which has a 0 beneath it. An asterisk is put *below* every L which has a 0 beneath it and below every M which has a 1 beneath it.

II *Rules for a new world*

A. If in a world w_i there occurs a formula $L\alpha$ with an asterisk *above* the L then in *every* world accessible to w_i (including w_i itself) α must be assigned 1.

B. If in a world w_i there occurs a formula $M\alpha$ with an asterisk above the M, then in *every* world accessible to w_i (including w_i itself), α must be assigned 0.

C. If in a world w_i there occurs a formula $L\alpha$ with an asterisk *below* the L then there must be a world accessible to w_i (w_i itself or some other) in which α is assigned 0.

D. If in a world w_i there occurs a formula $M\alpha$ with an asterisk below the M then there must be a world accessible to w_i (w_i itself or some other) in which α is assigned 1.

It should be clear that when we construct new worlds in accordance with these rules we do so in a way which complies with [V*L*] and [V*M*]. In terms of the diagrams a world, w_i, is represented by a rectangle with 'w_i' written beside it; and when a world, w_j, accessible to w_i but distinct from it, is required in order to satisfy C or D, we draw a rectangle labelled 'w_j', with an arrow to it from w_i to represent accessibility. Certain formulae will have to be written in w_j and certain values assigned to them as dictated by the rules in II (A–D). We shall refer to these values as the *initial* values in w_j; values which we then have to assign to various well-formed parts of the formulae in w_j in order to comply with the conditions for a T-assignment we shall call *consequential* values in w_j. Where the value of a modal operator with an asterisk beneath it is the same as the value of its argument then rules C and D will be satisfied without our needing to draw a new rectangle. (And indeed in such cases we may omit the asterisk.)

Third example

$$[3] \quad L(p \vee Mq) \supset (Lp \vee Mq)$$

If we suppose that $V([3], w_1) = 0$ then the PC rules give us the following values (in w_1):

```
       *                   *
w1 |   L(p ∨ Mq) ⊃ (Lp ∨ Mq)
       1           0  0  0 0
                      *
```

Rules A and B then give $V((p \vee Mq), w_1) = 1$ and $V(q, w_1) = 0$. Copying the value 0 for Mq under its other occurrences and applying [V∨], we get $V(p, w_1) = 1$. Hence we have (in w_1):

```
       *      *            *
w1 |   L(p ∨ Mq) ⊃ (Lp ∨ Mq)
       1 1 1 00  0  01 0 00
                    *
```

Since we now have Lp false but p true in w_1, rule C requires us to have a world, w_2, accessible to w_1, in which p is false; and rules

A and B require that in w_2 $(p \vee Mq)$ shall be true and q false. (All these are the initial values of w_2.) These values in w_2 are compatible with each other, but give $V(Mq, w_2) = 1$ and $V(q, w_2) = 0$. By rule D we therefore need a further world in the model, w_3, in which q is true. This, however, is the only restriction on the value-assignments in w_3, so no inconsistencies can arise. We have therefore found a falsifying T-model for [3], and this shows it not to be T-valid. The diagram is as follows:

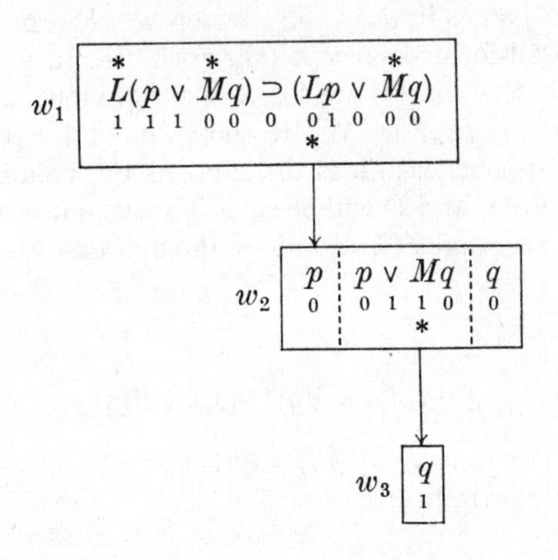

This diagram represents the following T-model: $W = \{w_1, w_2, w_3\}$. $w_1 R w_2$ and $w_2 R w_3$, but no other accessibility relations hold, apart from the accessibility of each world to itself. $V(p, w_1) = 1$; $V(q, w_1) = 0$; $V(p, w_2) = 0$; $V(q, w_2) = 0$; $V(q, w_3) = 1$. (The value of p in w_3 is irrelevant; it can be either 1 or 0.) [3] is invalid because in this T-model, $V([3], w_1) = 0$[49].

We can if we wish check that this is so by the following calculation. Each horizontal row represents the value-assignments in one of the worlds w_1, w_2 and w_3. The arrows indicate the accessibility relations. Given the initial assignments to p and q the calculations proceed in accordance with the rules for a T-

[49] Strictly we have here two falsifying T-models for [3], not one: for the model in which $V(p, w_3) = 1$ is different from the model in which $V(p, w_3) = 0$. Very often, as in this case, a rectangle specifies not one unique world but a set of worlds differing from one another only in ways that are irrelevant to the validity of the formula being tested.

assignment. To avoid confusion only the relevant values have been entered in the table.

	p	q	Mq	$p \vee Mq$	$L(p \vee Mq)$	Lp	Mq	$Lp \vee Mq$	$L(p \vee Mq) \supset (Lp \vee Mq)$
w_1	1	0	0	1	1	0	0	0	0
↓									
w_2	0	0	1	1					
↓									
w_3		1							

Fourth example

$$[4] \quad M(p \,.\, Mq) \supset (LMp \supset MLq)$$

By steps which should now be obvious we reach the following:

```
               *         *
w1   M(p . Mq) ⊃ (LMp ⊃ MLq)
     1         0  11  0 00
     *             *     *
```

We have, as yet, no definite values for p, q, or Mq in w_1. It is clear, however, that the value of $(p \,.\, Mq)$ in w_1 does not matter so long as there is some world (accessible to w_1) in which its value is 1. Similarly, all that is required in the case of p and q is that in some world (accessible to w_1) $V(p) = 1$, and that in some world (accessible to w_1) $V(q) = 0$.

In other words the fact that no further values have been assigned in w_1 does not in any way prevent the application of rules A–D. Continuing the procedure we get the following diagram which shows [4] to be invalid:

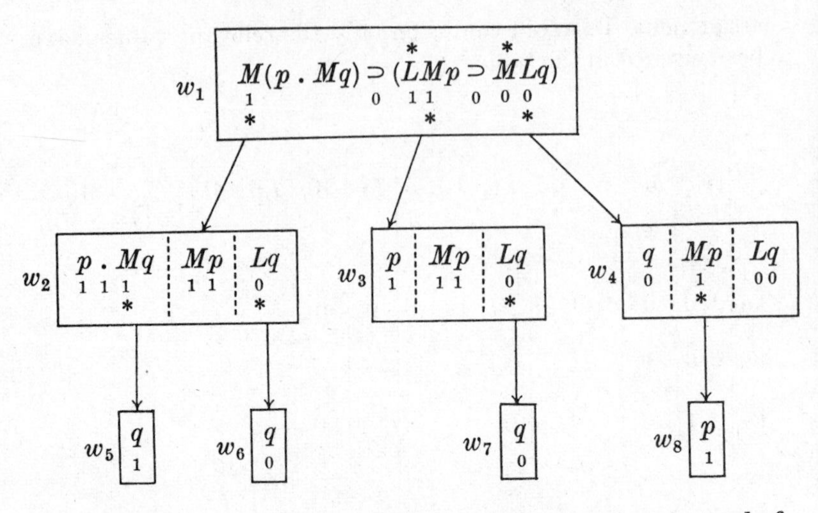

We can often shorten a diagram such as this, since instead of constructing a new rectangle whenever we need one we may find that a rectangle we have already constructed contains the values which are required in the new rectangle, or that it can be made to contain them by filling in values which although not required in the already existing rectangle, are compatible with it. In this way our present diagram can be shortened to the following:

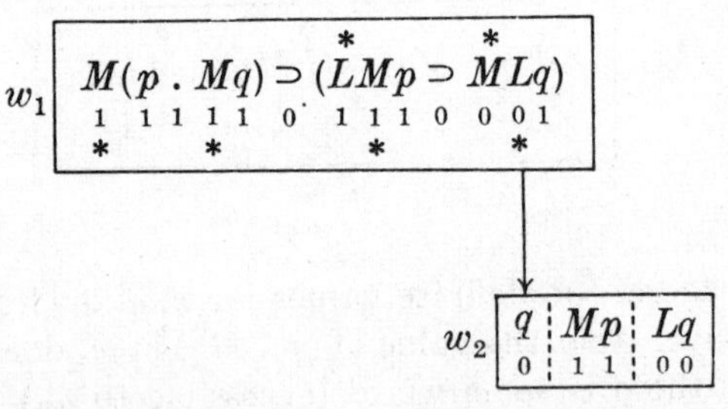

What has happened here is that we find that it is possible to let w_1 itself take over the functions for which we previously constructed many new rectangles and that only one other rectangle is required.

Although short cuts such as these can obviously save a lot of time in practice, we shall assume in our theoretical discussion of T-diagrams that no use has been made of them. In a T-diagram in which the short cuts are not used no values will occur in any rectangle unless they are explicitly required by the rules of the method.

Fifth example

$$[5] \quad L(Mp \equiv Mq) \supset L(p \equiv Lq)$$

The first rectangle in the diagram for [5] will be:

```
     *
w1   L(Mp ≡ Mq) ⊃ L(p ≡ Lq)
     1      1       0 0
            †         *
```

At the place marked by a † we have a situation which can also arise in the PC Reductio test (p.p 13f): a truth-functional operator has a value under it, but we cannot determine unambiguously the values of its arguments. We shall call such an operator, for brevity, a *†-operator*. In the present case, the first $\equiv$ in [5] is a †-operator in w_1: by [V$\equiv$], if $(Mp \equiv Mq)$ is to have the value 1 in w_1, then Mp and Mq must have the same value in w_1, but the assignment so far does not tell us which value this is. So we have two cases to consider, one in which Mp and Mq are both assigned 1, and one in which they are both assigned 0. We can represent these in this way:

```
         *
w1(i)    L(Mp ≡ Mq) ⊃ L(p ≡ Lq)
         1 1    1 1     0 0
           *      *       *
```

```
         * *      *
w1(ii)   L(Mp ≡ Mq) ⊃ L(p ≡ Lq)
         1 0    1 0     0 0
                          *
```

As in the parallel cases in the PC Reductio test, it is only if each of these assignments leads to an inconsistency that [5] is valid; i.e., if either of them leads to a falsifying model, [5] is invalid. Now neither $w_1(i)$ nor $w_1(ii)$ contains any †-operators, so we can begin a T-diagram with each of them by our earlier rules.

We take $w_1(ii)$ first, since it is the simpler. This does lead to an inconsistency, as the following diagram shows:

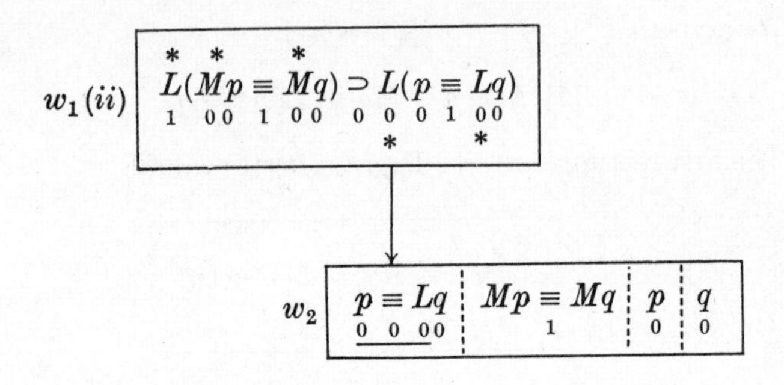

$w_1(i)$, however, gives us this:

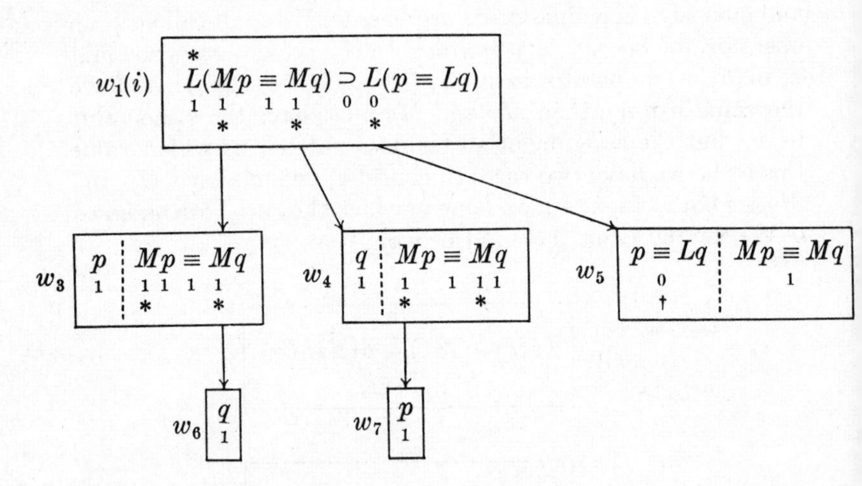

Here, in w_5, we find the same situation arising as in w_1, viz. the occurrence of a †-operator. (In fact in w_5 we have two †-operators, though we have only put a † under one of them, in accordance with a rule which we shall state shortly.) By [V≡], if $(p \equiv Lq)$ is to have the value 0 in w_5, p and Lq must have different values in w_5, but the assignments so far do not tell us what these values are to be. So we have two cases to consider for p and Lq in w_5, exactly as we had for Mp and Mq in w_1, and w_5 will count as containing an inconsistency iff each of these leads to an inconsistency. We represent the two cases as follows:

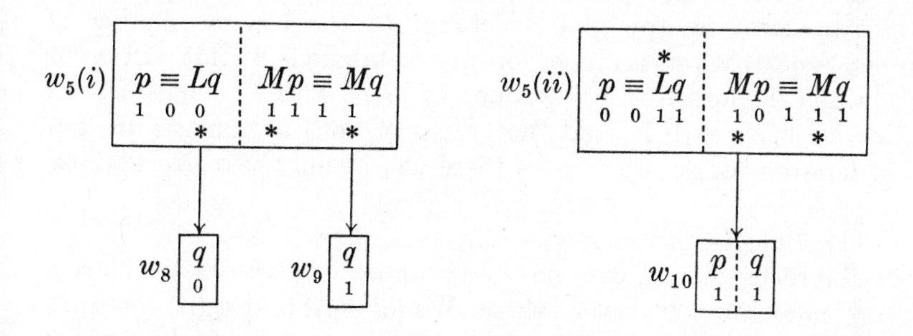

Neither $w_5(i)$ nor $w_5(ii)$ leads to an inconsistency, though of course in order to show that w_5 is not inconsistent it would have been sufficient for one of them not to lead to an inconsistency. So if we replaced w_5 by either $w_5(i)$ or $w_5(ii)$ in the diagram beginning with $w_1(i)$, we could use the diagram so obtained to construct a falsifying model for [5] and thus show it to be invalid. We leave the reader to verify this.

Note that in the present case neither $w_5(i)$ nor $w_5(ii)$ contains any †-operators, since in each case the assignments to p and Lq enable us to give definite values to Mp and Mq, the arguments of the other †-operator in w_5. But if this had not happened – if, e.g., in $w_5(i)$ we had not had definite values for Mp and Mq – we should have put a † under the $\equiv$ in $(Mp \equiv Mq)$ in that rectangle and constructed alternatives for it, which we should have called $(w_5(i))(i)$ and $(w_5(i))(ii)$. Quite in general, if †-operators appear, for whatever reason, in any rectangle, w_i, we put a † under one of them (let us say, for the sake of having a definite rule, the leftmost one) and construct alternatives in the way we have described. Since w_i can contain only a finite number of truth-functional operators, the task of constructing alternatives, alternatives of alternatives, and so on, of w_i is bound to be a finite one.

We shall now state a general rule for dealing with any †-operators that may occur in the construction of a T-diagram. It may be as well to re-state here what a †-operator is. A

†-operator in a rectangle, w_i, is a truth-functional operator[50] which has a value beneath it in w_i but whose arguments do not have their values determined unambiguously in w_i. Note that if we follow strictly the practice of putting a † under only one †-operator in any given rectangle[51], the largest number of alternatives we can have for any rectangle is 3: this will occur when the operator in question is $\vee$ or $\supset$ with 1 beneath it, or . with 0 beneath it, and the values of both arguments are undetermined. In other cases there will be only two alternatives.

III *Rule for alternatives*

If a rectangle, w_i, contains one or more †-operators, we place a † under the leftmost of them. We let $w_i(i)$ and $w_i(ii)$ (or $w_i(i)$, $w_i(ii)$ and $w_i(iii)$) be the two (or three) rectangles each of which reproduces w_i exactly and in addition contains a distinct one of the value-assignments to the arguments of the operator below which the † appears in w_i which are compatible with the value under that operator. We call these rectangles the *alternatives* of w_i, and beginning with each of them in turn we construct a fresh T-diagram. Iff each of these diagrams contains an inconsistency we regard w_i itself as inconsistent.

In each alternative of w_i the *initial* values are all the initial values in w_i together with the values assigned in that alternative to the arguments of the operator under which the † appears in w_i.

N.B. No arrows are drawn from a rectangle containing a †.

Summary of the method

We always begin by writing the formula to be tested, α, in a rectangle, w_1, and giving it the initial assignment 0. Within each rectangle, w_i, as we come to it, we work out all the consequential values in w_i before we proceed further (i.e., all the values of

[50] If a *modal* operator has a value under it but we cannot determine the value of its argument unambiguously, we handle the case by the rule for asterisks below operators, not by constructing alternative diagrams.

[51] Sometimes it is quicker in practice to put a † under more than one †-operator in a rectangle, list the n possible combinations of values of their arguments and construct n alternative diagrams, one for each of them. We leave the reader to work out the details for himself. In our exposition we assume that no such short cuts have been taken.

well-formed parts of formulae in w_i which are unambiguously determined by the initial assignments in w_i, the rules of PC and [V*L*] and [V*M*]). Rectangles other than w_1 are constructed in accordance with the following rules:

I. Rule for putting in asterisks (p. 86);
II. Rules for a new world (p. 86);
III. Rule for alternatives (p. 94).

When we have applied all these rules as often as we can, we say that we have a *complete system of T-diagrams for* α.

We now define an *inconsistent rectangle* in a complete system of T-diagrams as follows:

1. Any rectangle in which both 1 and 0 are assigned, initially or consequentially, to the same formula (or well-formed part of a formula) is an inconsistent rectangle. (We call these rectangles *explicitly inconsistent* rectangles; they are the ones which have underlined values in the diagrams we have drawn.)
2. Any rectangle to which an inconsistent rectangle is accessible is an inconsistent rectangle.
3. Any rectangle *all* of whose alternatives are inconsistent is an inconsistent rectangle.

A wff is T-valid iff the first rectangle in its complete system of T-diagrams, w_1, is inconsistent.

Semantic diagrams as a decision procedure for T

We have now to show that the method of semantic diagrams provides a decision procedure for T.

We shall say that a T-diagram (or system of T-diagrams) *terminates* when (i) for all rectangles containing a † the appropriate alternative diagrams have been constructed, and (ii) in all rectangles not containing a † every asterisk *below* a modal operator has been dealt with in accordance with rules C and D (p. 86); or when, before this stage is reached, the initial rectangle is shown to be inconsistent.

It is furthermore convenient to define the *modal degree of a rectangle* as the modal degree of the formula of highest modal degree in it.

Now every T-diagram (or system of T-diagrams) terminates in a finite number of steps, for the following reason. Whenever an arrow runs from a rectangle, w_i, to a rectangle, w_j, all the formulae in w_j will be the arguments of modal operators in w_i, and hence w_j will be of lower modal degree than w_i. It follows that, no matter what the modal degree of the initial formula in w_1 may be, we shall ultimately (unless the diagram terminates before this) reach rectangles containing only formulae of degree 0 (i.e., PC formulae); and since no asterisks can appear below PC formulae, the diagram will terminate then, if not before. Moreover, each alternative to a rectangle, w_i, is clearly of the same modal degree as w_i itself; and as we have already seen, there can be only a finite number of alternatives, alternatives of alternatives, and so on, of w_i.

Furthermore, the method can be applied to any wff of T whatsoever; for it explicitly specifies an initial value for the single formula in the first rectangle, and given this, it provides an effective way of proceeding until the diagram (or system of diagrams) terminates.

Finally, the method yields an unambiguous result in each case. For in the case of any wff, α, either the first rectangle, w_1, will be shown to be inconsistent, or else the complete system of T-diagrams for α will terminate without w_1 being shown to be inconsistent. In the former case we have shown that in any world in any T-model $V(\alpha)$ could be 0 only if that model contained some 'world' which could not be a world in a T-model at all, since it violates the conditions for V's being a T-assignment; and hence that in every world in every T-model, $V(\alpha) = 1$, i.e. that α is T-valid. And in the latter case, we can use the T-diagram to construct a falsifying T-model for α, and hence α is not T-valid.

The completeness of T [52]

We shall now prove that T is (weakly) complete; i.e., that every T-valid formula is a thesis of the axiomatic system T. We shall do this by showing that whenever a wff, α, is found to be valid by the method of semantic T-diagrams we can construct a proof of α in the axiomatic system.

To simplify what follows we shall assume that all modal

[52] A completeness proof based on semantic tableaux which is in many respects like ours is given in Kripke [1963a]. For proofs of a different kind, *vide* Chapters 9 and 17.

operators other than L have been eliminated by their definitions and that the decision-procedure has been applied as though we had no rules for M. This means that in each rectangle 0 will be initially assigned to exactly one formula, and 1 to all the others.

We confine ourselves to begin with to cases where the diagram contains no †-operators. We shall show later on how to deal with diagrams which do contain these.

Briefly, our proof will take the following form. We shall show how to construct, for each rectangle, w_j, a certain formula which we shall call $\mathbf{w_j}'$. The rules for constructing these formulae will ensure that $\mathbf{w_1}'$ is α (the initial formula in w_1) itself. We shall then prove two lemmas:

LEMMA 1

For any rectangle, w_j, if w_j is explicitly inconsistent then $\vdash_T \mathbf{w_j}'$[53].

LEMMA 2

For any rectangle, w_j, which is accessible to a rectangle, w_i, if $\vdash_T \mathbf{w_j}'$ then $\vdash_T \mathbf{w_i}'$.

Since w_1 can always be reached from any w_i in a finite number of steps, it will follow from these lemmas that if any rectangle in the T-diagram is explicitly inconsistent, then $\vdash_T \mathbf{w_1}'$ – i.e. $\vdash_T \alpha$.

Each $\mathbf{w_j}'$ is the disjunction of the formula in w_j whose initial assignment is 0 and the negations of all the formulae in w_j whose initial assignment is 1. Thus if w_j (with initial assignments) is:

$Lp \supset \sim Lq$	$p \supset L(r \supset q)$	r
0	1	1

then $\mathbf{w_j}'$ is:

$$(Lp \supset \sim Lq) \vee \sim(p \supset L(r \supset q)) \vee \sim r$$

It should be clear that assigning 0 to $\mathbf{w_j}'$ will result (by [V∨] and [V∼]) in giving precisely their initial assignments to all the

[53] Where α is any wff, $\vdash_T \alpha$ means that α is a thesis of T. We shall use other subscripts to $\vdash$ analogously.

formulae in w_j; and hence that we should have obtained exactly the same value-assignments to all the well-formed parts of the formulae in w_j if we had simply had in w_j the single formula $\mathbf{w_j}'$ and had assigned 0 to it. It should also be clear that since w_1 contains only the single formula α and α is assigned 0 in w_1 then $\mathbf{w_1}'$ is α itself.

As a step towards proving Lemma 1 we state and prove:

LEMMA 3

For any rectangle, w_j, and for any well-formed part β of any formula in w_j, if β is assigned 0 (initially or consequentially) in w_j then $\vdash_T(\beta \supset \mathbf{w_j}')$, and if β is assigned 1 then $\vdash_T({\sim}\beta \supset \mathbf{w_j}')$.

PROOF OF LEMMA 3

As we remarked above we can regard $\mathbf{w_j}'$ as the sole formula in w_j, with the initial assignment 0. Now where $\gamma_1, \ldots, \gamma_k$ are all the well-formed parts of formulae in w_j for which both they and $L\gamma_1, \ldots, L\gamma_k$ have been assigned a value we can show that:

(i) If β is assigned 0 then

$$((L\gamma_1 \supset \gamma_1) \cdot \ldots \cdot (L\gamma_k \supset \gamma_k)) \supset (\beta \supset \mathbf{w_j}')$$

is PC-valid (i.e. is a substitution-instance of a PC-tautology); and

(ii) If β is assigned 1 then

$$((L\gamma_1 \supset \gamma_1) \cdot \ldots \cdot (L\gamma_k \supset \gamma_k)) \supset ({\sim}\beta \supset \mathbf{w_j}')$$

is PC-valid.

We prove this as follows:

(i) In working out the consequential values in w_j, beginning from the assignment of 0 to $\mathbf{w_j}'$ and ending with the assignment of 0 to β, we showed that it would violate the rules of a *PC-assignment* if we were to give the value 0 to $\mathbf{w_j}'$, the value 1 to γ whenever we give 1 to $L\gamma$, the value 0 to $L\gamma$ whenever we give the value 0 to γ, and yet give the value 1 to β. But it is easy to check that this is the only kind of PC-assignment which could falsify

$$((L\gamma_1 \supset \gamma_1) \cdot \ldots \cdot (L\gamma_k \supset \gamma_k)) \supset (\beta \supset \mathbf{w_j}')$$

Hence since the only kind of PC-assignment which could falsify it has been shown to be impossible, the formula must be PC-valid.

(ii) The proof here is the same, except that we should have to be able to give the value 0 to β if the formula is to be falsified and that *this* is what is shown to violate the rules of a PC-assignment.

Now since T contains a basis which is known to be complete for PC, every substitution-instance of a PC-tautology is a theorem of T. Hence if β is assigned 0 in w_j, we have:

$$\vdash_T((L\gamma_1 \supset \gamma_1) \ldots . (L\gamma_k \supset \gamma_k)) \supset (\beta \supset \mathbf{w_j}')$$

and from this, by A5, Adjunction and MP, we obtain:

$$\vdash_T(\beta \supset \mathbf{w_j}')$$

Similarly, if β is assigned 1 in w_j, we have:

$$\vdash_T(\sim\beta \supset \mathbf{w_j}')$$ **Q.E.D.**

We are now in a position to prove Lemma 1, viz. that for any rectangle, w_j, if w_j is explicitly inconsistent then $\vdash_T \mathbf{w_j}'$.

PROOF OF LEMMA 1

If w_j is an explicitly inconsistent rectangle, then there is some well-formed part, β, of some formula in w_j such that β is assigned both 1 and 0 in w_j. Hence by Lemma 3, $\vdash_T(\sim\beta \supset \mathbf{w_j}')$ and $\vdash_T(\beta \supset \mathbf{w_j}')$; and hence, by $(\sim p \supset q) \supset ((p \supset q) \supset q)$ and MP, $\vdash_T \mathbf{w_j}'$.

We now prove Lemma 2, viz. that for any rectangle, w_j, which is accessible to a rectangle, w_i, if $\vdash_T \mathbf{w_j}'$ then $\vdash_T \mathbf{w_i}'$.

PROOF OF LEMMA 2

If w_i and w_j are the same rectangle the Lemma holds trivially.

If we have constructed a rectangle w_j from w_i according to the rules of the decision procedure, then the initial formulae of w_j will be the following:

(a) A set of formulae, $\beta_1, \ldots, \beta_k$, each assigned 1 in w_j and being all the formulae such that $L\beta_1, \ldots, L\beta_k$ are assigned 1 in w_i ($k \geqslant 0$, for there may be no such formulae in w_i);

(b) A formula, γ, assigned 0 in w_j, where $L\gamma$ is a formula assigned 0 in w_i[54].

Thus $\mathbf{w_j}'$ will be the formula $(\sim\beta_1 \vee \ldots \vee \sim\beta_k \vee \gamma)$, and so (by the hypothesis of the Lemma) we have:

$$\vdash_T(\sim\beta_1 \vee \ldots \vee \sim\beta_k \vee \gamma)$$

Hence by the rule of Necessitation:

$$\vdash_T L(\sim\beta_1 \vee \ldots \vee \sim\beta_k \vee \gamma)$$

i.e., by Def $\supset$:

$$\vdash_T L(\beta_1 \supset (\ldots \supset (\beta_k \supset \gamma) \ldots)$$

whence by repeated applications of A6:

$$\vdash_T L\beta_1 \supset (\ldots \supset (L\beta_k \supset L\gamma) \ldots)$$

i.e., by Def $\supset$:

$$(1) \quad \vdash_T(\sim L\beta_1 \vee \ldots \vee \sim L\beta_k \vee L\gamma)$$

Now in w_i each of $L\beta_1, \ldots, L\beta_k$ is assigned 1 and $L\gamma$ is assigned 0. Hence (by Lemma 3) we have:

$$\vdash_T(\sim L\beta_1 \supset \mathbf{w_i}'), \ldots, \vdash_T(\sim L\beta_k \supset \mathbf{w_i}'), \text{ and } \vdash_T(L\gamma \supset \mathbf{w_i}')$$

and hence (by PC):

$$\vdash_T((\sim L\beta_1 \vee \ldots \vee \sim L\beta_k \vee L\gamma) \supset \mathbf{w_i}')$$

which, together with (1), gives us by MP:

$$\vdash_T \mathbf{w_i}'$$

(If $k = 0$, we simply establish $\vdash_T L\gamma$ by N, and use Lemma 3 as before.)

This completes the proof of Lemma 2.

Lemmas 1 and 2 suffice to show that if α is T-valid then $\vdash_T\alpha$, except in the cases in which the T-diagram for α contains †-operators. We now show how to extend the proof to cover these cases.

It will simplify matters if we suppose that all the formulae in a rectangle, w_i, are written in primitive notation. Then the only case in which a † can occur below an operator in w_i is when we

54 As we mentioned earlier, we are assuming that M has been eliminated.

have a 1 under $(\beta \vee \gamma)$ but do not have determinate values for β and γ. Our procedure in this case is to construct three alternative diagrams each beginning with one of the rectangles $w_i(i)$, $w_i(ii)$, $w_i(iii)$. These rectangles reproduce the formulae and values in w_i, but in addition 1 is assigned both to β and to γ in $w_i(i)$, 1 is assigned to β and 0 to γ in $w_i(ii)$, and 0 is assigned to β and 1 to γ in $w_i(iii)$. These assignments to β and γ in the alternatives count as initial assignments in them, and so $\mathbf{w_i(i)}'$ is $(\mathbf{w_i}' \vee {\sim}\beta \vee {\sim}\gamma)$, $\mathbf{w_i(ii)}'$ is $(\mathbf{w_i}' \vee {\sim}\beta \vee \gamma)$, and $\mathbf{w_i(iii)}'$ is $(\mathbf{w_i}' \vee \beta \vee {\sim}\gamma)$. Now a rectangle containing a † counts as inconsistent if all its alternatives are inconsistent; so what we have to prove here is that if all of $\mathbf{w_i(i)}'$, $\mathbf{w_i(ii)}'$ and $\mathbf{w_i(iii)}'$ are theses of T, so is $\mathbf{w_i}'$.

LEMMA 4

$$\vdash_T \mathbf{w_i(i)}',\ \vdash_T \mathbf{w_i(ii)}',\ \vdash_T \mathbf{w_i(iii)}' \rightarrow \vdash_T \mathbf{w_i}'$$

PROOF

$(\beta \vee \gamma)$ is assigned 1 in w_i; hence by Lemma 3:

$$(1)\ \vdash_T {\sim}(\beta \vee \gamma) \supset \mathbf{w_i}'$$

But, as it is easy to verify, the following is a valid wff of PC, and hence a thesis of T:

$$({\sim}(p \vee q) \supset r) \supset (((r \vee {\sim}p \vee {\sim}q) \,.\, (r \vee {\sim}p \vee q) \,.\, (r \vee p \vee {\sim}q)) \supset r)$$

Hence by substitution:

$$\vdash_T({\sim}(\beta \vee \gamma) \supset \mathbf{w_i}') \supset (((\mathbf{w_i}' \vee {\sim}\beta \vee {\sim}\gamma) \,.\, (\mathbf{w_i}' \vee {\sim}\beta \vee \gamma) \,.\, (\mathbf{w_i}' \vee \beta \vee {\sim}\gamma)) \supset \mathbf{w_i}')$$

i.e. $$\vdash_T({\sim}(\beta \vee \gamma) \supset \mathbf{w_i}') \supset ((\mathbf{w_i(i)}' \,.\, \mathbf{w_i(ii)}' \,.\, \mathbf{w_i(iii)}') \supset \mathbf{w_i}')$$

From this and (1), by MP, we obtain:

$$\vdash_T(\mathbf{w_i(i)}' \,.\, \mathbf{w_i(ii)}' \,.\, \mathbf{w_i(iii)}') \supset \mathbf{w_i}'$$

But by the hypothesis of the Lemma, $\vdash_T \mathbf{w_i(i)}'$, $\vdash_T \mathbf{w_i(ii)}'$ and $\vdash_T \mathbf{w_i(iii)}'$. Hence by Adjunction and MP:

$$\vdash_T \mathbf{w_i}' \qquad \textbf{Q.E.D.}$$

We have thus shown that whenever a wff, α, is T-valid we can use its T-diagram to construct a proof of α in the axiomatic system T; i.e., that T is complete. Proofs constructed by this method will usually be very clumsy ones, but the point is that they always *can* be constructed.

As an illustration (both of the method of converting diagrams into deductive proofs and of the complexity of the proofs so obtained) take the formula $M(p \supset Lp)$. This becomes $\sim L\sim(p \supset Lp)$ and can be tested as follows:

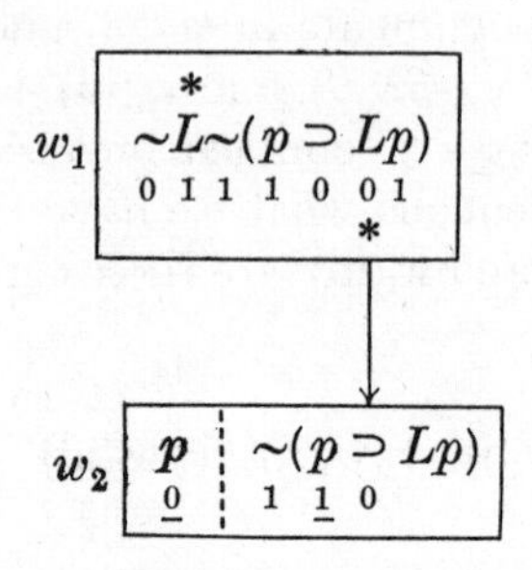

We reach an inconsistency in w_2 and thus the formula is shown to be T-valid. To construct a proof of it by the method just described we proceed as follows:

$\mathbf{w}_1'$ is $\sim L\sim(p \supset Lp)$

$\mathbf{w}_2'$ is $\sim\sim(p \supset Lp) \vee p$

Now (i) w_2 is explicitly inconsistent, so we must show by the method of Lemma 1 that $\vdash_T \mathbf{w}_2'$; and

(ii) w_2 is accessible to w_1, so (given that $\vdash_T \mathbf{w}_2'$) we must show by the method of Lemma 2 that $\vdash_T \mathbf{w}_1'$.

(i) The inconsistency in w_2 arises because p is there assigned both 1 and 0; i.e., p is the β of the proof of Lemma 1. We have therefore to show by the method of Lemma 3 that $\vdash_T(\sim p \supset \mathbf{w}_2')$ and that $\vdash_T(p \supset \mathbf{w}_2')$. Now no expression beginning with L is assigned a value in w_2[55] – i.e., no γ of Lemma 3 occurs in w_2 – so what we have to prove here is simply that $(\sim p \supset \mathbf{w}_2')$ and $(p \supset \mathbf{w}_2')$, i.e.

(1) $\sim p \supset (\sim\sim(p \supset Lp) \vee p)$

and (2) $p \supset (\sim\sim(p \supset Lp) \vee p)$

[55] We *could* have assigned a value (0) to Lp, but the inconsistency arises without our having to do so.

are PC-valid. That they are can be easily shown by applying truth-tables or some other method. (In doing so we of course treat Lp as if it had no connection with p; e.g., as if it were q.) Hence, since every PC-valid formula is a theorem of T), $\vdash_T(1)$ and $\vdash_T(2)$. Whence, by $(\sim p \supset q) \supset ((p \supset q) \supset q)$:

$$\vdash_T(\sim\sim(p \supset Lp) \vee p)$$

i.e.
$$\vdash_T \mathrm{w}_2'$$

(ii) We have just proved that $\vdash_T(\sim\sim(p \supset Lp) \vee p)$. Hence (following the method of Lemma 2), we have by N:

$$\vdash_T L(\sim\sim(p \supset Lp) \vee p)$$

i.e., by Def $\supset$: $\quad \vdash_T L(\sim(p \supset Lp) \supset p)$

whence, by A6: $\quad \vdash_T L\sim(p \supset Lp) \supset Lp$

i.e., by Def $\supset$: $\quad$ (3) $\vdash_T \sim L\sim(p \supset Lp) \vee Lp$

Now w_2 was constructed because in w_1 $L\sim(p \supset Lp)$ is assigned 1 and Lp is assigned 0. And in w_1 the γ's of Lemma 3 are $\sim(p \supset Lp)$ and p. Hence we should find that the following formulae are PC-valid:

(4) $((L\sim(p \supset Lp) \supset \sim(p \supset Lp)) \, . \, (Lp \supset p)) \supset$
$(\sim L\sim(p \supset Lp) \supset \sim L\sim(p \supset Lp))$

(5) $((L\sim(p \supset Lp) \supset \sim(p \supset Lp)) \, . \, (Lp \supset p)) \supset$
$(Lp \supset \sim L\sim(p \supset Lp))$

Again this can be checked by standard PC tests. ((4) is valid because its consequent is a substitution-instance of $p \supset p$. (5) is a substitution-instance of

$$((p \supset \sim(q \supset r)) \, . \, (r \supset q)) \supset (r \supset \sim p)$$

1 1 1 1 0 0 1 1 0 1 0 0 1

(we have written in the values which show validity by the Reductio test).)

(4) and (5) are therefore theses of T. Hence by A5, Adjunction and MP we have:

$$\vdash_T \sim L\sim(p \supset Lp) \supset \sim L\sim(p \supset Lp)$$

and
$$\vdash_T Lp \supset \sim L\sim(p \supset Lp)$$

whence by PC and (3):

$$\vdash_T \sim L \sim (p \supset Lp) \qquad [\mathbf{w_1}']$$

i.e. $$\vdash_T M(p \supset Lp)$$ **Q.E.D.**

It will be seen that if this were all written out as a proof in T it would be a very long proof indeed, even though in this case we had no †-operators to consider. Nevertheless it would clearly be a legitimate derivation from the axioms of T; and the completeness result assures us that a proof of this kind could always in theory be constructed for any T-valid formula.

Exercises – 5

5.1. Using the diagrams given on pp. 90 and 92f construct T-models which falsify examples [4] and [5].

5.2. Show that $\vdash \alpha \supset L(\beta \supset \gamma) \rightarrow \vdash \beta \supset L(\alpha \supset \gamma)$ is not validity-preserving in T. (*Hint*: use it to derive a formula which can be shown not to be T-valid.)

5.3. Test for validity in T:

(a) $M(Mp \,.\, \sim q) \vee L(p \supset Lq)$

(b) $(M(p \,.\, q) \vee M(p \,.\, r)) \supset Mp$

(c) $L(p \equiv q) \supset (Lp \equiv Lq)$

(d) $L(p \equiv q) \supset L(Lp \equiv Lq)$

5.4. Construct falsifying T-models for the invalid formulae in the previous exercise.

5.5. Where w_j is accessible to w_i prove in T: $L\mathbf{w_j}' \supset \mathbf{w_i}'$.

5.6. Where $w_i(i)$ and $w_i(ii)$ and $w_i(iii)$ are the alternatives of w_i, prove in T:

$$\mathbf{w_i}' \equiv (\mathbf{w_i(i)}' \vee \mathbf{w_i(ii)}' \vee \mathbf{w_i(iii)}')$$

CHAPTER SIX

S4 and S5: Decision Procedures and Completeness

S4-diagrams [56]

We shall now describe a decision procedure for S4 based, like the one we have given for T, on semantic diagrams. The reader is to assume that the rules of the method of T-diagrams hold throughout, except where we specify to the contrary.

The only difference between our definitions of T-validity and S4-validity is that in an S4-model the relation R must be transitive. Now R is the relation which holds between two worlds w_i and w_j when w_j is accessible to w_i. So the requirement that R be transitive can be expressed as the requirement that for any worlds, w_i, w_j and w_k, if w_j is accessible to w_i and w_k is accessible to w_j, then w_k is accessible to w_i.

Let us apply this to the diagrams. We shall say that in a semantic diagram a series of rectangles $w_i, \ldots, w_n$ form a *chain* if an arrow goes from each (except the last) to the next rectangle in the series. Thus in the diagram on p. 90, w_1, w_2, w_5 form a chain, and so do w_1, w_2, w_6, and so on. To take care of the transitivity requirement, an *S4-diagram* will differ from a T-diagram in the following way: an arrow must go from every rectangle to every other rectangle which occurs later in every chain to which the first belongs. This means that to satisfy rules A and B, whenever in any rectangle we have $L\alpha = 1$ (or $M\beta = 0$) we must now write α with a 1 under it (or β with 0 under it), not only in the next rectangle in the chain but in every subsequent one as well. When a rectangle, w_j, contains a †, then each alternative of w_j is regarded as belonging to the chain to which w_j belongs:

[56] Our procedure is again analogous to the method of semantic tableaux (cf. footnote 48). Other decision procedures for S4 are given in Von Wright [1951], Anderson [1954], (modified in Hanson [1966]), Ohnishi and Matsumoto [1957].

thus if an arrow goes from a rectangle, w_i, to w_j, arrows must be drawn from w_i to each of $w_j(i)$, $w_j(ii)$ and $w_j(iii)$.

Clearly the transitivity requirement will make no difference in the case of a diagram in which no chain is more than two rectangles long. Hence, since the T-diagram for a formula of first modal degree never contains chains more than two rectangles long, we have the result that a first degree formula is S4-valid iff it is T-valid.

When, however, the T-diagram for a formula contains any longer chain this this, there will .be a difference between its T-diagram and its S4-diagram. Consider, e.g., the formula:

$$(1)\ L(p \,.\, q) \supset LL(Mp \supset Mq)$$

Its T-diagram is

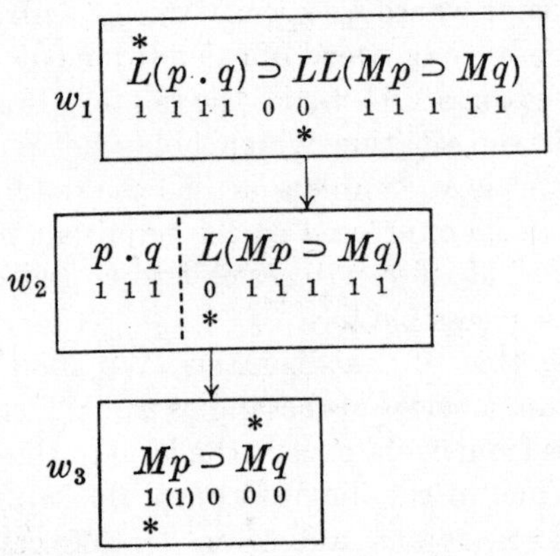

But its S4-diagram is:

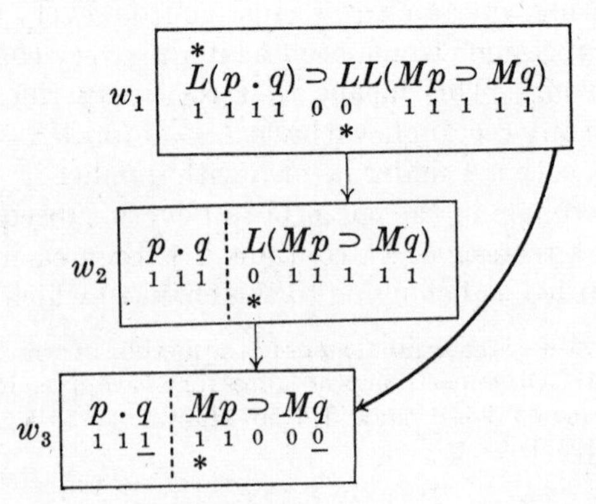

Here the assignment of the value 1 to $L(p \,.\, q)$ in w_1 requires the presence of $p \,.\, q$ (= 1) not merely in w_2 but in w_3 as well. This kind of addition to the contents of rectangles, of course, creates new possibilities of inconsistencies arising in the diagrams. When we find an inconsistency in the S4-diagram of a formula but not in its T-diagram, that formula is S4-valid but not T-valid. (1), in fact, is a case in point, as the diagrams show.

In order to show that the method of semantic diagrams provides a decision procedure for S4, we have to show that for every wff an S4-diagram[57] of finite length can be constructed; or more exactly, that for every wff, α, we can construct an S4-diagram which will in a finite number of steps either (a) show α to be valid (by containing an inconsistency in some rectangle), or (b) enable us to construct a falsifying S4-model for α.

Now it was easy to show that every T-diagram is finite. For in every chain in a T-diagram the modal degree of the rectangles is constantly diminishing. Hence every chain must at worst lead us to a rectangle containing nothing but PC formulae, and such formulae never generate further rectangles.

This does not, however, apply to S4-diagrams. For in an S4-diagram, if rectangle w_1 contains say the assignment of 1 to $L\beta$ then β will appear in every rectangle in the diagram; hence if β is of modal degree n, every rectangle will contain at least one formula of modal degree n. So we do not have the same simple proof as we have for T to show that every S4-diagram is finite. In fact in S4 the following tantalizing situation can arise. Consider the formula:

$$(2) \quad LMp \supset MLp$$

This is not S4-valid[58]; but its S4-diagram (by our present rules) goes like this:

[57] Or a complete system of S4-diagrams; but for simplicity we omit reference to alternatives here, since they cannot affect the finiteness of the procedure for the same reason as we gave in the case of T on p. 93.

[58] Perhaps the easiest way to convince oneself of this is to notice that by the reduction laws of S5, (2) is equivalent *in S5* to $Mp \supset Lp$, which is clearly not S5-valid. Hence (2) is not S5-valid, and if a formula is not S5-valid it cannot be S4-valid either.

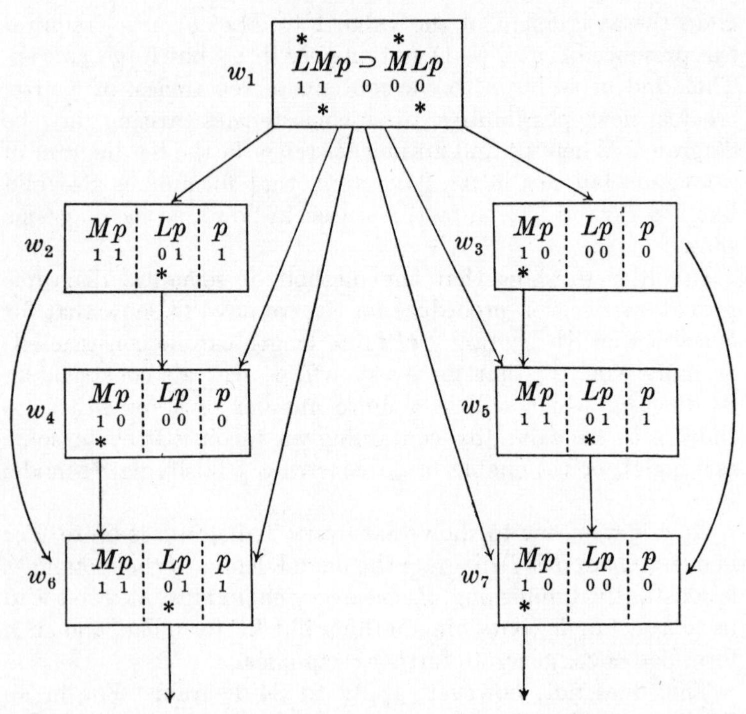

Here rectangle w_6 is needed because of the asterisk in w_4. But an arrow goes from w_1 to w_6 as well as from w_4 to w_6, and as a result the contents of w_6 turn out to be identical with those of w_2; hence we need yet another rectangle below w_6 whose contents will turn out to be the same as those of w_4, and so on for ever. And the same situation obtains on the right hand side of the diagram. Thus a falsifying model for (2) always seems to be within our grasp at the next step, but once we take that step it seems to be one step further on still. Yet we never strike an inconsistency in the diagram either [59].

Clearly this diagram is not giving us a decision for (2). A simple modification of it, however, will do so: we delete rectangle w_6 altogether, and run the arrow from w_4 upwards to w_2 instead;

[59] Adapting a phrase from Kripke [1963a] (p. 71), we might call diagrams constructed in accordance with our present rules 'tree' diagrams. Thus $LMp \supset MLp$ cannot be falsified in a finite tree diagram in S4. In T, however, every invalid formula can be falsified in a finite tree diagram.

and we treat the other side of the diagram in the same way. We then have a five-world falsifying model for (2), as we can easily check by a truth-table. The diagram will be this:

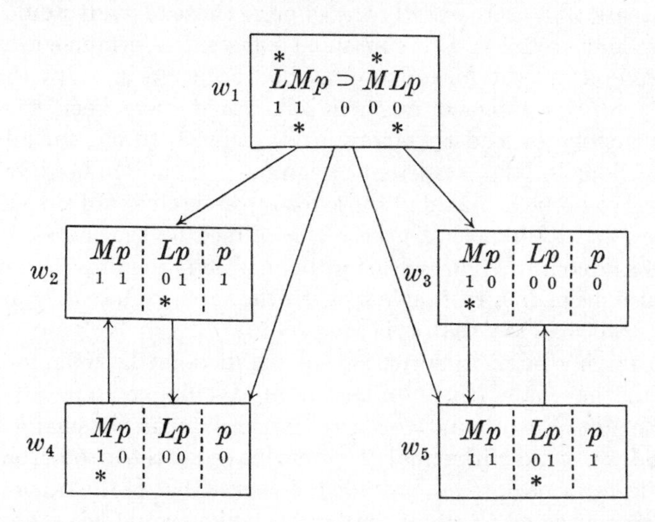

That this diagram fulfils the conditions which previously looked as if they would lead us to an infinite diagram, can be seen as follows:

1. We needed a world w_6 (accessible to w_4) in the first diagram to enable the initial values to be consistently assigned to the formulae in w_4. Since the formulae in w_2 and the values assigned to them there are exactly the same as those in w_6, making w_2 accessible to w_4 is equally satisfactory.

2. The conditions for the assignment of the initial values in w_6 were that it in turn should be succeeded by a further world in which certain value-assignments should obtain. But since w_6 is identical with w_2, these are precisely the conditions for the initial value-assignments in w_2, and we have already provided for their fulfilment in making w_4 accessible to w_2.

In short, instead of the endless chain, $w_1, w_2, w_4, w_6, \ldots$, we have w_1 followed by w_2 and w_4 in endless alternation; and for this we only need three worlds. Exactly the same considerations apply to the right hand side of the diagram [60].

[60] In the case of the present example we can in fact do better than this. Since w_3 and w_4 are identical, and w_2 and w_5 are too, we could abandon

We now show how to generalize this procedure to avoid infinite diagrams in all cases.

We note first of all that although in the above example the contents of w_2 were exactly the same as those of w_6, it would not have mattered if w_2 had contained some extra formulae as well. So long as all the formulae in w_6 had occurred in w_2 (with the same values assigned to them), it would have been equally satisfactory to lead the arrow back from w_4 to w_2; for all the conditions for the consistent assignment of the required values in w_6 would be included in those for the assignment of the values in w_2, and by hypothesis these are fulfilled by the successors of w_2 in the chain. When all the formulae which occur in a rectangle, w_j also occur in a rectangle w_i with the same values assigned to them we shall say that w_j is *contained in* w_i.

A further point to notice is that the distance between w_2 and w_6 in the chain (the number of intervening rectangles) was irrelevant. Even had w_6 occurred much later in the chain than it did, we could with equal propriety have led the arrow from its predecessor back to w_2, provided of course that at the same time we also directed to w_2 all the arrows which would have gone to w_6.

We now state the following additional rule for S4-diagrams.

w_4 and w_5 altogether and instead draw arrows from w_2 to w_3, and from w_3 to w_2. We then obtain a three-world falsifying model for (2). Indeed by using 'optional' values in w_1 we can do better still and produce the following diagram which gives a two-world falsifying model:

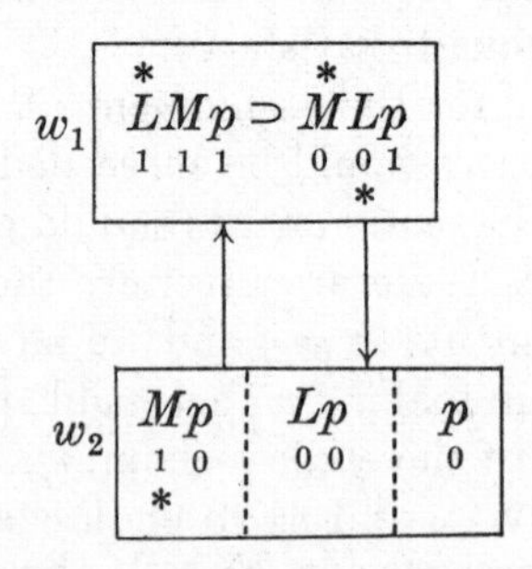

(Note that this, as will appear later, is an S5-diagram as well as an S4-diagram, and shows that (2) is invalid in S5, not only in S4.) But these possibilities depend on special features of (2) and we cannot generalize from them.

Rule of repeating chains.

Whenever in any chain in an S4-diagram a rectangle, w_j, is contained in a rectangle, w_i, which occurs *earlier* in that chain, we delete w_j and lead every arrow which would have gone to w_j to w_i instead.

We shall call a chain to which we have applied this rule, a *repeating* chain.

Observing the rule of repeating chains will guarantee that every chain in an S4-diagram is of finite length, and hence that every diagram contains only a finite number of rectangles, for the following reason. It is clear from the rules for constructing the diagrams that in the diagram for a wff, α, every formula which occurs in any rectangle must be a well-formed part of α itself. Now α has only a finite number of well-formed parts; and hence there can be only a finite number of sets of formulae selected from these, and of course only a finite number of ways of assigning values to the formulae in any such set. So while the contents of the rectangles in a chain can vary a great deal, they cannot vary *indefinitely*; therefore in an infinite chain we must sooner or later come across a rectangle which is contained in an earlier rectangle, and to which we can therefore apply the rule of repeating chains. Once we have done so, of course, the chain will only contain a finite number of rectangles.

Each chain in an S4-diagram, then, is finite. Now an S4-diagram consists of a set of chains each beginning with w_1. Each rectangle (apart from w_1) is generated, in accordance with rules C and D on p. 86, by a modal operator below which an asterisk appears in the immediately preceding rectangle in the chain; and since each rectangle only contains a finite number of modal operators, there can only be a finite number of chains in any S4-diagram. Hence every S4-diagram contains a finite number of rectangles. The presence of †-operators and the consequent construction of alternatives cannot affect this result, for the same reason as in the case of T.

The one remaining modification we have to make in the method of T-diagrams concerns the definition of an inconsistent rectangle. We said on p. 95 that any rectangle to which an inconsistent rectangle is accessible is itself inconsistent. In a T-diagram this means that a rectangle is inconsistent if a rectangle immediately below it (i.e., connected by an arrow) is inconsistent. But of course in an S4-diagram arrows go not only to the rectangles immediately below but to the ones below *them* and so on; and indeed arrows even go to alternatives (which they do not in a T-diagram). So we shall replace clause 2 of the definition of an inconsistent rectangle on p. 95 by:

2′. Any rectangle which is the immediate predecessor of an inconsistent rectangle is an inconsistent rectangle.

We define the *immediate predecessor* of a rectangle as follows. If the construction of a rectangle, w_j, accessible to but distinct from w_i, is required by rule C or rule D (p. 86) in order to satisfy the value-assignments in w_i, then w_i is the immediate predecessor of w_j. (Note that in a T-diagram, constructed without short cuts, whenever w_j is accessible to w_i, w_i is the immediate predecessor of w_j; but this of course does not hold for S4-diagrams.)

A wff will be S4-valid iff the first rectangle in its complete system of S4-rectangles, w_1, is inconsistent.

The completeness of S4

To prove the completeness of S4 we have to show that every wff which is S4-valid is a thesis of the axiomatic system S4. Our proof will follow the general pattern of the completeness proof we gave for T in the previous chapter. As we did there, we shall first of all consider only cases where no †-operators occur. And we shall again assume that M has been replaced everywhere by $\sim L \sim$. Every rectangle will then contain exactly one formula, γ, initially assigned 0, and a set of formulae, $\beta_1, \ldots, \beta_k (k \geqslant 0)$, each initially assigned 1 (cf. p. 97).

In proving the completeness of T we defined for each rectangle, w_j, a formula, $\mathbf{w_j}'$, which was:

$$\sim\beta_1 \vee \ldots \vee \sim\beta_k \vee \gamma$$

and we proved (i) that if w_j is explicitly inconsistent, then $\vdash_T \mathbf{w_j}'$, and (ii) that if w_j is accessible to w_i, then $\vdash_T \mathbf{w_j}' \rightarrow \vdash_T \mathbf{w_i}'$ (Lemmas 1 and 2, p. 97). To prove the completeness of S4 we define, for each rectangle, w_j, in an S4-diagram a formula, $\mathbf{w_j}''$, which is

$$\sim L\beta_1 \vee \ldots \vee \sim L\beta_k \vee \gamma$$

where $\beta_1, \ldots, \beta_k$ and γ are as before; and we now state two corresponding lemmas for S4:

LEMMA 5

If any rectangle, w_j, is explicitly inconsistent, then $\vdash_{S4} \mathbf{w_j}''$.

LEMMA 6

If any rectangle, w_i, is the immediate predecessor of a rectangle, w_j, then if $\vdash_{S4}\mathbf{w_j}''$, $\vdash_{S4}\mathbf{w_i}''$.

Since w_1 can always be reached from any w_j in a finite number of steps, it will follow from these two lemmas that if any rectangle in an S4-diagram is explicitly inconsistent, then $\vdash_{S4}\mathbf{w_1}''$; and since α is the only formula in w_1 and is assigned 0 there, $\mathbf{w_1}''$ is α itself.

As a step towards proving Lemmas 5 and 6 we first state:

LEMMA 7

For any rectangle, w_j, $\vdash_{S4}\mathbf{w_j}' \supset \mathbf{w_j}''$.

The proof is by $\sim p \supset \sim Lp$ (a simple derivative of A5) and PC.

We also note that Lemmas 1 and 3 (pp. 97, 98) hold for S4 – i.e. they hold when for '$\vdash_T$' we read '$\vdash_{S4}$' – since their proofs depend only on features which T-diagrams and S4-diagrams have in common, and every thesis of T is also a thesis of S4[60a].

Lemma 5 can now be seen to be an immediate consequence of Lemmas 1 and 7 by MP. Lemma 6 will, however, require a more complicated proof.

PROOF OF LEMMA 6

Of the formulae placed in w_j and initially assigned values there, one (viz. γ) will be assigned 0 in w_j because $L\gamma$ is assigned 0 (initially or consequentially) in w_i, and some others, say $\beta_1, \ldots, \beta_h$ ($k \geqslant h \geqslant 0$) will be assigned 1 in w_j because $L\beta_1, \ldots, L\beta_h$ are assigned 1 (initially or consequentially) in w_i. Hence by Lemma 3, as on p. 100:

$$\vdash_{S4}(\sim L\beta_1 \vee \ldots \vee \sim L\beta_h \vee L\gamma) \supset \mathbf{w_i}'$$

and hence by Lemma 7 and Syll:

$$(1)\ \vdash_{S4}(\sim L\beta_1 \vee \ldots \vee \sim L\beta_h \vee L\gamma) \supset \mathbf{w_i}''$$

[60a] Lemma 2, however, does not hold for S4. Its proof assumes that whenever β is initially assigned 1 in w_j, $L\beta$ is assigned 1 in the immediate predecessor of w_j, viz. w_i; and this is not necessarily so in an S4-diagram since, by the transitivity requirement, β may be in w_j because $L\beta$ is assigned 1 in some predecessor of w_i, though not in w_i itself. It is this fact which makes it more difficult to prove a corresponding lemma for S4, and which requires us to formulate such a lemma (viz. Lemma 6) in terms of a more complicated kind of formula ($\mathbf{w_j}''$) than the kind ($\mathbf{w_j}'$) which sufficed in the completeness proof for T.

Any remaining formulae in w_j will be some formulae $\beta_{h+1}, \ldots,$ β_k, each assigned 1 in w_j because $L\beta_{h+1}, \ldots, L\beta_k$ have each been assigned 1 in some rectangle further up the chain than w_i. In this case, by Rule A (p. 86), $\beta_{h+1}, \ldots, \beta_k$ will all have been *initially* assigned 1 in w_i (as well as in w_j), and so $\sim L\beta_{h+1}, \ldots, \sim L\beta_k$ will be among the disjuncts in $\mathbf{w_i}''$. Hence by p ⊃ (p ∨ q):

$$(2)\ \vdash_{S4}(\sim L\beta_{h+1} \vee \ldots \vee \sim L\beta_k) \supset \mathbf{w_i}''$$

From (1) and (2) we have (by PC):

$$\vdash_{S4}(\sim L\beta_1 \vee \ldots \vee \sim L\beta_h \vee \sim L\beta_{h+1} \vee \ldots \vee \sim L\beta_k \vee L\gamma) \supset \mathbf{w_i}''$$

i.e. $$(3)\ \vdash_{S4}(\sim L\beta_1 \vee \ldots \vee \sim L\beta_k \vee L\gamma) \supset \mathbf{w_i}''$$

Now by hypothesis, $\vdash_{S4}\mathbf{w_j}''$, i.e.

$$\vdash_{S4}(\sim L\beta_1 \vee \ldots \vee \sim L\beta_k \vee \gamma)$$

From this, by N, Def ⊃ and A6, as on p. 100, we have:

$$\vdash_{S4}(\sim LL\beta_1 \vee \ldots \vee \sim LL\beta_k \vee L\gamma)$$

whence by R4 ($Lp \equiv LLp$):

$$\vdash_{S4}(\sim L\beta_1 \vee \ldots \vee \sim L\beta_k \vee L\gamma)$$

whence by (3) and MP:

$$\vdash_{S4}\mathbf{w_i}''$$ **Q.E.D.**

We have finally to deal with †-operators. Where w_i, $w_i(i)$, $w_i(ii)$, and $w_i(iii)$ are as on pp. 100f, we define $\mathbf{w_i(i)}''$, $\mathbf{w_i(ii)}''$ and $\mathbf{w_i(iii)}''$ respectively as $(\mathbf{w_i}'' \vee \sim\beta \vee \sim\gamma)$, $(\mathbf{w_i}'' \vee \sim\beta \vee \gamma)$ and $(\mathbf{w_i}'' \vee \beta \vee \sim\gamma)$. Simple modifications of the proofs will show that for these definitions Lemmas 5 and 6 still hold. It remains to prove an analogue of Lemma 4 (p. 101), viz.

$$\vdash_{S4}\mathbf{w_i(i)}'',\quad \vdash_{S4}\mathbf{w_i(ii)}'',\quad \vdash_{S4}\mathbf{w_i(iii)}'' \rightarrow \vdash_{S4}\mathbf{w_i}''$$

The proof of this is the same as for Lemma 4 except that (a) we replace the formula numbered (1) by

$$\vdash_{S4}\sim(\beta \vee \gamma) \supset \mathbf{w_i}''$$

which is a simple consequence of (1) and Lemma 7, and (b) in the remainder of the proof we replace $\mathbf{w_i}'$ throughout by $\mathbf{w_i}''$ and consequently $\mathbf{w_i}(\mathbf{i})'$, $\mathbf{w_i}(\mathbf{ii})'$ and $\mathbf{w_i}(\mathbf{iii})'$ by $\mathbf{w_i}(\mathbf{i})''$, $\mathbf{w_i}(\mathbf{ii})''$ and $\mathbf{w_i}(\mathbf{iii})''$ respectively.

We can, therefore, as in the case of T, use the S4-diagram of any S4-valid formula to construct a proof of that formula in the axiomatic system S4.

S5-diagrams

The method of diagrams can be extended to provide a Decision Procedure for S5. In an S5-model every world stands in the relation R to every other world[61]. The extra rule that has to be observed in constructing an S5-diagram is therefore that an arrow must go from every rectangle to every other one. This means that whenever we add a new rectangle in constructing an S5-diagram we must draw an arrow from it to every rectangle already in the diagram, as well as from all other rectangles to it, and then enter in these rectangles any formulae which the new arrows make necessary. In this way the possibilities of inconsistencies arising in rectangles are increased – as, of course, we should expect, since a formula can be S5-valid without being S4-valid.

As an illustration take the following formula (the immediate values are written in):

$$\underset{1}{MMp} \underset{0}{\supset} \underset{0}{LMp}$$

The S5-diagram is given below. For simplicity we write the two arrows going from w_1 to w_2 and from w_2 to w_1 respectively as a single double-headed arrow, and analogously in the other cases.

Here rectangles w_2 (with $Mp = 1$) and w_3 (with $Mp = 0$) are required by the asterisks in w_1 by the ordinary rules for T-diagrams. But the S5 arrows from w_3 to w_1 and w_2 lead to p being assigned 0 in each of these, with the result that we cannot give p the value 1 in w_2 but must have another rectangle w_4 in

[61] Alternatively, R must be symmetrical as well as transitive and reflexive; but we have already (p. 67n) shown the coincidence between this condition and the one stated in the text.

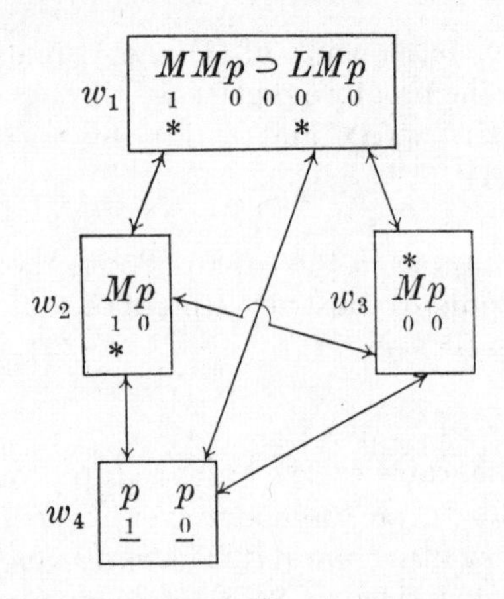

which p shall have the value 1. However an arrow must also go from w_3 to w_4, and the asterisk above the M in w_3 gives the value 0 to p in w_4, and thus an inconsistency there. Hence the formula in w_1 is S5-valid.

It would be possible to set out a completeness proof for S5 based on semantic diagrams, analogous to the ones we have given for T and S4 [62]. In the case of S5, however, it is also possible to give a completeness proof based on reduction to modal conjunctive normal form (MCNF) and this is what we shall do instead. To do so we first show how reduction to MCNF yields a decision procedure for S5.

An alternative decision procedure for S5 [63]

A wff, α, can be tested for S5-validity by the following method:

1. Reduce α to a wff, α', in MCNF by the method explained on p. 55.

2. Within each conjunct, C_i, in α' arrange the disjuncts in the following order: first, all unmodalized disjuncts (PC formulae); next, all disjuncts beginning with L; finally, all disjuncts beginning with M. Since a disjunction of PC formulae is itself a PC formula, each C_i will now have the form:

$$\beta \vee L\gamma_1 \vee \ldots \vee L\gamma_n \vee M\delta_1 \vee \ldots \vee M\delta_m$$

[62] For such a proof *vide*, e.g., Kripke [1963a].

[63] The decision procedure described here is due essentially to Carnap [1946]. A slightly more elaborate method is found in Wajsberg [1933].

where $\beta, \gamma_1, \ldots, \gamma_n, \delta_1, \ldots, \delta_m$ are all PC formulae, $n \geqslant 0, m \geqslant 0$, and it may be that there is no unmodalized disjunct β.

3. Use the law of M-distribution to replace $M\delta_1 \vee \ldots \vee M\delta_m$ by $M(\delta_1 \vee \ldots \vee \delta_m)$. Since $\delta_1, \ldots, \delta_m$ are all PC formulae, their disjunction is a PC formula, and can be referred to simply as δ. Each C_i is therefore now of the form:

$$\beta \vee L\gamma_1 \vee \ldots \vee L\gamma_n \vee M\delta$$

(When each C_i is in this form we shall say that the formula is in *ordered* MCNF.)

4. Test each C_i by the following rule. Form $n + 1$ disjunctions, each of which has δ as one disjunct and a distinct one of β, $\gamma_1, \ldots, \gamma_n$ as the other; i.e. form the $n + 1$ PC formulae, $(\beta \vee \delta)$, $(\gamma_1 \vee \delta), \ldots, (\gamma_n \vee \delta)$[64]. Iff at least one of these is PC-valid, C_i is S5-valid.

5. Iff each C_i in α' is S5-valid, α is S5-valid.

As examples consider the two formulae we reduced to MCNF on p. 56. In MCNF these are:

EXAMPLE 1

$$(Mp \vee L{\sim}p \vee Lp) \,.\, (M{\sim}p \vee L{\sim}p \vee Lp)$$

This will be valid if each conjunct is. Take the first conjunct, $Mp \vee L{\sim}p \vee Lp$. By step 2 this becomes $L{\sim}p \vee Lp \vee Mp$. (Step 3 is inapplicable.) Here we have no β (for all disjuncts are modalized). γ_1 is ${\sim}p$, γ_2 is p and δ is p. Now this conjunct will be valid iff either $(\gamma_1 \vee \delta)$ or $(\gamma_2 \vee \delta)$ is PC-valid; i.e. iff either ${\sim}p \vee p$ or $p \vee p$ is PC-valid. And ${\sim}p \vee p$ *is* PC-valid, hence the whole conjunct is valid.

The second conjunct becomes $L{\sim}p \vee Lp \vee M{\sim}p$ which will be valid iff either ${\sim}p \vee {\sim}p$ or $p \vee {\sim}p$ is valid, and $p \vee {\sim}p$ *is* valid.

Thus both conjuncts are valid and hence the whole formula is valid. Therefore the original formula is valid too (since any formula is equivalent in S5 to its MCNF).

EXAMPLE 2

$$(Mp \vee M(p \,.\, {\sim}q) \vee L({\sim}p \vee q) \vee Mr)$$
$$.\, (L{\sim}r \vee M(p \,.\, {\sim}q) \vee L({\sim}p \vee q) \vee Mr)$$

[64] Of course if C_i contains no unmodalized disjunct, β, there will be only the n disjunctions, $(\gamma_1 \vee \delta), \ldots, (\gamma_n \vee \delta)$, to consider; and if there is no δ we simply test $\beta, \gamma_1, \ldots, \gamma_n$.

By steps 2 and 3 this becomes:

$$(L(\sim p \vee q) \vee M(p \vee (p \,.\, \sim q) \vee r)) \\ .\, (L{\sim}r \vee L(\sim p \vee q) \vee M((p \,.\, \sim q) \vee r))$$

The first conjunct will be valid iff $(\sim p \vee q) \vee (p \vee (p \,.\, \sim q) \vee r)$ is valid (which it is); the second will be valid iff either $\sim r \vee ((p \,.\, \sim q) \vee r)$ or $(\sim p \vee q) \vee ((p \,.\, \sim q) \vee r)$ is valid, and in fact both are valid. So once more the original formula is valid.

Neither of these examples has any unmodalized formulae; but the following (assumed already reduced to MCNF) does:

EXAMPLE 3

$$(Lq \vee M{\sim}p \vee r \vee L{\sim}(p \,.\, r) \vee ((p \,.\, q) \supset r)) \,.\, (Mp \vee L{\sim}p)$$

By step 2 (step 3 is inapplicable) the first conjunct becomes:

$$(r \vee ((p \,.\, q) \supset r)) \vee Lq \vee L{\sim}(p \,.\, r) \vee M{\sim}p$$

Here β is $(r \vee ((p \,.\, q) \supset r))$, γ_1 is q, γ_2 is $\sim(p \,.\, r)$, and δ is $\sim p$.

This conjunct will be valid iff one of $(\beta \vee \delta)$, $(\gamma_1 \vee \delta)$, $(\gamma_2 \vee \delta)$ is.

I.e. iff $\quad r \vee ((p \,.\, q) \supset r) \vee \sim p$

or $\quad q \vee \sim p$

or $\quad \sim(p \,.\, r) \vee \sim p$

is valid, and none of them are. So the conjunct is not valid and hence the whole formula is not valid (we do not need to test the other conjunct). Therefore if Example 3 is the result of reducing a formula to MCNF, the original formula is not valid.

We have now to show that the procedure we have described is an effective test for S5-validity.

We have already shown on pp. 55f that there is an effective procedure for finding, for every wff, α, a wff, α', which is in ordered MCNF and is such that $\vdash_{S5}(\alpha \equiv \alpha')$. (The equivalences used in steps 2 and 3 of the decision procedure are among those used in Chapter 3 for reduction to MCNF.) We have also shown on p. 69 that every thesis of S5 is S5-valid. Hence $(\alpha \equiv \alpha')$ is S5-valid, from which it follows that α is S5-valid iff α' is. Now a conjunction is valid iff each of its conjuncts is valid, and each conjunct in α' is of the form:

$$\beta \vee L\gamma_1 \vee \ldots \vee L\gamma_n \vee M\delta$$

What remains therefore is to prove that the test described in step 4 of the decision procedure is an effective test of S5-validity for formulae of this form.

(a) We prove that if any of $(\beta \vee \delta)$, $(\gamma_1 \vee \delta)$, . . . , $(\gamma_n \vee \delta)$ is PC-valid, then

$$(1)\ \beta \vee L\gamma_1 \vee \ldots \vee L\gamma_n \vee M\delta$$

is S5-valid.

Suppose that $(\beta \vee \delta)$ is PC-valid. Then since S5 contains a complete basis for PC, $\vdash_{S5}(\beta \vee \delta)$. By T1, $\vdash_{S5}(\delta \supset M\delta)$. Hence by $(q \supset r) \supset ((p \vee q) \supset (p \vee r))$ and MP, $\vdash_{S5}(\beta \vee M\delta)$; and hence by $p \supset (q \vee p)$ and Comm, $\vdash_{S5}(1)$. But every thesis of S5 is S5-valid, therefore (1) is S5-valid.

Suppose now that one of $(\gamma_1 \vee \delta)$, . . . , $(\gamma_n \vee \delta)$, say $(\gamma_j \vee \delta)$, is PC-valid. Then as before, $\vdash_{S5}(\gamma_j \vee \delta)$, and by N, $\vdash_{S5}L(\gamma_j \vee \delta)$. Hence by T28 (p. 51), $\vdash_{S5}(L\gamma_j \vee M\delta)$. From this it follows as before that $\vdash_{S5}(1)$, and therefore that (1) is S5-valid.

(b) We prove that if none of $(\beta \vee \delta)$, $(\gamma_1 \vee \delta)$, . . . , $(\gamma_n \vee \delta)$ is PC-valid, then (1) is not S5-valid.

To do so we construct the following S5-model $\langle W, V \rangle$. $W = \{w_0, w_1, \ldots, w_n\}$. I.e., there are exactly $n+1$ worlds in the model. With w_0 we associate $(\beta \vee \delta)$, and with each of $w_1, \ldots, w_n$ we associate a distinct one of $(\gamma_1 \vee \delta)$, . . . , $(\gamma_n \vee \delta)$, as indicated by the subscript to the γ. We define V as follows. In w_0, V makes some value-assignment to the variables which will give $V((\beta \vee \delta), w_0) = 0$; and in each w_j among $w_1, \ldots, w_n$, V makes some assignment to the variables which will give $V((\gamma_j \vee \delta), w_j) = 0$. (Since each of $(\beta \vee \delta)$, $(\gamma_1 \vee \delta)$, . . . , $(\gamma_n \vee \delta)$ is by hypothesis an invalid PC formula, there will be for each of them a *PC-assignment* which falsifies it. We simply let the V in our S5-model give in w_0 the values given by the particular PC-assignment which falsifies $(\beta \vee \delta)$, and in each w_j the values given by the particular PC-assignment which falsifies $(\gamma_j \vee \delta)$.)

We now show that in such a model, $V((1), w_0) = 0$, and therefore that (1) is not S5-valid.

By the definition of V, $V((\beta \vee \delta), w_0) = 0$. Hence by [Vv], $V(\beta, w_0) = 0$ and $V(\delta, w_0) = 0$. Similarly, for each w_j among $w_1, \ldots, w_n$, $V(\gamma_j, w_j) = 0$ and $V(\delta, w_j) = 0$. Thus for every $w_i \in W$, $V(\delta, w_i) = 0$, and hence by [V*M*], $V(M\delta, w_0) = 0$. Moreover, for each γ_j among $\gamma_1, \ldots, \gamma_n$, there is some $w_i \in W$ (viz. w_j) such that $V(\gamma_j, w_i) = 0$; and hence by [V*L*], $V(L\gamma_j, w_0) = 0$ in each case. Therefore each disjunct in (1) has the value 0 in w_0, and so by [Vv], $V((1), w_0) = 0$.

(If (1) contains no unmodalized disjunct β, we omit w_0 from the model. We can then prove by the same method that for any $w_i \in W$ whatever, $V((1), w_i) = 0$.)

We can illustrate the construction of a falsifying S5-model in a particular case by the first conjunct in our Example 3, which turned out to be invalid. This is:

$$(r \vee ((p \cdot q) \supset r)) \vee Lq \vee L{\sim}(p \cdot r) \vee M{\sim}p$$

Here β is $(r \vee ((p \cdot q) \supset r))$, γ_1 is q, γ_2 is ${\sim}(p \cdot r)$, and δ is ${\sim}p$. Now,

1. $(((r \vee ((p \cdot q) \supset r) \vee {\sim}p)$ is not PC-valid and is falsified by the following PC-assignment V_1:

$$V_1(p) = 1,\ V_1(q) = 1,\ V_1(r) = 0$$

2. $q \vee {\sim}p$ is not PC-valid and is falsified by the assignment:

$$V_2(p) = 1,\ V_2(q) = 0,\ V_2(r) = 1$$

3. ${\sim}(p \cdot r) \vee {\sim}p$ is not PC-valid and is falsified by the assignment:

$$V_3(p) = 1,\ V_3(q) = 1,\ V_3(r) = 1$$

We construct the following S5-model:

Let $$W = \{w_0, w_1, w_2\}$$

Let

$V(p, w_0) = V_1(p) = 1$, $V(q, w_0) = V_1(q) = 1$, $V(r, w_0) = V_1(r) = 0$
$V(p, w_1) = V_2(p) = 1$, $V(q, w_1) = V_2(q) = 0$, $V(r, w_1) = V_2(r) = 1$
$V(p, w_2) = V_3(p) = 1$, $V(q, w_2) = V_3(q) = 1$, $V(r, w_2) = V_3(r) = 1$

From this it is easy to show that $V((r \vee ((p \cdot q) \supset r)), w_0) = 0$; $V(q, w_1) = 0$, and hence $V(Lq, w_0) = 0$; $V({\sim}(p \cdot r), w_2) = 0$ and hence $V(L{\sim}(p \cdot r), w_0) = 0$; $V({\sim}p, w_0) = V({\sim}p, w_1) = V({\sim}p, w_2) = 0$ hence $V(M{\sim}p, w_0) = 0$; and so $V((r \vee ((p \cdot q) \supset r) \vee Lq \vee L{\sim}(p \cdot r) \vee M{\sim}p), w_0) = 0$. Hence the conjunct (and therefore the whole conjunction) is invalid.

The completeness of S5

It is now a simple matter to prove that S5 is complete, i.e. that if α is S5-valid, then $\vdash_{S5}\alpha$.

Let α' be the ordered MCNF of α. We have already shown that α is S5-valid iff each conjunct, C_i, in α' is S5-valid. Now each C_i is of the form

$$(1)\ \beta \vee L\gamma_1 \vee \ldots \vee L\gamma_n \vee M\delta$$

and we proved at (b) above that if (1) is S5-valid then at least one of $(\beta \vee \delta)$, $(\gamma_1 \vee \delta)$, . . . , $(\gamma_n \vee \delta)$ is PC-valid. But we also proved in the course of (a) above that if any of these is PC-valid, then $\vdash_{S5}(1)$. Hence if α is S5-valid, then for each C_i in α', $\vdash_{S5}C_i$; and therefore by Adjunction, $\vdash_{S5}\alpha'$. But we have also shown that $\vdash_{S5}(\alpha \equiv \alpha')$; hence $\vdash_{S5}\alpha$.

This completeness proof, like those we have given for T and S4, shows how we can transform a validity test into a proof in the axiomatic system whenever the formula is valid. As an example of the method of proof we can take our Example 1 (p. 56 and p. 117).

We prove the first conjunct, viz.

$$L{\sim}p \vee Lp \vee Mp$$

This was shown to be valid since ${\sim}p \vee p$ is PC-valid;

hence: $\vdash_{S5}{\sim}p \vee p$

hence by N: $\vdash_{S5}L({\sim}p \vee p)$

hence by T28: $\vdash_{S5}L{\sim}p \vee Mp$

and hence: $\vdash_{S5}L{\sim}p \vee Lp \vee Mp$

For the other conjunct we go from $\vdash_{S5}p \vee {\sim}p$ to $\vdash_{S5}Lp \vee M{\sim}p$ and hence to $\vdash_{S5}L{\sim}p \vee Lp \vee M{\sim}p$.

With both conjuncts proved we use Adjunction to prove the MCNF and the various S5 equivalences and Eq to prove the original formula.

Exercises – 6

6.1. Test for validity in S4:

(a) $M(Mp \,.\, {\sim}q) \vee L(p \supset Lq)$

(b) $MLp \supset LMp$

(c) $MLp \equiv LMp$

(d) $L(Lp \supset q) \vee L(Lq \supset p)$

(e) $L(Lp \supset q) \supset L(Lp \supset Lq)$

(f) $L(p \equiv q) \supset L(Lp \equiv Lq)$

6.2. For the formulae of 6.1:

(a) Construct falsifying S4-models for those which are invalid.

(b) For those which are S4-valid but not T-valid construct falsifying T-models and show why they cannot be S4-models.

6.3. Test for validity in S5 all the formulae of Exercise 3.5.

6.4. Reduce the following formulae to MCNF and test them for S5-validity:

(a) $(p \strictif (q \strictif r)) \supset (q \supset (p \strictif r))$

(b) $((p = q) \strictif Mq) \supset ((p = q) \strictif q)$

(c) $((p \strictif Lp) \strictif Lp) \supset (MLp \supset Lp)$

6.5. For the formulae of exercises 6.3 and 6.4 give a falsifying S5-model if they are invalid and a sketch proof if they are valid.

CHAPTER SEVEN

Some Alternative Bases for T, S4 and S5

We can define an axiomatic system in either of two ways: with reference to its basis, or with reference to its theses. For many purposes it does not matter much which course we adopt, but each will lead us to speak in somewhat different ways. Suppose we have two systems, A and B, with precisely the same theses but different bases. If we speak in the first way – the way we have been speaking up to now – we shall say that A and B are different systems, but that they are deductively equivalent. If we speak in the second way, we shall say that they are the same system but with different bases, or that B provides an alternative basis for, or an alternative axiomatization of, A. Sometimes it is more convenient to speak in this second way; and so long as no confusion arises, no harm is done.

In this chapter we consider some alternative bases for the systems T, S4 and S5.

Alternative bases for T

The first new basis for T we shall consider is the following:

The primitive symbols, formation rules and definitions are the same as for T (p. 30).

The axioms however differ. They consist of some complete set for PC[65], such as those of PM, plus

A5 $Lp \supset p$ (as before)
A9 $(Lp \, . \, Lq) \equiv L(p \, . \, q)$
A10 $L(p \supset p)$

[65] In Part III we consider axiomatic bases which are not formed by making additions to PC.

As transformation rules we again have Substitution and Detachment, but in place of TR3 we have

TR4 $\vdash(\alpha \equiv \beta) \rightarrow \vdash(L\alpha \equiv L\beta)$

We shall call the system derived from this basis, T′. We shall now prove that it is deductively equivalent to T, i.e. that T′ contains T and T contains T′.

To prove that T′ contains T it is sufficient to derive from the basis of T′ those items in the basis of T which are not in the basis of T′. I.e. we have to show that A6 is a theorem of T′ and that TR3 is a derived transformation rule of T′.

A6 $L(p \supset q) \supset (Lp \supset Lq)$

PROOF IN T′

A9 $[p \supset q/q]$:	(1) $(Lp \,.\, L(p \supset q)) \equiv L(p \,.\, (p \supset q))$
PC[66]:	(2) $(p \,.\, (p \supset q)) \equiv (p \,.\, q)$
(2) × TR4:	(3) $L(p \,.\, (p \supset q)) \equiv L(p \,.\, q)$
PC:	(4) $(p \equiv q) \supset ((q \equiv r) \supset (p \equiv r))$
(1), (3) × (4):	(5) $(Lp \,.\, L(p \supset q)) \equiv L(p \,.\, q)$
PC:	(6) $(p \equiv q) \supset ((r \equiv q) \supset (p \equiv r))$
(5), A9 × (6):	(7) $(Lp \,.\, L(p \supset q)) \equiv (Lp \,.\, Lq)$
(7) × PC:	(8) $(Lp \,.\, L(p \supset q)) \supset (Lp \,.\, Lq)$
PC:	(9) $((p \,.\, q) \supset (p \,.\, r)) \supset (q \supset (p \supset r))$
(8) × (9):	(10) $L(p \supset q) \supset (Lp \supset Lq)$ **Q.E.D.**

TR3 (N) $\vdash\alpha \rightarrow \vdash L\alpha$

DERIVATION IN T′

Given:	(1) α
PC $[\alpha/q]$:	(2) $\alpha \supset ((p \supset p) \equiv \alpha)$
(1), (2) × MP:	(3) $(p \supset p) \equiv \alpha$
(3) × TR4:	(4) $L(p \supset p) \equiv L\alpha$
(4) × PC:	(5) $L(p \supset p) \supset L\alpha$
A10, (5) × MP:	(6) $L\alpha$ **Q.E.D.**[67]

[66] From now on, relying on the known completeness of a basis such as that of PM, we simply write 'PC' to justify the citing of any valid PC formula, whether it occurs in the lists in Chapter 1 or not.

[67] The replaceability of TR3 by TR4 and A10 (which is what this derivation shows) was noted by Sobociński [1953].

We have now shown that T′ contains T. To show that T contains T′ we have merely to refer to the appropriate items in Chapter 2. A9 is T3 (p. 34); TR4 is DR2 (p. 34); and A10 follows immediately from the PC thesis $p \supset p$ by N.

Therefore T′ is deductively equivalent to T.

T and T′ are L-based systems in the sense explained on p. 26. An M-based system deductively equivalent to T is von Wright's system M[68]. To obtain M we make the following additions to a complete basis for PC:

M as primitive.

[Def L] $L\alpha =_{\mathrm{Df}} \sim M \sim \alpha$

Axioms

A11 $p \supset Mp$

A12 $M(p \vee q) \equiv (Mp \vee Mq)$

Transformation rules

TR3 $\vdash \alpha \rightarrow \vdash L\alpha$

TR5 $\vdash (\alpha \equiv \beta) \rightarrow \vdash (M\alpha \equiv M\beta)$

It is easy to show that T contains M: A11 and A12 are T1 and T7 respectively (in Chapter 3); TR5 is easily derived from DR3; and T5, by Eq, can be made to do the work of Def L. The proof that M contains T is left as an exercise.

In view of the deductive equivalence of T, T′ and M, we shall usually refer to them indifferently as T, and we shall use whichever axiomatization happens to be most convenient.

Alternative bases for S4

In Chapter 4 we obtained the system S4 by adding to the basis of T the extra axiom A7: $Lp \supset LLp$. An alternative way is to *replace* A6 ($L(p \supset q) \supset (Lp \supset Lq)$) by the stronger axiom:

A13 $L(p \supset q) \supset L(Lp \supset Lq)$[69]

[68] Von Wright [1951], pp. 84–5. The deductive equivalence of M and T was shown by Sobociński [1953], pp. 173–175.

[69] Cf. Parry [1939], especially Theorem 32.1, p. 147; and Lemmon [1957], p. 178.

We call the system so obtained, S4′. We now prove it is deductively equivalent to S4.

To prove that S4′ contains S4 we have to derive A6 and A7 in S4′. Clearly A6 follows from A13 by A5 and Syll. (Hence S4′ contains T.) For A7 we proceed as follows.

A7 $Lp \supset LLp$

PROOF IN S4′

PC:	(1) $p \supset ((p \supset p) \supset p)$
(1) × DR1:	(2) $Lp \supset L((p \supset p) \supset p)$
A13 $[p \supset p/p, p/q]$:	(3) $L((p \supset p) \supset p) \supset L(L(p \supset p) \supset Lp)$
A6 $[L(p \supset p)/p, Lp/q]$:	(4) $L(L(p \supset p) \supset Lp) \supset (LL(p \supset p) \supset LLp)$
(2), (3), (4) × Syll:	(5) $Lp \supset (LL(p \supset p) \supset LLp)$
(5) × Perm:	(6) $LL(p \supset p) \supset (Lp \supset LLp)$
Id × TR3 (twice):	(7) $LL(p \supset p)$
(7), (6) × MP:	(8) $Lp \supset LLp$ **Q.E.D.**

To prove that S4 contains S4′ we have simply to derive A13 in S4.

A13 $L(p \supset q) \supset L(Lp \supset Lq)$

PROOF IN S4

A7 $[p \supset q/p]$:	(1) $L(p \supset q) \supset LL(p \supset q)$
A6 × DR1:	(2) $LL(p \supset q) \supset L(Lp \supset Lq)$
(1), (2) × Syll:	(3) $L(p \supset q) \supset L(Lp \supset Lq)$ **Q.E.D.**

Von Wright's system M′ is obtained by adding to the system M the extra axiom

A14 $MMp \supset Mp$

A14 is T18 in Chapter 3; and the derivation from it of the characteristic S4 axiom ($Lp \supset LLp$) is very straightforward. Hence M′ is an *M*-based system deductively equivalent to S4[70].

[70] Von Wright [1951], pp. 84–5. On pp. 89–90 von Wright proves the equivalence of M′ and S4 – a less trivial result in that context than might appear from our present treatment, since he was not regarding S4 as an extension of T or of any equivalent system. (T was not one of the original Lewis systems – *vide infra*, pp. 236f.)

Alternative bases for S5

In Chapter 3 we obtained S5 by adding A8 ($Mp \supset LMp$) to T. This was part of our programme of displaying T, S4 and S5 as progressively stronger systems, and it had the obvious advantage of enabling us to study what can be proved without using the various reduction laws. If, however, we want to construct S5 directly, without starting from T, we can do so more economically as follows.

We start as before with a complete basis for PC, L as primitive, and the usual definitions of the other modal operators. To this we add no new axioms, but the following two transformation rules:

TR6 $\vdash(\alpha \supset \beta) \rightarrow \vdash(L\alpha \supset \beta)$

TR7 If α is fully modalised (i.e., if every propositional variable in α occurs within the scope of a modal operator), then $\vdash(\alpha \supset \beta) \rightarrow \vdash(\alpha \supset L\beta)$[71].

A minor variant of this basis is obtained by using the familiar A5 ($Lp \supset p$) in place of TR6. It is easy to show that A5 and TR6 can be derived from each other. If we have TR6, then since, by PC, $\vdash(p \supset p)$, we immediately obtain $\vdash(Lp \supset p)$, i.e. A5. And if we have A5 as an axiom we can derive TR6 as follows:

Given:	(1) $\alpha \supset \beta$
A5 $[\alpha/p]$:	(2) $L\alpha \supset \alpha$
(2), (1) × Syll:	(3) $L\alpha \supset \beta$

Q.E.D.

We shall call the system derived from the above basis (whether using TR6 or A5), S5′. We now prove it is deductively equivalent to S5.

To prove that S5′ contains S5 we have to derive A6, A8 and TR3 in S5′. We do so as follows

A6 $L(p \supset q) \supset (Lp \supset Lq)$

PROOF IN S5′

A5 $[p \supset q/p]$:	(1) $L(p \supset q) \supset (p \supset q)$
(1) × Perm:	(2) $p \supset (L(p \supset q) \supset q)$
(2) × TR6:	(3) $Lp \supset (L(p \supset q) \supset q)$

[71] Cf. Prior [1955], p. 202. Note that the 1st edition of Prior's book contains an error: in fact *every* variable in α must be modalized, not merely those also occurring in β, as is stated there.

(3) × Imp: (4) $(Lp \,.\, L(p \supset q)) \supset q$
(4) × TR7: (5) $(Lp \,.\, L(p \supset q)) \supset Lq$
(5) × Comm, × Exp: (6) $L(p \supset q) \supset (Lp \supset Lq)$ **Q.E.D.**

A8 $Mp \supset LMp$

PROOF IN S5′
Id $[Mp/p]$: (1) $Mp \supset Mp$
(1) × TR7: (2) $Mp \supset LMp$ **Q.E.D.**

TR3 $\vdash \alpha \rightarrow \vdash L\alpha$

DERIVATION IN S5′
Given: (1) α
PC: (2) $p \supset (q \supset p)$
(1) × (2) $[\alpha/p,\ Lp \supset Lp/q]$: (3) $(Lp \supset Lp) \supset \alpha$
(3) × TR7: (4) $(Lp \supset Lp) \supset L\alpha$
Id $[Lp/p]$: (5) $Lp \supset Lp$
(5), (4) × MP: (6) $L\alpha$ **Q.E.D.**

To prove that S5 contains S5′ we have only to derive TR7 in S5[72].

We first prove the following Lemma:
If α is fully modalized, then $\vdash_{S5}(\alpha \equiv L\alpha)$

PROOF
We shall assume that α is written in primitive notation (i.e., with $\sim$, $\vee$ and L as the only operators). Then α is either of the form $L\beta$ (where β is some wff), or else is constructed by $\sim$ and/or $\vee$, by the formation rules, from one or more components each of which is of the form $L\beta$. We show (i) that the Lemma holds for every wff of the form $L\beta$; and then (ii) that if it holds for a certain set of formulae, it also holds for any wff which can be constructed from these formulae by $\sim$ and $\vee$. From this by induction on the construction of α it follows that the Lemma holds for every fully modalized formula.

(i) If α is $L\beta$, then $L\alpha$ is $LL\beta$. But by R4 (p. 49), $\vdash(L\beta \equiv LL\beta)$ in S5 (and S4). Hence the Lemma holds in this case.

[72] Lemmon [1956].

(ii) Suppose γ is some formula for which the Lemma holds. Then we have:

Given: (1) $\gamma \equiv L\gamma$
(1) × PC: (2) $\sim\gamma \equiv \sim L\gamma$
R1 $[\sim\gamma/p]$: (3) $M\sim\gamma \equiv LM\sim\gamma$
(3) × LMI: (4) $\sim L\gamma \equiv L\sim L\gamma$
(2), (4) × Eq: (5) $\sim\gamma \equiv L\sim L\gamma$
(5), (1) × Eq: (6) $\sim\gamma \equiv L\sim\gamma$

Therefore if the Lemma holds for γ, it holds for $\sim\gamma$.

Suppose now that γ and δ are two formulae for which the Lemma holds. Then we have:

Given: (1) $\gamma \equiv L\gamma$
(2) $\delta \equiv L\delta$
PC: (3) $(p \equiv q) \supset ((r \equiv s) \supset ((p \vee r) \equiv (q \vee s)))$
(1), (2) × (3): (4) $(\gamma \vee \delta) \equiv (L\gamma \vee L\delta)$
T29 (p. 51) $[\gamma/p, \delta/q]$: (5) $L(\gamma \vee L\delta) \equiv (L\gamma \vee L\delta)$
(4), (5) × Eq: (6) $(\gamma \vee \delta) \equiv L(\gamma \vee L\delta)$
(6), (2) × Eq: (7) $(\gamma \vee \delta) \equiv L(\gamma \vee \delta)$

Therefore if the Lemma holds for γ and for δ, it holds for $(\gamma \vee \delta)$.

This completes the proof of the Lemma.

The derivation of TR7 in S5 is now straightforward. Assuming that α is fully modalized, we have:

Given: (1) $\alpha \supset \beta$
By Lemma: (2) $\alpha \equiv L\alpha$
(1) × DR1: (3) $L\alpha \supset L\beta$
(3), (2) × Eq: (4) $\alpha \supset L\beta$ **Q.E.D.**

This completes the proof that S5′ is deductively equivalent to S5.

An M-based system deductively equivalent to S5 is von Wright's M″[73], which is obtained by adding to M the extra axiom:

[73] Von Wright [1951], pp. 84–5, 90.

A15 $M{\sim}Mp \supset {\sim}Mp$

A15 yields the characteristic S5 axiom ($Mp \supset LMp$) by Transp and Def L, and is itself derivable from the S5 theorem T25 (p. 49) by $[{\sim}p/p]$ and LMI.

The bases considered in this chapter are only a few among many possible alternatives. But we should have said enough by now to give some indication of what alternative bases can be like.

Exercises – 7

7.1. (a) T″ is T′ with DR1 ($\vdash \alpha \supset \beta \rightarrow \vdash L\alpha \supset L\beta$) as a primitive rule in place of TR4. Show that T″ and T′ are deductively equivalent.

(b) Show that the omission of A10 from T′ would give a weaker system (i.e., show that A10 is independent).

7.2. (a) Show that the system M contains the system T.

(b) M* is M but with A10 replacing TR3. Show that M* and M are deductively equivalent.

7.3. T* is formed from T by replacing A6 and TR3 by A10 and the rule $\vdash (\alpha \,.\, \beta) \supset \gamma \rightarrow \vdash (L\alpha \,.\, L\beta) \supset L\gamma$; and S4* is formed from T by replacing A6 and TR3 by A10 and the rule

$$\vdash (L\alpha \,.\, L\beta) \supset \gamma \rightarrow \vdash (L\alpha \,.\, L\beta) \supset L\gamma$$

(a) Show that T and T* are deductively equivalent.

(b) Show that S4 and S4* are deductively equivalent.

7.4. (a) Show that the addition to T as an axiom of the formula (B″) $p \supset LLMp$ gives S5.

(b) Show that TR3 is not independent in M″ (i.e., that it can be derived from the rest of the basis).

PART TWO

Modal Predicate Logic

CHAPTER EIGHT

The Lower Predicate Calculus

The non-modal lower predicate calculus

We first do for Part II what we did in Chapter 1 for Part I. That is to say, we give an outline of what has to be known about the 'ordinary', non-modal Lower Predicate Calculus if the subsequent modal developments are to be understood. We normally abbreviate 'Lower Predicate Calculus' to 'LPC' [74]. For simplicity we adopt a form of LPC which does not contain either propositional variables or individual constants or predicate constants, since these are not essential for expounding the distinctive features of modal predicate logic.

Primitive symbols of LPC

1. A set of *individual-variables*: $x, y, z, \ldots$ (with or without numerical subscripts).

2. A set of *predicate variables*: $\phi, \psi, \chi, \ldots$

We suppose ourselves to have an unlimited number of each type of variable.

3. The symbols $\sim, \vee, \forall$, (,).

Formation rules

FR1 An expression consisting of a predicate variable followed by any number of individual-variables is a wff (an *atomic* wff).

(In such a wff, the individual-variables are the arguments of the predicate variable. A predicate variable with n arguments [75] is said to be of *degree* n, or to be an n-adic variable.)

[74] The Predicate Calculus is also known as the Functional Calculus. Instead of 'Lower' the term 'First-order' is often used. What distinguishes LPC from calculi of higher order is that in LPC the only variables that are ever quantified over are individual-variables. A standard development is found in Church [1956], Chapters III and IV. Other accounts will be found in Quine [1940] and [1950].

[75] For the purposes of this definition the number of arguments is not the number of *distinct* variables but the total number of variables whether distinct or not. Thus in both ϕxyz and ϕxxz, ϕ is of degree 3.

FR2 If α is a wff, so is $\sim\alpha$.

FR3 If α and β are wffs, so is $(\alpha \vee \beta)$, *provided that* any predicate variable which occurs more than once in $(\alpha \vee \beta)$ is of the same degree on each occurrence.

FR4 If α is a wff and $\mathbf{a}$[76] is any individual-variable, $(\forall\mathbf{a})\alpha$ is a wff. (Where no confusion is likely to arise, we normally omit the symbol $\forall$ in wffs of this form; e.g. we write $(x)\phi x$ instead of $(\forall x)\phi x$.)

In a wff of the kind just described, $(\forall\mathbf{a})$ (or $(\mathbf{a})$) is called a *universal quantifier*, and α is said to be the *scope* of that quantifier. An occurrence of a variable, $\mathbf{a}$, in a wff, α (not as part of a quantifier) is said to be either *free* or *bound* in α. If it does not lie within the scope of any quantifier in α which contains $\mathbf{a}$, it is said to be free in α. Otherwise it is said to be bound in α; and if $\mathbf{a}$ is free in α, it is said to be *bound by* $(\mathbf{a})$ in $(\mathbf{a})\alpha$. Thus in the formula

$$(1) \quad (x)(\phi x \vee \psi y)$$

the x occurring immediately after ϕ is bound (by the quantifier (x) at the beginning), but y is free. Note, however, that even where we are speaking of (1) we say that x is free in $(\phi x \vee \psi y)$, since no quantifier containing x occurs in *that* expression, though of course x is not free in (1) itself. Note also that it is *occurrences* of variables that are bound or free, and that the same variable may occur both bound and free in the same formula, as, e.g., x does in $(x)\phi x \supset \phi x$.

We adopt, as we did for PC, the convention that the outermost brackets round a formula may be omitted. We also sometimes insert brackets not required by the formation rules, for ease of reading.

We introduce the operators $.$, $\supset$ and $\equiv$ by the same definitions as in PC. We also introduce the symbol $\exists$ by the definition:

Where $\mathbf{a}$ is any individual-variable and α is any wff,

$$(\exists\mathbf{a})\alpha =_{\mathrm{Df}} \sim(\forall\mathbf{a})\sim\alpha$$

In a wff of the form $(\exists\mathbf{a})\alpha$, $(\exists\mathbf{a})$ is called an *existential quantifier*. The remarks made above about the scope of quantifiers and about free and bound variables apply also in the case of existential quantifiers. (As with L and M, a parallel development of LPC with $\exists$ as primitive and $\forall$ defined is possible.)

[76] We use $\mathbf{a}$ and $\mathbf{b}$ as meta-logical variables ranging over individual-variables.

Interpretation

The intended interpretation of the formulae of LPC can be sufficiently indicated as follows. The individual-variables are to be thought of as ranging over objects or individuals, in the widest sense. Predicate variables of degree 1 range over properties of objects, again in the widest sense; those of degree n ($n > 1$) range over n-adic relations. Atomic formulae express (the forms of) propositions to the effect that a certain object has a certain property, or that a certain group of objects are related in a certain way: thus 'Smith is blue-eyed' is of the form ϕx, 'Tom is the father of Bill' is of the form ϕxy, '24 is between 19 and 32' is of the form ϕxyz, and so on. The operators $\sim$, $\vee$, $.$, $\supset$ and $\equiv$ are to be understood as in PC. The universal and existential quantifiers express the notions *all* and *some* (i.e., *at least one*) respectively; more exactly, $(\mathbf{a})\alpha$ is to mean 'for every value of $\mathbf{a}$, α', and $(\exists\mathbf{a})\alpha$ is to mean 'for at least one value of $\mathbf{a}$, α'. Thus, $(x)\phi x$ means that no matter what value we assign to x, ϕx is true, or more succinctly that everything has the property ϕ; and similarly $(\exists x)\phi x$ means that something has the property ϕ. It is easy to see that the definition of the existential in terms of the universal quantifier is appropriate to this interpretation.

Validity in LPC

We shall define validity in LPC, as we did for previous systems, in terms of value-assignments and models. For an LPC value-assignment it is not sufficient to think of assigning a truth-value to certain formulae and then giving rules for calculating the truth-values of the remainder, since here we have individual-variables and predicate variables to consider, and we cannot assign truth-values to these. To form an LPC-model we first of all assume some non-empty domain or set, D, of individuals, $\{u_1, u_2, \ldots, u_i, \ldots\}$. An LPC-model $\langle D, V \rangle$ then consists of this domain together with a value-assignment, V, satisfying the following conditions:

To each individual-variable V assigns some member of D. We write '$V(x) = u$' to mean 'V assigns u to x', and so forth.

For predicate variables we proceed as follows: If ϕ is a monadic predicate it is convenient for our present purposes to think of the property ϕ as being determined by the set of things that possess it; e.g. if ϕx means 'x is human' we can, instead of thinking of ϕ as the abstract property of humanity, think of it in terms of the set of human beings. For the case of a dyadic predicate, if ϕxy means 'x is the father of y', we can think of ϕ in terms of the set of ordered pairs such that in each the first is the father of the second; and we can treat predicates of higher degree analogously. This way of looking at things suggests that we could let V assign to each n-adic predicate variable a set of ordered n-tuples of members of D; and this is what we shall do. More precisely, the condition to be satisfied by V as far as predicate variables are concerned is this: for each n-adic variable, ϕ, $V(\phi)$ is some set of ordered n-tuples

$$\{\langle u_{i1}, \ldots, u_{in}\rangle, \langle u_{j1}, \ldots, u_{jn}\rangle, \ldots\}$$

of members of D.

Given $V(x_i)$ and $V(\phi)$ for each individual-variable x_i and each predicate variable ϕ, we evaluate all formulae as follows:

1. For an atomic formula – i.e. one consisting of an n-adic variable ϕ followed by n individual variables $x_1, \ldots, x_n$ $(n \geqslant 1)$ – $V(\phi(x_1, \ldots, x_n)) = 1$ iff $\langle V(x_1), \ldots, V(x_n)\rangle \in V(\phi)$; otherwise $V(\phi(x_1, \ldots, x_n)) = 0$. (I.e. the formula has the value 1 or 0 according as the n-tuple of those members of D assigned by V to $x_1, \ldots, x_n$ respectively is or is not one of those assigned by V to ϕ.)

For $\sim$ and $\vee$ the rules are the same as for PC, viz:

2. [V$\sim$]. If α is a wff, then $V(\sim\alpha) = 1$ if $V(\alpha) = 0$; otherwise $V(\sim\alpha) = 0$.

3. [V$\vee$]. If α and β are wffs, then $V(\alpha \vee \beta) = 1$ if either $V(\alpha) = 1$ or $V(\beta) = 1$; otherwise $V(\alpha \vee \beta) = 0$.

For quantification the rule is:

4. [V$\forall$]. If α is a wff and **a** is an individual-variable, then $V((\mathbf{a})\alpha) = 1$ if for every LPC assignment V' which gives to all variables except **a** the same values which V gives to them, $V'(\alpha) = 1$; otherwise $V((\mathbf{a})\alpha) = 0$.

(This last rule can be expressed intuitively by saying that $(\mathbf{a})\alpha$ is true in the model – has the value 1 – iff α is true no matter what value is assigned to **a**; but since α may contain other free variables than **a** we have to assume that they are held constant, and thus we require that V' be the same as V except in respect of its assignment to **a**.)

Having now defined an LPC-model we can define validity in LPC by saying that a wff α is LPC-valid iff for every LPC-model $\langle D, V\rangle$, $V(\alpha) = 1$.

Although it is possible to give a perfectly clear account of what constitutes validity in LPC, yet LPC is not a decidable system; i.e. it is not possible to give a general decision procedure for LPC-validity[77]. The same is true of the modal extensions of LPC which we shall study shortly. In this respect LPC differs in an important way from all the systems we have considered up to now.

Axiomatization

Our style of axiomatization for LPC will be different from the one we have used in dealing with the propositional systems. In place of axioms in the sense of Part I we shall use axiom-schemata[78]; and parallel to axiom-schemata we shall have theorem-schemata, i.e. general principles to the effect that any wff of a certain form is a theorem. For the most part we shall prove such theorem-schemata rather than particular theorems.

It is usual, and very convenient, to axiomatize LPC as an extension of PC, and there are several ways of doing this. The way we shall choose is to have an axiom-schema which gives as an axiom any LPC substitution-instance of any valid wff of PC[79]. (An LPC substitution-instance of a PC thesis is a formula obtained from a valid PC wff by uniformly replacing every variable in it by some wff of LPC.)

The axiom-schemata we shall use for LPC are the following:

PC If α is an LPC substitution-instance of a valid PC wff then α is an axiom.

[77] This was proved in Church [1936]. There are, however, a number of decidable fragments of LPC (cf. Church [1956], pp. 247–257).

[78] *Vide* p. 17n. The elimination of primitive rules of uniform substitution made possible by the use of axiom-schemata instead of ordinary axioms is a great convenience in LPC, where such rules, though formulable, are very complicated both to state and to apply.

[79] An alternative method would be to incorporate in the basis for LPC some complete set of axioms (or axiom-schemata) for PC with appropriate transformation rules, but since PC is decidable there is an effective method for deciding of any wff whether it is an axiom in virtue of the axiom-schema in question or not.

∀1 If **a** is any individual-variable, α any wff, and β any wff differing from α only in having some individual-variable **b** replacing every free occurrence of **a** in α, then

$$(\mathbf{a})\alpha \supset \beta$$

is an axiom, provided that **a** does not occur within the scope of any occurrence of a quantifier containing **b**[80].

The primitive transformation rules are:

∀2 If **a** is any individual-variable and α and β are any wffs, then

$$\vdash \alpha \supset \beta \rightarrow \vdash \alpha \supset (\mathbf{a})\beta$$

provided that **a** does not occur free in α[81].

MP $\vdash \alpha,\ \vdash(\alpha \supset \beta) \rightarrow \vdash \beta$

This basis can be proved to be (weakly) complete; i.e. it can be shown that, for every wff α, if α is LPC-valid, then α is derivable from the above basis (*vide* pp. 159–164).

Some theorems of LPC

We list here some theorems of LPC (in the form of schemata) which will be useful in later developments. We omit proofs,

80 Another way of expressing this proviso is by saying that we may not replace **a** by any variable which would become bound in the process, but only by one which would also be free. The need for such a restriction can be seen from a simple example. Let **a** be x, let α be $(\exists y)\phi xy$, and let β (in defiance of the proviso) be $(\exists y)\phi yy$. Then ∀1 would, but for the proviso, give us as an axiom:

$$(1)\ \ (x)(\exists y)\phi xy \supset (\exists y)\phi yy$$

But (1) is not valid; for if ϕxy means 'x is a child of y', then the antecedent means 'everyone is a child of someone' (which is true), but the consequent means 'someone is a child of himself' (which is false). The fact that we cannot replace x in $(\exists y)\phi xy$ by y (but could replace it, e.g. by z) is sometimes expressed by saying that although x is free in $(\exists y)\phi xy$, it is not *free for* y (though it is free for z).

81 ∀2 is often replaced by alternative rules and axiom-schemata; e.g. by the rule UG. $\vdash \alpha \rightarrow \vdash(\mathbf{a})\alpha$ and the axiom-schema $(\mathbf{a})(\alpha \supset \beta) \supset (\alpha \supset (\mathbf{a})\beta)$ (provided **a** is not free in α), or by the rule UG, and the two axiom-schemata, $\alpha \equiv (\mathbf{a})\alpha$ (provided **a** is not free in α) and $(\mathbf{a})(\alpha \supset \beta) \supset ((\mathbf{a})\alpha \supset (\mathbf{a})\beta)$. The bases so obtained can easily be proved equivalent to the one we are using.

which are in any case quite standard. In each theorem-schema, α and β are any wffs and **a** and **b** are any individual-variables.

LPC1 $(\mathbf{a})(\alpha \supset \beta) \supset ((\mathbf{a})\alpha \supset (\mathbf{a})\beta)$

LPC2 $(\mathbf{a})(\alpha \equiv \beta) \supset ((\mathbf{a})\alpha \equiv (\mathbf{a})\beta)$

LPC3 $(\mathbf{a})\alpha \equiv \sim(\exists\mathbf{a})\sim\alpha$

LPC4 $(\exists\mathbf{b})((\exists\mathbf{a})\alpha \supset \beta)$, where β is α with free **b** replacing free **a** everywhere, provided that **b** is not free in α[82].

LPC5 $(\exists\mathbf{a})(\alpha \,.\, \beta) \equiv (\alpha \,.\, (\exists\mathbf{a})\beta)$, provided that **a** does not occur free in α.

LPC6 $(\exists\mathbf{a})(\alpha \supset \beta) \equiv (\alpha \supset (\exists\mathbf{a})\beta)$, provided that **a** does not occur free in α.

LPC7 $(\mathbf{a})\alpha \equiv (\mathbf{b})\beta$, where α and β differ only in that α has free **a** where and only where β has free **b**.

When α and β differ only in the way just described, we shall say that each is an *alphabetic variant* of the other, and that $(\mathbf{a})\alpha$ and $(\mathbf{b})\beta$ (and likewise $(\exists\mathbf{a})\alpha$ and $(\exists\mathbf{b})\beta$) are *bound alphabetic variants* (of each other). We shall also say that two wffs, α and β, are bound alphabetic variants if α differs from β only in having some well-formed part, γ, where β has δ, and γ is a bound alphabetic variant of δ[83].

LPC8 $\alpha \equiv \beta$, where α is a bound alphabetic variant of β.

As a special case of LPC8 we have:

LPC9 $(\mathbf{a})\alpha \equiv (\mathbf{a})\beta$, where α is a bound alphabetic variant of β.

[82] E.g., if $(\exists\mathbf{a})\alpha$ is $(\exists x)(\exists y)\phi xyz$, β might be $(\exists y)\phi wyz$ ($\mathbf{b} = w$); but not $(\exists y)\phi yyz$ ($\mathbf{b} = y$), since y is not free in this formula, and not $(\exists y)\phi zyz$ ($\mathbf{b} = z$), since z *is* free in $(\exists y)\phi xyz$. In other words,

$$(\exists w)((\exists x)(\exists y)\phi xyz \supset (\exists y)\phi wyz)$$

is a theorem, but

$$(\exists y)((\exists x)(\exists y)\phi xyz \supset (\exists y)\phi yyz) \quad \text{and} \quad (\exists z)((\exists x)(\exists y)\phi xyz \supset (\exists y)\phi zyz)$$

are not.

[83] What this comes to is that we form a bound alphabetic variant of a wff, α, by taking some quantifier in α and replacing the individual-variable in it, together with all occurrences of that variable within its scope, by some other individual variable. (This process is often called *re-lettering* a bound variable.) Intuitively, bound alphabetic variants 'say the same thing'; in the simplest case, $(x)\phi x$ and $(y)\phi y$ each mean simply that everything is ϕ. If α and β are bound alphabetic variants, any LPC value-assignment will assign the same value to each; i.e., if V is any LPC value-assignment, $V(\alpha) = V(\beta)$.

Derived transformation rules

The following derived rules are useful. In them α and β are again any wffs and **a** is any individual-variable.

1. The rule of universal generalization:

UG $\vdash\alpha \rightarrow \vdash(\mathbf{a})\alpha$

2. Where α is any wff containing free **a**, and β is α with free **b** replacing every occurrence of free **a**, then

$$\vdash\alpha \rightarrow \vdash\beta$$

(the proof is by UG and ∀1.)

3. $\vdash(\alpha \equiv \beta) \rightarrow \vdash((\mathbf{a})\alpha \equiv (\mathbf{a})\beta)$

From 3 and the rules of PC we can derive the following rule for substitution of proved equivalents:

Eq If $\vdash\alpha$, and β differs from α only in having some wff, δ, at one or more places where α has γ, then if $\vdash(\gamma \equiv \delta)$, $\vdash\beta$.

We now prove that every thesis derivable from our basis for LPC is LPC-valid.

The axiom-schema PC requires no comment, since all theses given by it are valid for the same reasons as in the Propositional Calculus itself.

∀1 gives $(\mathbf{a})\alpha \supset \beta$ as a thesis whenever β differs from α only in having free **b** wherever α has free **a**. Now in any LPC-model, if $V((\mathbf{a})\alpha) = 1$ then (by [V∀]) $V'(\alpha) = 1$ for every V' differing from V only in assignment to **a**. Among these V's there will be one which assigns to **a** the same individual which V assigns to **b**, and thus assigns to α the same value which V assigns to β. Hence for this V', $V'(\alpha) = V(\beta)$. But $V'(\alpha) = 1$; therefore $V(\beta) = 1$. I.e., whenever $V((\mathbf{a})\alpha) = 1$, $V(\beta) = 1$, and hence $(\mathbf{a})\alpha \supset \beta$ is valid.

The transformation rule ∀2, gives $\alpha \supset (\mathbf{a})\beta$ as a thesis whenever $\vdash(\alpha \supset \beta)$ and **a** is not free in α. We show that this rule is validity-preserving. It is sufficient to prove that if $\alpha \supset \beta$ is valid so is $\alpha \supset (\mathbf{a})\beta$, provided that **a** is not free in α.

Suppose that in a given LPC-model $\langle D, V\rangle$, $V(\alpha) = 1$. Since by hypothesis $\alpha \supset \beta$ is valid, it has the value 1 in every model; and in particular in every model $\langle D, V'\rangle$ differing from $\langle D, V\rangle$ only in the assignment V' makes to **a**. Since **a** does not occur free in α, we have $V'(\alpha) = V(\alpha) = 1$. Hence since $V'(\alpha \supset \beta) = 1$, $V'(\beta) = 1$ (for every V' differing from V only in assignment to **a**);

hence $V((\mathbf{a})\beta) = 1$. I.e., whenever $V(\alpha) = 1$, $V((\mathbf{a})\beta) = 1$, and so $\alpha \supset (\mathbf{a})\beta$ is valid.

MP is clearly validity-preserving for the same reasons as in PC.

Therefore every thesis of LPC is LPC-valid.

Modal lower predicate calculi

We can graft modal elements on to LPC and thus produce predicate calculi analogous to the propositional systems T, S4 and S5.

The system we shall call LPC + T can be obtained by adding to the basis for LPC given above (or any equivalent basis):

1. The primitive operator L, with the appropriate formation rule: if α is a wff, so is $L\alpha$.
2. Definitions of M, ⥽ and $=$ as in T.
3. Axiom-schemata corresponding to the modal axioms of T, viz.

A5[84] $L\alpha \supset \alpha$

A6 $L(\alpha \supset \beta) \supset (L\alpha \supset L\beta)$

where α and β are any wffs of LPC + T.

4. The rule of Necessitation (N): $\vdash \alpha \rightarrow \vdash L\alpha$.

LPC + S4 is LPC + T together with the axiom-schema:

A7 $L\alpha \supset LL\alpha$

and LPC + S5 is LPC + T together with the axiom-schema:

A8 $M\alpha \supset LM\alpha$

Clearly our derived rules of T, S4 and S5 hold in the corresponding predicate calculi; and so do all their theorems, when restated as theorem-schemata.

We shall not give samples of theorems at the outset, as we did with the propositional modal systems; but we shall prove theorems for special purposes from time to time, and these will sufficiently illustrate the techniques of proof.

In the case of the propositional systems discussed in Part I, T appeared to contain all the modal formulae which we should intuitively regard as valid, apart from those which depend on

[84] We keep the same numbering as for the corresponding axioms of T (p. 31), since the schemata do the same work as is done by these axioms together with the rule of substitution.

the various reduction laws for iterated modalities. It may, however, be doubted whether LPC + T is as successful as this for predicate logic. For example, one of the theorem-schemata of LPC is

LPC1 $(\mathbf{a})(\alpha \supset \beta) \supset ((\mathbf{a})\alpha \supset (\mathbf{a})\beta)$

and its analogue for strict implication, viz.

$$(1)\ (\mathbf{a})(\alpha \prec \beta) \supset ((\mathbf{a})\alpha \prec (\mathbf{a})\beta)$$

seems intuitively to be valid. Yet (1) is not a theorem of LPC + T – or of LPC + S4 either, though it is of LPC + S5.

Let us elaborate this a little. Expanding (1) by Def $\prec$ we have:

$$(2)\ (\mathbf{a})L(\alpha \supset \beta) \supset L((\mathbf{a})\alpha \supset (\mathbf{a})\beta)$$

This looks like a combination of LPC1 and A6; but what we notice about it is that in the antecedent the modal operator is within the scope of the quantifier, but in the consequent the quantifiers are within the scope of the modal operator; and it is this feature which prevents it from being derivable in LPC + T. What we can prove quite easily in LPC + T is:

T1 $L(\mathbf{a})(\alpha \supset \beta) \supset L((\mathbf{a})\alpha \supset (\mathbf{a})\beta)$

(The proof is by LPC1 and DR1 (p. 33).) But T1 is not (2) – in the antecedent the modal operator and the quantifier are in the wrong order. We could, however, derive (2) from T1 by Syll if we also had:

$$(3)\ (\mathbf{a})L(\alpha \supset \beta) \supset L(\mathbf{a})(\alpha \supset \beta)$$

Now (3) is simply a special case of a more general formula which is often called the *Barcan formula*, viz.

BF $(\mathbf{a})L\alpha \supset L(\mathbf{a})\alpha$

So if LPC + T contained BF, it would also contain (2); but it does not, nor does LPC + S4[85].

BF is a formula of some importance, which has given rise to some philosophical controversy[86]. Its name derives from Ruth C. Barcan[87], who called attention to it – or more accurately

[85] *Vide* Lemmon [1960b] and *infra* p. 173.

[86] The interpretation of this formula is further discussed in Chapter 10.

[87] Later Mrs J. A. Marcus. *Vide* her [1946], where (4) is axiom 11 on p. 2.

to a formula which can easily be seen to be interdeducible with it, viz.

$$(4)\ M(\exists \mathbf{a})\alpha \prec (\exists \mathbf{a})M\alpha$$

((4) follows from BF by Def M, Def $\exists$, standard PC equivalences and N, and the reverse derivation is equally straightforward; in view of this, we call each indifferently the *Barcan* formula[88].)

We have shown that if the Barcan formula were a thesis of LPC + T, (2) would be one too. And the converse also holds. We show this by deriving BF on the supposition that (2) is a thesis, as follows:

If (2) holds for all wffs α and β, then as a special case we have:

	(i)	$(\mathbf{a})L((\alpha \supset \alpha) \supset \alpha) \supset L((\mathbf{a})(\alpha \supset \alpha) \supset (\mathbf{a})\alpha)$
PC:	(ii)	$((\alpha \supset \alpha) \supset \alpha) \equiv \alpha$
(ii) × DR2:	(iii)	$L((\alpha \supset \alpha) \supset \alpha) \equiv L\alpha$
PC, UG:	(iv)	$(\mathbf{a})(\alpha \supset \alpha)$
PC:	(v)	$(\mathbf{a})(\alpha \supset \alpha) \supset [((\mathbf{a})(\alpha \supset \alpha) \supset (\mathbf{a})\alpha) \equiv (\mathbf{a})\alpha]$
(iv), (v) × MP:	(vi)	$((\mathbf{a})(\alpha \supset \alpha) \supset (\mathbf{a})\alpha) \equiv (\mathbf{a})\alpha$
(i), (iii), (vi) × Eq:	(vii)	$(\mathbf{a})L\alpha \supset L(\mathbf{a})\alpha$ **Q.E.D.**

Hence as additions to LPC + T, the Barcan formula and (2) – the strict implication analogue of LPC1 – stand or fall together.

The converse of the Barcan formula (i.e. $L(\mathbf{a})\alpha \supset (\mathbf{a})L\alpha$) is easily provable in LPC + T as it stands:

T2 $L(\mathbf{a})\alpha \supset (\mathbf{a})L\alpha$

PROOF

$\forall$1:	(1)	$(\mathbf{a})\alpha \supset \alpha$
(1) × DR1:	(2)	$L(\mathbf{a})\alpha \supset L\alpha$
(2) × $\forall$2:	(3)	$L(\mathbf{a})\alpha \supset (\mathbf{a})L\alpha$ **Q.E.D.**

(Step (3) is legitimate since clearly **a** cannot be free in $L(\mathbf{a})\alpha$.)

Hence if we had the Barcan formula we could easily derive:

$$L(\mathbf{a})\alpha \equiv (\mathbf{a})L\alpha$$

[88] And clearly, in systems containing N and A5, it does not matter whether the main operator of a thesis is $\supset$ or $\prec$.

and from this, by Def M, Def $\exists$ and PC equivalences,

$$M(\exists \mathbf{a})\alpha \equiv (\exists \mathbf{a})M\alpha$$

Two related theorems which can be proved without the Barcan formula are:

T3 $M(\mathbf{a})\alpha \supset (\mathbf{a})M\alpha$

(whose proof is as for T2, but using DR3 (p. 37) instead of DR1), and its straightforward derivative:

T4 $(\exists \mathbf{a})L\alpha \supset L(\exists \mathbf{a})\alpha$

The converses of T3 and T4, however, are not provable even with the Barcan formula. To see that under the intended interpretation the converse of T4 is not valid, consider the simple substitution-instance $L(\exists x)\phi x \supset (\exists x)L\phi x$, and let ϕx be 'x is the number of the planets'. Then the antecedent is true, for there must be some number which is the number of the planets (even if there were no planets at all there would still be such a number, viz. 0): but the consequent is false, for since it is a contingent matter how many planets there are, there is no number which *must* be the number of the planets. It is equally easy to see that the converse of T3 is not valid. (See, however, pp. 196–198.)

As we have already said, the Barcan formula is not a thesis of LPC + T or of LPC + S4. It is, however, consistent with each of these systems, and it could be added to either without strengthening it to LPC + S5. It is convenient to think of ourselves as having two versions of each of these systems, one without and one with the Barcan formula as an axiom. We shall call the former, as we have already been doing, LPC + T and LPC + S4; and we shall call the latter T + BF and S4 + BF[89]. In LPC + S5, however, the Barcan formula is a theorem[90]. The proof is as follows:

BF $(\mathbf{a})L\alpha \supset L(\mathbf{a})\alpha$

[89] Later on we shall be considering other variants of all three systems. When we want to refer to the modal predicate calculi without specifying a particular version we shall speak of 'quantificational T (S4, S5)', or (where no confusion with the propositional systems can arise) simply of T, S4 or S5.

[90] Prior [1956]. But *vide infra*, pp. 178–182, for a version of LPC for which BF is not a theorem of quantificational S5.

PROOF IN LPC + S5

$\forall 1$:	(1) $(\mathbf{a})L\alpha \supset L\alpha$
(1) $\times$ DR3:	(2) $M(\mathbf{a})L\alpha \supset ML\alpha$
(2), T34 $\times$ Syll:	(3) $M(\mathbf{a})L\alpha \supset \alpha$
(3) $\times \forall 2$:	(4) $M(\mathbf{a})L\alpha \supset (\mathbf{a})\alpha$
(4) $\times$ DR5:	(5) $(\mathbf{a})L\alpha \supset L(\mathbf{a})\alpha$ **Q.E.D.**

Note that all the modal theses and rules used in the above proof are in the Brouwerian system (*vide* pp. 57f); the Barcan formula is therefore a thesis of LPC + B (i.e. the system which results from adding the axiom-schema $\alpha \supset LM\alpha$ to LPC + T)[91].

The consistency of modal LPC

It is not difficult to show that the systems we have been considering are consistent. In outline the proof is as follows; the details are left to the reader.

For any wff, α, of modal LPC form its PC-transform, α', in the following way:

1. Eliminate all occurrences of $\prec$ and $=$ by using the definitions.
2. Delete all modal operators, quantifiers and individual variables.
3. Replace each distinct predicate variable by a distinct propositional variable.

The resulting expression, α', will be a wff of PC. A survey will show that the PC-transform of every axiom of T + BF, S4 + BF and LPC + S5 is a *valid* wff of PC. Moreover, if β is obtained by MP from two wffs, α and $\alpha \supset \beta$, each of which has a valid PC-transform, then the PC-transform of β will be valid too; if $L\alpha$ is obtained by N from a wff α with a valid PC-transform, then the PC-transform of $L\alpha$, being identical with that of α, will obviously be valid too; and for $\forall 2$ the situation is similar. Hence the PC-transform of every thesis of T + BF, S4 + BF or LPC + S5 is valid; and from this the consistency of these systems follows in the way that was explained for T on pp. 41f.

Validity in modal LPC

We shall now give definitions of validity in T + BF, S4 + BF

[91] Prior [1967] p. 146 credits the first proof of this to E. J. Lemmon.

and LPC + S5[92]. Essentially our procedure will be to combine the definitions of validity for the propositional systems given in Chapter 4 with that for LPC given earlier in this chapter. Later on we shall give alternative definitions for systems without the Barcan formula.

We begin by defining a T + BF-model. Such a model is an ordered quadruple ⟨W, R, D, V⟩, where W is a set of 'worlds' (in the sense of Chapter 4), R is a reflexive relation over the members of W, D is a set (or domain) of individuals, and V is a value-assignment satisfying conditions which we shall now set out and explain.

We want any particular value-assignment, V, to enable us to calculate a value (1 or 0) for any wff of modal LPC in any given world. And just as in the propositional systems of Part I we allowed the propositional variables to take different values in different worlds, so here we want to allow atomic formulae to take different values in different worlds. We can achieve this as follows.

As in the case of non-modal LPC, we let V assign some member, u, of D to each individual-variable. (We again write '$V(x) = u$' for 'V assigns u to x', and so forth.) This time, however, we cannot let the value of an n-adic predicate variable, ϕ, be simply a set of ordered n-tuples of members of D, and then define the value of an atomic formula as before, since this would result in every atomic formula having the same value in every world. What we shall do instead is, for every n-adic ϕ, to let $V(\phi)$ be a set of ordered $(n + 1)$-tuples, each of the form $\langle u_1, \ldots, u_n, w_i \rangle$, where each of $u_1, \ldots, u_n \in D$ and $w_i \in W$. Given $V(x_k)$ for every individual-variable x_k and $V(\phi)$ for every predicate variable ϕ, we can then calculate $V(\alpha, w_i)$ for every wff α and every $w_i \in W$ by the following rules, which are obviously adaptations of those for propositional T and for LPC.

1. For atomic formulae: If ϕ is any n-adic predicate variable, then $V(\phi(x_1, \ldots, x_n), w_i) = 1$ if $\langle V(x_1), \ldots, V(x_n), w_i \rangle \in V(\phi)$. Otherwise $V(\phi(x_1, \ldots, x_n), w_i) = 0$.

[92] The definitions which follow are adapted from Kripke [1963b]. Kripke, however, assumes a different domain of individuals for each world, and as a result obtains a semantics which does not, as ours does, validate the Barcan formula; *vide infra*, pp. 178–183, *vide* also Hintikka [1961] and Bayart [1958].

2. [V~]. For any wff α and any $w_i \in W$, $V(\sim\alpha, w_i) = 1$ if $V(\alpha, w_i) = 0$. Otherwise $V(\sim\alpha, w_i) = 0$.

3. [Vv]. For any wffs α and β and any $w_i \in W$, $V((\alpha \vee \beta), w_i) = 1$ if either $V(\alpha, w_i) = 1$ or $V(\beta, w_i) = 1$. Otherwise $V((\alpha \vee \beta), w_i) = 0$.

4. [V∀]. For any wff α, any individual-variable **a**, and any $w_i \in W$, $V((\mathbf{a})\alpha, w_i) = 1$ if for every T + BF-assignment V′ which makes the same assignment as V does to all variables other than **a**, $V'(\alpha, w_i) = 1$. Otherwise $V((\mathbf{a})\alpha, w_i) = 0$.

5. [VL]. For any wff α and any $w_i \in W$, $V(L\alpha, w_i) = 1$ if for every $w_j \in W$ such that $w_i R w_j$, $V(\alpha, w_j) = 1$. Otherwise $V(L\alpha, w_i) = 0$.

We can now define validity for T + BF by saying that a wff, α, is T + BF-valid iff for every T + BF-model $\langle W, R, D, V\rangle$, $V(\alpha, w_i) = 1$ for every $w_i \in W$.

For S4 + BF we make the additional requirement that R be transitive, and for LPC + S5 that it be transitive and symmetrical. The definitions of models and validity are otherwise the same as for T + BF.

All the theses of the axiomatic systems T + BF, S4 + BF and LPC + S5 are valid by these definitions. The proof of this follows the standard method of showing that all the axioms are valid and that the rules preserve validity.

That PC and the modal axiom-schemata A5–A8 are valid[93] can be seen by going through the analogous proofs for the propositional systems on pp. 68f, but with LPC formulae in mind. The same holds for the validity-preservingness of the rules MP and N.

The validity of ∀1 and ∀2 can be shown by straightforward adaptations of the proofs for non-modal LPC on pp. 140f.

This leaves only the Barcan formula, $(\mathbf{a})L\alpha \supset L(\mathbf{a})\alpha$. Suppose that for some world, w_i, in a T + BF-model, $V((\mathbf{a})L\alpha, w_i) = 1$. Let V′ differ from V only in assignment to **a**, and let w_j be any world such that $w_i R w_j$.

Then (by [V∀]):

$$V'(L\alpha, w_i) = 1, \text{ for every } V'.$$

[93] To say that an axiom-*schema* is valid is to say that every wff which is an instance of it is valid.

Hence (by [V*L*]):

$$V'(\alpha, w_j) = 1, \text{ for every } V' \text{ and every } w_j.$$

Hence (by [V∀]):

$$V((\mathbf{a})\alpha, w_j) = 1, \text{ for every } w_j.$$

Hence finally (by [V*L*]):

$$V(L((\mathbf{a})\alpha), w_i) = 1.$$

Therefore the Barcan formula is T + BF-valid (and hence of course S4 + BF-valid and LPC + S5-valid as well).

In the next chapter we shall prove that the axiomatic systems are also complete relative to our definitions of validity; i.e., that for any wff, α, if α is T + BF (S4 + BF, LPC + S5)-valid, then α is a thesis of T + BF (S4 + BF, LPC + S5).

The monadic modal LPC

We mentioned on p. 137n that although non-modal LPC is not a decidable system, yet a number of fragments of it are decidable. Of these the best-known is the *monadic Lower Predicate Calculus*, i.e. that fragment of LPC in which all predicate variables are monadic ones. In view of this we might expect that the monadic modal LPC would likewise be a decidable system. Kripke, however, has shown [94] that it is not. His proof depends on the standard result for non-modal LPC that the fragment which contains no predicate variables except a single dyadic one (say ϕ) is undecidable [95]. Let us call this fragment the *single dyadic LPC*. Briefly, Kripke's method is to prove that if we take any wff, α, of this calculus and in it replace every expression of the form $\phi\mathbf{ab}$ (where **a** and **b** are any individual-variables) by $M(\psi\mathbf{a} \,.\, \chi\mathbf{b})$, we obtain a wff of the monadic modal LPC, α', which is such that α is valid iff α' is valid. From this it follows that if the single dyadic LPC is undecidable, so is the monadic modal LPC – and indeed even that still smaller fragment of modal LPC in which not more than two monadic predicate variables occur. This result of Kripke's holds for LPC + S5 and for all systems of modal LPC contained therein.

[94] The proof appears in Kripke [1962], which also contains a discussion of a purported decision procedure in Poliferno [1961].

[95] Kalmar [1936]. (*Vide* Church [1956] pp. 272–279.)

CHAPTER NINE

The Completeness of Modal LPC

In 1949 Leon Henkin[96] published a completeness proof for LPC which, though fairly complicated, was simpler than any which had previously appeared. We shall call any completeness proof (whether for LPC or for any other system) which uses the same general principles as Henkin's does, a *Henkin proof*. Later in this chapter we shall use proofs of this kind to show the completeness of T + BF, S4 + BF and LPC + S5.

Henkin proofs can be applied to PC and to propositional modal systems such as T, S4 and S5[97], and the most straightforward introduction to the topic is to show first of all how they work for these simpler systems. The reader will then be in a better position to approach the more complicated Predicate Calculus cases.

The general structure of a Henkin proof

The first notion we shall introduce is that of the *consistency of a formula* with respect to a given system. We have already met the notion of the consistency of a *system*. The present notion, though related, is somewhat different. A formula α of a system S is said to be *consistent with respect to S* (or, where no confusion is likely

[96] Henkin [1949].

[97] Henkin himself mentions this as far as PC is concerned. He also remarks, however, that the standard completeness proofs for PC are to be preferred since, unlike his own, they start from a decision procedure and show how to construct, for any wff α which turns out valid by that decision procedure, a derivation of α from the axiomatic basis. Our own completeness proofs for T, S4 and S5 in Chapters 5 and 6 have also been of this kind. Such proofs are indeed to be preferred where they may be had; but for LPC (and its modal extensions) they are not to be had, since, as Church proved in [1936], LPC is not a decidable system. Completeness proofs for LPC systems can in fact be based on the methods of Chapters 5 and 6 (using semantic tableaux), though these methods do not of course yield a decision procedure in such cases.

to arise, simply *consistent*) iff $\sim\alpha$ is not a thesis of S[98]. What this amounts to is that the negation of a thesis counts as inconsistent (i.e. not consistent), but every other formula counts as consistent. By extension we say that a finite set $\{\alpha_1, \ldots, \alpha_n\}$ of formulae of S is consistent iff

$$\dashv_S \sim(\alpha_1 \cdot \ldots \cdot \alpha_n)$$ [99]

(Here '$\dashv_S \alpha$' is a convenient abbreviation for 'α is not a thesis of S'.) Finally, if Λ is an infinite set of formulae we shall say that Λ is consistent iff it contains no inconsistent finite subset of formulae, i.e. no subset $\{\alpha_1, \ldots, \alpha_n\}$ such that

$$\vdash_S \sim(\alpha_1 \cdot \ldots \cdot \alpha_n)$$

The connection between consistency in this sense and completeness is as follows. To prove that a system, S, is complete is to prove that if any formula α is S-valid, then $\vdash_S \alpha$; or, what comes to the same thing, that if $\dashv_S \alpha$, then α is not S-valid. Now $\dashv_S \alpha$ iff $\sim\alpha$ is consistent (with respect to S); and one method of showing that α is not S-valid is to construct a falsifying S-model for α (i.e. one in which $V(\alpha) = 0$) – which will of course be a verifying S-model for $\sim\alpha$ (i.e. one in which $V(\sim\alpha) = 1$)[100]. So an alternative way of proving that S is complete is to show that for every consistent formula of S there is a verifying S-model.

And this is basically what a Henkin proof proves about a system. It does so by showing (a) that if α is any *consistent* formula of a system S, we can construct a special kind of set of formulae of S, called a *maximal consistent set*, which contains α; and (b) that we can then form an S-model which verifies every formula in the set, and hence *inter alia* α itself.

Maximal consistent sets

We have defined the term 'consistent' as applied to a set of

[98] This account of consistency presupposes that S contains an identifiable negation operator – a requirement complied with by all the systems we shall consider here. The reason for using the term 'consistent' should be sufficiently clear.

[99] This presupposes that S contains an identifiable conjunction operator.

[100] In the case of the modal systems, a falsifying model for α will be one in which for some $w_i \in W$, $V(\alpha, w_i) = 0$, and a verifying model for $\sim\alpha$ one in which for some $w_i \in W$, $V(\sim\alpha, w_i) = 1$.

formulae. But obviously there will be many consistent sets of formulae which are in a certain sense incomplete – incomplete in the sense that there are formulae not already in them which could be consistently added to them. Consider, for example, the set $\{p, p \supset q, r\}$. We could not consistently add $\sim p$ to this set (assuming that S contains PC), for then it would become $\{p, \sim p, p \supset q, r\}$, and since $\vdash \sim(p \,.\, \sim p \,.\, (p \supset q) \,.\, r)$, this set would be inconsistent. But we *could* consistently add, say, s or $\sim s$ (though not both) or $(p \,.\, q)$ or $(p \,.\, r)$ or an infinite number of others if properly selected. Such formulae, we shall say, are *consistent with* the original set, though not *in* it. Now a *maximal consistent set* of formulae is a consistent set which is not incomplete in the sense we have explained; i.e. it is one for which any formula not already in it would, if added, make the set inconsistent. In other words, a set of formulae of S is maximal consistent iff it is consistent and every formula of S which is not in it is inconsistent with it.

What we have first to show is that if we are given a consistent formula, α (or a consistent set of formulae, Λ) of S, we can always construct a maximal consistent set, Γ, which contains α (or Λ).

To do this we assume that all the formulae of S are arranged in some kind of determinate order, so that we can speak of the 1st, 2nd, . . . , nth, . . . etc. formula of S.[101]

Let us call these formulae, $\alpha_1, \alpha_2, \ldots, \alpha_n, \ldots$, respectively. We then form Γ by starting with α[102] (or Λ) and then running through all the formulae of S in the order we have chosen, adding each in turn iff we can consistently do so at that point.

To express this more strictly: We construct Γ by forming a series of sets, $\Gamma_0, \Gamma_1, \ldots, \Gamma_n, \ldots$, in the following way. Γ_0 is $\{\alpha\}$[103]. To form Γ_1 we take α_1 (the first formula of S in the chosen ordering), and if α_1 is consistent with Γ_0 (i.e. if $\dashv_S \sim(\alpha \,.\, \alpha_1)$) then we let Γ_1 be $\{\alpha, \alpha_1\}$, but if it is not then we let Γ_1 simply be Γ_0 itself. We form each subsequent set in the series analogously: given Γ_n, we form Γ_{n+1} by taking α_{n+1} and if $\Gamma_n \cup \{\alpha_{n+1}\}$ – i.e. the set consisting of all the members of Γ_n together with α_{n+1} – is consistent, then we let Γ_{n+1} be this set, but if it is not then we

[101] It should be clear that something rather analogous to the alphabetical ordering of a dictionary could be used for this purpose.

[102] More accurately, $\{\alpha\}$ – the set whose only member is α.

[103] Or Λ – and similarly in what follows.

let Γ_{n+1} be Γ_n. Finally we let Γ be the set of all formulae which are in any set in the series $\Gamma_0, \Gamma_1, \ldots, \Gamma_n, \ldots$

Now clearly each set in this series is consistent; for Γ_0 ($\{\alpha\}$) is consistent by hypothesis, and each Γ_{n+1} is consistent if Γ_n is. It follows that Γ itself is consistent. For let Λ be any finite subset of Γ, and let α_n be the last formula (in the chosen ordering) in Λ. Then Λ is a subset of Γ_n. But Γ_n is consistent, and any subset of a consistent set must also be consistent. Hence Γ contains no finite inconsistent subsets, and is therefore itself consistent.

Moreover Γ is maximal. For consider any formula which is consistent with Γ. Let this be α_m – i.e. the *m*th in the chosen ordering. Since α_m is consistent with Γ it is consistent with any subset of Γ, and in particular with Γ_{m-1}. Therefore in the construction of Γ it will have been added to Γ_{m-1} to form Γ_m, and hence will be in Γ.

Thus for any consistent formula α of S we can construct a maximal consistent set Γ containing α.

Some properties of maximal consistent sets

Let us suppose that our system S contains some adequate axiomatic version of PC, such as PM (pp. 17f). All the systems with which we shall be dealing do in fact have this feature. We can then prove the following lemmas.

LEMMA 1

If Γ is maximal consistent relative to S, then for any wff α, α and $\sim\alpha$ are not both in Γ.

PROOF

If α and $\sim\alpha$ are both in Γ, then $\{\alpha, \sim\alpha\}$ is a subset of Γ. Hence since Γ is consistent, $\{\alpha, \sim\alpha\}$ is consistent, i.e.

$$\dashv_S \sim(\alpha \,.\, \sim\alpha)$$

But this is false, for, since S contains PC

$$\vdash_S \sim(\alpha \,.\, \sim\alpha)$$

for every wff α. Hence α and $\sim\alpha$ are not both in Γ.

LEMMA 2

If Γ is maximal consistent relative to S, then for any wff α, either $\alpha \in \Gamma$ or $\sim\alpha \in \Gamma$.

PROOF

Suppose that neither $\alpha \in \Gamma$ nor $\sim\alpha \in \Gamma$. Then since Γ is maximal, for some finite subset Γ_i of Γ, $\Gamma_i \cup \{\alpha\}$ is inconsistent, and for some finite subset Γ_j of Γ, $\Gamma_j \cup \{\sim\alpha\}$ is inconsistent; i.e., writing 'β' for the conjunction of all the wffs in Γ_i and 'γ' for the conjunction of all the wffs in Γ_j,

$$\vdash_S \sim(\beta \, . \, \alpha)$$

and

$$\vdash_S \sim(\gamma \, . \sim\alpha)$$

whence by $\vdash \sim(p \, . \, q) \supset (\sim(r \, . \sim q) \supset \sim(p \, . \, r))$ (since S contains PC),

$$\vdash_S \sim(\beta \, . \, \gamma)$$

But this means that $\Gamma_i \cup \Gamma_j$ is inconsistent, and therefore, since $\Gamma_i \cup \Gamma_j$ is a finite subset of Γ, that Γ itself is inconsistent, contrary to the hypothesis.

It is a simple corollary of Lemma 2 that every thesis is in every maximal consistent set. For if $\vdash\alpha$, then $\sim\alpha$ is inconsistent and hence cannot be in any consistent set. So if Γ is maximal consistent, $\sim\alpha$ is not in Γ, and therefore (by Lemma 2) α is.

LEMMA 3

If Γ is maximal consistent relative to S, then for any wffs α and β, if $\alpha \in \Gamma$ and $(\alpha \supset \beta) \in \Gamma$, $\beta \in \Gamma$.

PROOF

Suppose α and $(\alpha \supset \beta)$ were in Γ but β were not. Then (by Lemma 2) $\sim\beta$ would be in Γ, and hence $\{\alpha, (\alpha \supset \beta), \sim\beta\}$ would be a subset of Γ. But, since S contains PC, we have:

$$\vdash_S \sim(\alpha \, . \, (\alpha \supset \beta) \, . \sim\beta)$$

Hence the subset in question, and therefore Γ itself, would be inconsistent, which by hypothesis it is not.

Since, by the corollary of Lemma 2, if $\vdash(\alpha \supset \beta)$ then $(\alpha \supset \beta)$ is in every maximal consistent set, we have the following corollary of Lemma 3, whose proof is obvious:

If Γ is maximal consistent relative to S, then for any wffs α and β, if $\alpha \in \Gamma$ and $\vdash_S(\alpha \supset \beta)$, $\beta \in \Gamma$.

This corollary, though weaker than Lemma 3, is often more useful in practice than the Lemma itself. We shall refer both to the Lemma and to the corollary simply as *Lemma 3*.

The completeness of PM

Suppose we wish to prove that the axiomatic system PM is complete for PC. In the light of the foregoing, this will be done if we can prove that for every wff, α, which is consistent with respect to PM, there is a PC assignment, V, such that $V(\alpha) = 1$.

We have shown how, given that α is consistent, we can construct a maximal consistent set, Γ, containing α. In the present case, we assume α to be consistent with respect to PM and we make Γ maximal consistent also with respect to PM. Having constructed Γ, we make the following PC value-assignment, V: to each propositional variable p_i we assign 1 if it is one of the wffs in Γ and 0 if it is not; the values assigned to all other wffs can be calculated by the standard PC rules for $\sim$ and $\vee$ given on p. 10 ([V$\sim$] and [V$\vee$]).

We now state the following theorem:

THEOREM 1

Given V *as defined above, for every PC wff,* β, $V(\beta) = 1$ *or* 0 *according as* $\beta \in \Gamma$ *or not.*

(Less formally expressed, what Theorem 1 states is that if Γ is maximal consistent with respect to PM, then assigning 1 to all the *variables* in Γ (and 0 to all the others) results in assigning 1 to all the *wffs* without exception in Γ (and 0 to all the others).)

Clearly, since by hypothesis our original consistent formula, α, is in Γ, it will follow from Theorem 1 that $V(\alpha) = 1$, and the completeness of PM will thus be established.

We prove Theorem 1 by induction in the construction of PC formulae.

1. If β is a propositional variable, the theorem holds for β by the initial value-assignment to propositional variables.

2. If the theorem holds for a wff β (induction hypothesis), then it also holds for $\sim\beta$. For (a) if $\sim\beta \in \Gamma$, then (by Lemma 1)[104]

[104] In proving the completeness of PM we need of course to satisfy ourselves that the PC theses required to prove Lemmas 1–3 actually *are* derivable in PM.

$\beta \notin \Gamma$ (i.e. β is not in Γ); hence (by induction hypothesis) $V(\beta) = 0$; and hence $V(\sim\beta) = 1$. And (b) if $\sim\beta \notin \Gamma$, then (by Lemma 2) $\beta \in \Gamma$; hence (by induction hypothesis) $V(\beta) = 1$; and hence $V(\sim\beta) = 0$.

3. If the theorem holds for β and for γ (induction hypothesis), then it also holds for $(\beta \vee \gamma)$.

(a) Suppose that $(\beta \vee \gamma) \in \Gamma$. In that case either $\beta \in \Gamma$ or $\gamma \in \Gamma$, for the following reason: if neither $\beta \in \Gamma$ nor $\gamma \in \Gamma$, then (by Lemma 2) $\sim\beta \in \Gamma$ and $\sim\gamma \in \Gamma$, and hence (by Lemma 3), since $\vdash_{PM}(\sim\beta \supset (\sim\gamma \supset \sim(\beta \vee \gamma)))$, $\sim(\beta \vee \gamma) \in \Gamma$; but then (by Lemma 1), $(\beta \vee \gamma) \notin \Gamma$, which contradicts the original supposition. Hence if $(\beta \vee \gamma) \in \Gamma$, then either $\beta \in \Gamma$ or $\gamma \in \Gamma$. Now if $\beta \in \Gamma$, then (by induction hypothesis) $V(\beta) = 1$, and therefore $V(\beta \vee \gamma) = 1$; and likewise if $\gamma \in \Gamma$, $V(\gamma) = 1$, and therefore once more $V(\beta \vee \gamma) = 1$.

(b) Suppose that $(\beta \vee \gamma) \notin \Gamma$. Then (by Lemma 2) $\sim(\beta \vee \gamma) \in \Gamma$; hence (by Lemma 3), since $\vdash_{PM}(\sim(\beta \vee \gamma) \supset \sim\beta)$ and $\vdash_{PM}(\sim(\beta \vee \gamma) \supset \sim\gamma)$, $\sim\beta \in \Gamma$ and $\sim\gamma \in \Gamma$; and hence in turn (by Lemma 1) $\beta \notin \Gamma$ and $\gamma \notin \Gamma$. By the induction hypothesis we then have $V(\beta) = 0$ and $V(\gamma) = 0$, and therefore $V(\beta \vee \gamma) = 0$.

Since every wff of PC (in primitive notation) is either a propositional variable or else formed from propositional variables by $\sim$ and $\vee$ in accordance with the formation rules, Theorem 1 holds for every wff of PC.

The completeness of T, S4 and S5

We now show how to give a Henkin proof of the completeness of T. The general principle of the proof is the same as before: we show that if α is any formula which is consistent with respect to T, we can construct a verifying T-model for α, i.e. a T-model $\langle W, R, V \rangle$ in which for some $w_i \in W$, $V(\alpha, w_i) = 1$[105].

As a preliminary we prove the following lemma, which holds for all systems containing T.

LEMMA 4

Where $\beta, \gamma_1, \ldots, \gamma_n$ are any wffs, if $\{L\gamma_1, \ldots, L\gamma_n, M\beta\}$ is consistent, then $\{\gamma_1, \ldots, \gamma_n, \beta\}$ is consistent.

PROOF

The proof proceeds by showing that if $\{\gamma_1, \ldots, \gamma_n, \beta\}$ is inconsistent, then so is $\{L\gamma_1, \ldots, L\gamma_n, M\beta\}$.

[105] Cf. Kaplan [1966], p. 122 (review of Kripke [1963a]), and Makinson [1966c].

Suppose that $\{\gamma_1, \ldots, \gamma_n, \beta\}$ is inconsistent, i.e. that $\vdash \sim(\gamma_1 \ldots . \gamma_n . \beta)$. Then by N and LMI:

$$(1) \vdash \sim M(\gamma_1 \ldots . \gamma_n . \beta)$$

But substitution in T17 (× Imp) gives:

$$(2) \vdash (M\beta . L(\gamma_1 \ldots . \gamma_n)) \supset M(\beta . \gamma_1 \ldots . \gamma_n)$$

(1) and (2) by PC give:

$$\vdash \sim(M\beta . L(\gamma_1 \ldots . \gamma_n))$$

Hence by T3:

$$\vdash \sim(M\beta . L\gamma_1 \ldots . L\gamma_n)$$

and therefore $\{L\gamma_1, \ldots, L\gamma_n, M\beta\}$ is inconsistent. **Q.E.D.**

For a Henkin proof of the completeness of T we have to show how to construct, not a single maximal consistent set, but a whole system of such sets, starting with a given consistent formula α. We use the symbol Γ this time to refer to this whole system, and the maximal consistent sets composing it we call $\Gamma_1, \ldots, \Gamma_i, \ldots$

We first construct Γ_1 by taking $\{\alpha\}$ as the initial set and making Γ_1 maximal consistent (with respect to T) in the way already described. Having constructed Γ_1 we then construct for every Γ_i in Γ (including Γ_1 itself) further maximal consistent sets in accordance with the following plan. For every wff of the form $M\beta$[106] in Γ_i we construct a maximal consistent set Γ_j, by taking as Γ_{j0} (i.e. the initial set for Γ_j) the set consisting of β together with every wff γ such that $L\gamma \in \Gamma_i$. That is, we start Γ_j by taking the arguments of *one* occurrence of M and of *every* occurrence of L in Γ_i, and then make Γ_j maximal consistent by running through the wffs of T in the way we described earlier. This assumes that Γ_{j0} is itself consistent, but this assumption is easily justified as follows. Let $\{\gamma_1, \ldots, \gamma_n, \beta\}$ be any finite subset[107] of Γ_{j0}. Now $L\gamma_1, \ldots, L\gamma_n$ and $M\beta$ are all in Γ_i; and since Γ_i is consistent by hypothesis, $\{L\gamma_1, \ldots, L\gamma_n, M\beta\}$ is consistent. Hence by Lemma 4, $\{\gamma_1, \ldots, \gamma_n, \beta\}$ is consistent. Thus every finite subset of Γ_{j0} is consistent; i.e. Γ_{j0} is consistent.

[106] I.e. (unabbreviated) $\sim L \sim \beta$.

[107] Of course β will not be in every subset of Γ_{j0}; but clearly if $\{\gamma_1, \ldots, \gamma_n, \beta\}$ is consistent, so is $\{\gamma_1, \ldots, \gamma_n\}$.

If Γ_j is formed from Γ_i in the way we have just described, we say that Γ_j is a *subordinate* of Γ_i.

Γ, then, is the system of maximal consistent sets formed by beginning with some maximal consistent set Γ_1 and constructing a subordinate set for every wff of the form $M\beta$ in every set in Γ. (Note that Γ will contain an infinite number of sets.)

To summarise: the whole system Γ has the following properties: for each Γ_i in Γ,

1. Γ_i is maximal consistent.
2. For every $M\beta \in \Gamma_i$ there is a subordinate maximal consistent set, Γ_j, such that $\beta \in \Gamma_j$ and for every $L\gamma \in \Gamma_i$, $\gamma \in \Gamma_j$.

We can easily form a T-model $\langle$W, R, V$\rangle$ on the basis of Γ. With each Γ_i in Γ we associate a 'world' in our model. (For simplicity we can call the worlds associated with Γ_i, Γ_j etc., w_i, w_j etc. respectively.) Let W be the set of all such worlds. Let R be the relation such that $w_i \mathrm{R} w_j$ iff either Γ_j is a subordinate of Γ_i or else Γ_j is Γ_i itself. Let V be the following value-assignment: for every propositional variable p_k and every $w_i \in \mathrm{W}$, $\mathrm{V}(p_k, w_i) = 1$ or 0 according as $p_k \in \Gamma_i$ or not; and for $\sim$, $\vee$ and L, V complies with the standard conditions for a T-assignment, viz. [V$\sim$], [V$\vee$] and [VL]. It is then clear that $\langle$W, R, V$\rangle$ is a T-model.

We now state a theorem which stands to T as Theorem 1 does to PM.

THEOREM 2

Given W, R *and* V *as defined above, for every wff*, β, *of* T *and for every* $w_i \in \mathrm{W}$, $\mathrm{V}(\beta, w_i) = 1$ *or* 0 *according as* $\beta \in \Gamma_i$ *or not.*

Since our initial consistent formula, α, is in Γ_1, it will follow from Theorem 2 that there is some $w_i \in \mathrm{W}$ (viz. w_1) for which $\mathrm{V}(\alpha, w_i) = 1$, and the completeness of T will thus be established.

Theorem 2 is proved by induction on the construction of wffs of T. For variables, $\sim$ and $\vee$ the proof is as for Theorem 1 with obvious modifications. It remains therefore to prove:

If Theorem 2 holds for a wff β (induction hypothesis), then it holds for $L\beta$.

PROOF

(a) Suppose $L\beta \in \Gamma_i$. Then for every Γ_j subordinate to Γ_i, $\beta \in \Gamma_j$ (by construction of Γ_j); hence (by induction hypothesis)

$V(\beta, w_j) = 1$. Furthermore (by Lemma 3), since $\vdash_T(L\beta \supset \beta)$, $\beta \in \Gamma_i$, and so (by induction hypothesis) $V(\beta, w_i) = 1$. Thus for every w_j such that $w_i R w_j$, $V(\beta, w_j) = 1$; hence $V(L\beta, w_i) = 1$.

(b) Suppose $L\beta \notin \Gamma_i$. Then (by Lemma 2) $\sim L\beta \in \Gamma_i$, and hence, since $\vdash_T(\sim L\beta \supset M\sim\beta)$, $M\sim\beta \in \Gamma_i$. Therefore (by construction of Γ) there is some Γ_j subordinate to Γ_i such that $\sim\beta \in \Gamma_j$, and therefore (by Lemma 1) $\beta \notin \Gamma_j$. Hence (by induction hypothesis) $V(\beta, w_j) = 0$. I.e. for some w_j such that $w_i R w_j$, $V(\beta, w_j) = 0$, and so $V(L\beta, w_i) = 0$.

Therefore Theorem 2 holds for every wff of T.

To prove the completeness of S4 by this method we proceed as for T, with the following exceptions. Every Γ_i in Γ is now to be maximal consistent with respect to S4; and we modify the account of R so that $w_i R w_j$ iff Γ_j is a subordinate$_*$[108] of Γ_i (this takes care of the transitivity of R in an S4-model).

The only change this requires in the proof is that in showing that Theorem 2 holds for L we have to show that if $L\beta$ is in Γ_i then β is not only in Γ_i and in every subordinate of Γ_i, but in every subordinate$_*$ of Γ_i. And this is proved quite easily. Let Γ_j be a subordinate of Γ_i and Γ_k a subordinate of Γ_j. If $L\beta \in \Gamma_i$, then since $\vdash_{S4}(L\beta \supset LL\beta)$ $LL\beta \in \Gamma_i$. Hence $L\beta \in \Gamma_j$, and hence $\beta \in \Gamma_k$ (by construction of Γ_j and Γ_k). By induction on subordination the theorem therefore holds for any subordinate$_*$ of Γ_i.

To deal with S5 we must also require that R be symmetrical. We therefore have to say that whenever Γ_j is a subordinate of Γ_i, then not merely $w_i R w_j$ but $w_j R w_i$ as well. This means that we have to add to the S4 proof that Theorem 2 holds for L a proof that if Γ_j is subordinate to Γ_i, then if $L\beta \in \Gamma_j$, $\beta \in \Gamma_i$. (Γ_i and Γ_j are of course assumed to be maximal consistent with respect to S5.)

We prove this by showing that if β is not in Γ_i, neither is $L\beta$ in Γ_j. We do this as follows. If $\beta \notin \Gamma_i$, then (by Lemma 2)

108 A *subordinate$_*$* of Γ_i is a set which is either Γ_i itself or a subordinate of Γ_i or a subordinate of a subordinate of Γ_i or . . . Our notation here is derived from that used in Whitehead and Russell [1910] for an 'ancestral' relation. (We shall where convenient use 'accessible$_*$' and 'R$_*$' in an analogous way.)

$\sim\beta \in \Gamma_i$. Hence, since $\vdash(\sim\beta \supset \sim L\beta)$ and $\vdash_{S5}(\sim L\beta \supset L{\sim}L\beta)$, $L{\sim}L\beta \in \Gamma_i$. Therefore (by construction of Γ_j) $\sim L\beta \in \Gamma_j$; and so (by Lemma 1) $L\beta \notin \Gamma_j$.

The completeness of (non-modal) LPC[109]

We now show by a Henkin proof that the basis we gave for LPC in Chapter 8 is complete for LPC. Following our previous pattern, what we shall show is that for any wff which is consistent with respect to this basis we can always construct a verifying model.

Beginning with any consistent formula, α, we construct, as in the case of PC, a single maximal consistent set, Γ. Γ, of course, is to be maximal consistent with respect to LPC (i.e. with respect to the particular basis for LPC we have given). We also require that Γ shall have what we shall call the *E-property*. A set, Λ, is said to have the E-property iff for every wff of the form $(\exists \mathbf{a})\beta$[110] in Λ there is also in Λ some wff, for which we shall use the notation $\beta[\mathbf{b}/\mathbf{a}]$, which differs from β only in that wherever β has free $\mathbf{a}$, $\beta[\mathbf{b}/\mathbf{a}]$ has some individual-variable, $\mathbf{b}$, which is free in $\beta[\mathbf{b}/\mathbf{a}]$ but not free in β.

We now show how to ensure that Γ has the E-property. We begin with some definitions.

1. Any wff of the form $(\exists \mathbf{a})\beta \supset \beta[\mathbf{b}/\mathbf{a}]$ – where $\beta[\mathbf{b}/\mathbf{a}]$ is understood in the way just explained – we shall call an *E-formula with respect to* **b**. In such a formula we shall refer to **b** as the *replacement variable.*

2. All E-formulae which differ only in that each is an E-formula with respect to a different variable will be said to have the same *E-form.* (An E-form can be thought of as the set of all E-formulae with the same antecedent.)

Clearly the E-forms are enumerable.

3. A set of wffs will be said to have the *E′-property* iff it contains at least one E-formula of every E-form.

It is easy to show that if a *maximal consistent* set, Λ, has the E′-property, it also has the E-property. For suppose $(\exists \mathbf{a})\beta$ is in Λ. Since Λ has the E′-property, there is some variable **b** such that $(\exists \mathbf{a})\beta \supset \beta[\mathbf{b}/\mathbf{a}]$ is also in Λ. Therefore (by Lemma 3), $\beta[\mathbf{b}/\mathbf{a}]$ is in Λ.

[109] This section contains a modified version of the original proof in Henkin [1949].

[110] I.e. (unabbreviated) $\sim(\mathbf{a})\sim\beta$.

For a given consistent wff, α, we construct Γ as follows. We begin with $\{\alpha\}$. We then suppose that all the E-forms are arranged in some standard order, and for each one of them we add, in that order, an E-formula with respect to some variable which does not occur anywhere else in that E-formula, or in any preceding E-formula, or in α[111]. (Since we have an unlimited supply of individual-variables at our disposal, and since at each stage only a finite number of formulae are already in the set, there will always be a fresh variable available for this purpose.) Finally we increase the set to a maximal consistent one in the standard way[112]. Clearly Γ, as thus constructed, has the E′-property, and therefore the E-property.

We need to prove at this point that the set obtained when all the E-formulae are added is consistent, since otherwise the final stage in the construction could not even be begun. Since $\{\alpha\}$ is by hypothesis consistent, it is sufficient to prove the following lemma:

LEMMA 5

If Λ is a consistent set of wffs, then $\Lambda \cup \{(\exists \mathbf{a})\beta \supset \beta[\mathbf{b}/\mathbf{a}]\}$ is consistent, where **b** does not occur in any wff in Λ or in $(\exists \mathbf{a})\beta$.

PROOF

Let Λ' be any finite sub-set of Λ; then we have to prove that $\Lambda' \cup \{(\exists \mathbf{a})\beta \supset \beta[\mathbf{b}/\mathbf{a}]\}$ is consistent. Suppose it is not; i.e. (writing 'γ' for the conjunction of the wffs in Λ'),

$$\vdash \sim(\gamma \,.\, ((\exists \mathbf{a})\beta \supset \beta[\mathbf{b}/\mathbf{a}])$$

Then, by $\vdash(\sim(p \,.\, q) \supset (p \supset \sim q))$,

$$\vdash(\gamma \supset \sim((\exists \mathbf{a})\beta \supset \beta[\mathbf{b}/\mathbf{a}])$$

111 Equivalently we assume that the wffs of the form $(\exists \mathbf{a})\beta$ are arranged in a standard order and then add for each an E-formula with it as antecedent, etc.

112 More formally expressed, the construction of Γ is as follows. Γ_{00} is $\{\alpha\}$. Each Γ_{0i} ($i > 0$) is $\Gamma_{0(i-1)} \cup \{(\exists \mathbf{a})\beta \supset \beta[\mathbf{b}/\mathbf{a}]\}$ where $(\exists \mathbf{a})\beta$ is the ith wff of that form (in the standard ordering) and **b** does not occur in $(\exists \mathbf{a})\beta$ or in any member of $\Gamma_{0(i-1)}$. Γ_0 is the union of all Γ_{0j}'s ($j \geqslant 0$). Γ is formed from Γ_0 and the wffs of LPC in a standard ordering in the way explained earlier for the general case. Clearly Γ_0 will be consistent if each Γ_{0i} is.

Since **b** is not free in γ, we then have, by $\forall 2$,

$$\vdash(\gamma \supset (\mathbf{b}){\sim}((\exists \mathbf{a})\beta \supset \beta[\mathbf{b}/\mathbf{a}])$$

and hence by Transp and Def $\exists$,

$$\vdash((\exists \mathbf{b})((\exists \mathbf{a})\beta \supset \beta[\mathbf{b}/\mathbf{a}]) \supset {\sim}\gamma)$$

But by LPC4, since **b** is not free in β,

$$\vdash(\exists \mathbf{b})((\exists \mathbf{a})\beta \supset \beta[\mathbf{b}/\mathbf{a}])$$

Hence by MP:

$$\vdash{\sim}\gamma$$

i.e. Λ' is inconsistent, contrary to the original hypothesis of the consistency of Λ.

It follows that the set obtained when we have added all the E-formulae specified in the construction of Γ is consistent, and hence Γ itself is consistent (and of course maximal) for the reasons given earlier for the general case.

We now construct the LPC-model $\langle$D, V$\rangle$. D will have to be some infinite domain of individuals; for simplicity let it be the set of all the individual-variables themselves, considered as objects (letters, typographical entities). Let V assign to each individual-variable (considered as a variable) itself (considered as an object). Thus V(x) – the value of the *variable* x – is to be the *typographical entity* 'x', and so on[113]. Where ϕ is any n-adic predicate variable, let V(ϕ) be the set of just those ordered n-tuples $\{\langle x_1, \ldots, x_n\rangle, \langle y_1, \ldots, y_n\rangle, \ldots\}$ such that $\phi(x_1, \ldots, x_n)$, $\phi(y_1, \ldots, y_n) \ldots$ are in Γ. Then for any n-adic ϕ and any individual-variables $z_1, \ldots, z_n$, we shall have V($\phi(z_1, \ldots, z_n)$) = 1 or 0 according as $\phi(z_1, \ldots, z_n) \in \Gamma$ or not. I.e., if β is an atomic formula, V(β) = 1 or 0 according as $\beta \in \Gamma$ or not. All other wffs can then be evaluated by the standard rules for an LPC assignment ([V$\sim$], [V$\vee$], [V$\forall$]).

We can now state the crucial theorem for LPC:

THEOREM 3
Given D *and* V *as defined above, for every wff*, β, *of LPC*, V(β) = 1 *or* 0 *according as* $\beta \in \Gamma$ *or not.*

[113] Clearly any 1–1 correspondence between variables and a set of objects would do, but this is the simplest one to specify.

The completeness result follows from this theorem as in the previous cases.

The proof of the theorem is as before by induction on the construction of a wff (of LPC).

For atomic formulae the theorem holds by the initial assignment to the variables. For $\sim$ and $\vee$ the proof is as for PC (pp. 154f). It remains therefore to prove that the theorem holds for quantification.

Now if β is an atomic wff, the theorem holds not merely for β but for every substitution-instance (for individual-variables)[114] of β, since every such substitution-instance is itself an atomic wff. Moreover the proof for $\sim$ could easily be re-cast to show that if the theorem holds for every substitution-instance of β, then it holds for every substitution-instance of $\sim\beta$; and the position is similar for $\vee$. Hence for quantification we can take it as our induction hypothesis that the theorem holds for β and for every substitution-instance (for individual-variables) thereof, and from this we have to prove that it holds for $(\mathbf{a})\beta$ and for every such substitution-instance thereof.

(A) Suppose that $(\mathbf{a})\beta \in \Gamma$. We have to show that in that case $V((\mathbf{a})\beta) = 1$, i.e. that for every V′ differing from V only in the assignment it makes to **a**, $V'(\beta) = 1$. Consider any such V′. The value it assigns to **a** will of course still be some member of D, i.e. some individual-variable. Let this be **b**.

Now suppose (i) that **b** is a variable for which **a** is free in β[115]. Then since V′ makes the same assignments as V does except that whereas $V(\mathbf{a}) = \mathbf{a}$, $V'(\mathbf{a}) = \mathbf{b}$, the value which V′ assigns to β will be the same as V assigns to the formula which we expressed earlier as $\beta[\mathbf{b}/\mathbf{a}]$; i.e. $V'(\beta) = V(\beta[\mathbf{b}/\mathbf{a}])$. But by ∀1,

$$\vdash (\mathbf{a})\beta \supset \beta[\mathbf{b}/\mathbf{a}]$$

Therefore (by Lemma 3), since $(\mathbf{a})\beta \in \Gamma$, $\beta[\mathbf{b}/\mathbf{a}] \in \Gamma$; and so (by

[114] γ is a substitution-instance (for individual-variables) of β iff it results from β by the uniform replacement of any free individual-variable in β by any other individual-variable, provided that that individual-variable does not become bound in γ. By extension we also count substitution-instances of bound alphabetic variants of β (*vide* p. 139) as substitution-instances of β.

[115] I.e. that in β **a** does not occur within the scope of any quantifier containing **b**.

induction hypothesis), since $\beta[\mathbf{b}/\mathbf{a}]$ is a substitution-instance of β, $V(\beta[\mathbf{b}/\mathbf{a}]) = 1$. Hence $V'(\beta) = 1$.

Suppose however (ii) that **b** is a variable for which **a** is not free in β – i.e. that in β **a** does occur within the scope of (**b**) or ($\exists$**b**). In this case we form a bound alphabetic variant of β, in the sense explained on p. 139, in which neither (**b**) nor ($\exists$**b**) occurs. Let us call the resulting formula, γ. We now replace free **a** everywhere in γ by **b**; and since no quantifier containing **b** occurs in γ, the result is $\gamma[\mathbf{b}/\mathbf{a}]$. Since, however, β and γ are bound alphabetic variants, $\vdash(\mathbf{a})\beta \equiv (\mathbf{a})\gamma$ (by LPC9); hence, since $(\mathbf{a})\beta \in \Gamma$, $(\mathbf{a})\gamma \in \Gamma$. But by $\forall 1$, $\vdash(\mathbf{a})\gamma \supset \gamma[\mathbf{b}/\mathbf{a}]$; hence $\gamma[\mathbf{b}/\mathbf{a}] \in \Gamma$. Now $\gamma[\mathbf{b}/\mathbf{a}]$ is a substitution-instance of γ; and so (by induction hypothesis), $V(\gamma[\mathbf{b}/\mathbf{a}]) = 1$. Therefore by the same proof as in (i), $V'(\gamma) = V(\gamma[\mathbf{b}/\mathbf{a}]) = 1$. But since γ is a bound alphabetic variant of β, any LPC value-assignment will give the same value to each; so we have $V'(\beta) = V'(\gamma)$, and hence in this case too, $V'(\beta) = 1$[116].

(B) Suppose $(\mathbf{a})\beta \notin \Gamma$. We have to prove that $V((\mathbf{a})\beta) = 0$, i.e. that for some V′ differing from V only in assignment to **a**, $V'(\beta) = 0$. Since $(\mathbf{a})\beta \notin \Gamma$, $\sim(\mathbf{a})\beta \in \Gamma$ (by Lemma 2); hence by LPC3 and Lemma 3, $(\exists\mathbf{a})\sim\beta \in \Gamma$. Therefore since Γ has the

[116] An example may make this clearer. Let **a** be x and let β be $(\exists y)\phi xy$. Then $(\mathbf{a})\beta$ is $(x)(\exists y)\phi xy$. We suppose that $(x)(\exists y)\phi xy \in \Gamma$, and we want to prove that in that case $V((x)(\exists y)\phi xy) = 1$; i.e. that for every V′ differing from V only in assignment to x, $V'((\exists y)\phi xy) = 1$. The induction hypothesis is that every substitution-instance of $(\exists y)\phi xy$ has the value 1 iff it is in Γ. Now the x in $(\exists y)\phi xy$ is free for every variable except y; and V′ must be such that $V'(x) = y$, or else such that $V'(x)$ is some other variable, say z. If the latter, we have case (i); $\beta[\mathbf{b}/\mathbf{a}]$ will then be $(\exists y)\phi zy$ (**b** is just $V'(x)$ in each case); and it should be clear that in that case the value V′ gives to $(\exists y)\phi xy$ will be the same as V (which assigns every variable to itself) gives to this formula $(\exists y)\phi zy$. I.e., $V'((\exists y)\phi xy) = V((\exists y)\phi zy)$. But $\forall 1$ ensures that $(\exists y)\phi zy$ is in Γ, and moreover $(\exists y)\phi zy$ is a substitution-instance of $(\exists y)\phi xy$, so the induction hypothesis applies to it (i.e. $V((\exists y)\phi zy) = 1$). Therefore for this V′, and for any except one which makes $V'(x) = y$, $V'((\exists y)\phi xy) = 1$.

Turning now to the more awkward case when $V'(x) = y$, we choose some variable other than y, say w, and form a bound alphabetic variant of $(\exists y)\phi xy$, viz. $(\exists w)\phi xw$ (this is the γ of case (ii)). In *this* we can replace x by y without binding y, and hence obtain a substitution-instance of $(\exists y)\phi xy$, viz. $(\exists w)\phi yw$. LPC9 and $\forall 1$ assure us that if $(x)(\exists y)\phi xy$ is in Γ, so is $(\exists w)\phi yw$, and since $(\exists y)\phi xy$ and $(\exists w)\phi xw$ must have the same value, the proof then proceeds as in (i). So for *every* V′ that we have to consider, $V'((\exists y)\phi xy) = 1$, which is what we had to prove.

E-property, there is also in Γ some wff $\sim\beta[\mathbf{b}/\mathbf{a}]$; and hence (by Lemma 1) $\beta[\mathbf{b}/\mathbf{a}] \notin \Gamma$. Now $\beta[\mathbf{b}/\mathbf{a}]$ is a substitution-instance of β, and so (by the induction hypothesis) $V(\beta[\mathbf{b}/\mathbf{a}]) = 0$. Let V′ differ from V only in that $V'(\mathbf{a}) = \mathbf{b}$. Then for this V′, $V'(\beta) = V(\beta[\mathbf{b}/\mathbf{a}])$, and hence $V'(\beta) = 0$.

For any substitution-instance of $(\mathbf{a})\beta$ an analogous proof can clearly be constructed from the same induction hypothesis; hence Theorem 3 holds for quantification; and this completes the proof that it holds for all wffs of LPC.

The completeness of modal LPC

We are now in a position to deal with the modal predicate calculi. We shall first give a Henkin proof of the completeness of T + BF, and then indicate the modifications required for S4 + BF and LPC + S5. In broad terms our methods will be to combine the proofs given for the propositional modal systems and for LPC in the previous two sections[117].

As before, we prove completeness by showing that for any consistent formula of the system with which we are dealing we can construct a verifying model. Given a formula, α, which is consistent with respect to T + BF, we show how to construct, beginning with α, a system of maximal consistent sets, Γ, in which the sets are related to each other as they are in the case of T, and in which each set has the E-property. To be more explicit, Γ is to have the following features:

1. $\alpha \in \Gamma_1$.
2. For every Γ_i in Γ and for every wff of the form $M\beta$ in Γ_i there is a subordinate set Γ_j such that (i) $\beta \in \Gamma_j$ and (ii) for every wff of the form $L\gamma$ in Γ_i, $\gamma \in \Gamma_j$.
3. For every Γ_i in Γ and for every wff of the form $(\exists\mathbf{a})\beta$ in Γ_i there is in Γ_i a wff $\beta[\mathbf{b}/\mathbf{a}]$, differing from β only in that free **b** replaces free **a** everywhere.

The construction will not, however, be as straightforward as in the previous cases, for the following reason. Suppose we are

[117] Other completeness proofs exist for some modal predicate calculi. Thus Kripke [1959] bases a proof for LPC + S5 on semantic tableaux. Bayart [1959] has also proved the completeness of LPC + S5 by a method similar to ours, though a little simpler since S5 does not require a relation R. Bayart (op. cit.) also proves the completeness of second-order quantificational S5 (in the sense of completeness defined for higher-order calculi in Henkin [1950]).

to construct a maximal consistent set Γ_j for a wff $M\beta$ in Γ_i. The natural plan might seem to be this:

(i) Start with $\{\beta\}$.
(ii) Add every wff γ such that $L\gamma \in \Gamma_i$.
(iii) Give the set the E′-property in the way described for LPC.
(iv) Finally increase the set to a maximal consistent one.

But here we strike a difficulty. While it is easy to prove (in the way we did for T and LPC respectively) that, having started with $\{\beta\}$ we can consistently do *either* (ii) *or* (iii), we have no guarantee that we can consistently do both. For in proving, for LPC, that we could consistently give the set the E′-property, it was essential that the replacement variable **b** in each successive E-formula was one which had not appeared up to that point; and the availability of these fresh variables depended on the fact that each E-formula was added to a *finite* set of wffs, in which therefore only a finite number of variables could have occurred. But in the present case we should have to add these E-formulae to an infinite set, viz. the γ's mentioned in (ii), and we therefore could not be sure that fresh variables were available. A similar difficulty would arise if we were to reverse the order and do (iii) before (ii): we could not be sure that the γ's were consistent with all the E-formulae.

This difficulty can be overcome in the following way[118]. We begin by extending the notions defined on p. 159:

1. *An E_M-formula with respect to* **b** is defined as follows:

(i) Any wff of the form $(\exists \mathbf{a})\beta \supset \beta[\mathbf{b}/\mathbf{a}]$ is an E_M-formula with respect to **b**.

(ii) If γ is an E_M-formula with respect to **b** and δ is any wff not containing free **b**, then $M\delta \supset M(\delta \,.\, \gamma)$ is an E_M-formula with respect to **b**.

In virtue of (i), every E-formula with respect to **b** is also an E_M-formula with respect to **b**.

[118] The solution which follows is the simplest one we have been able to discover; we do not guarantee that no simpler one is possible. It is not of course the mere fact that the set obtained after (i) and (ii) is infinite that causes the difficulty. It is rather the fact that the set depends on an already established maximal consistent set (Γ_i) of wffs of LPC and that we have no way of ensuring that all the individual-variables of LPC do not occur in the set obtained when the γ's are introduced.

2. All E_M-formulae which differ only in that each is an E_M-formula with respect to a different variable will be said to have the same E_M-form. (Clearly the E_M-forms are enumerable.)

3. A set of wffs will be said to have the *E_M'-property* iff it contains at least one E_M-formula of every E_M-form.

Obviously a set which has the E_M'-property has the E′-property; and hence, since we proved on p. 159 that a maximal consistent set with the E′-property has the E-property, if a maximal consistent set has the E_M'-property it also has the E-property[119].

We next prove two lemmas, the first of which is a preliminary to the second.

LEMMA 6

If γ is an E_M-formula with respect to **b**, then $\vdash(\exists \mathbf{b})\gamma$.

The proof is by induction on the construction of an E_M-formula:

1. If γ is of the form $(\exists \mathbf{a})\beta \supset \beta[\mathbf{b}/\mathbf{a}]$, then by LPC4, $\vdash(\exists \mathbf{b})((\exists \mathbf{a})\beta \supset \beta[\mathbf{b}/\mathbf{a}])$, i.e. $\vdash(\exists \mathbf{b})\gamma$.

2. Suppose (induction hypothesis) that γ is an E_M-formula with respect to **b** and that $\vdash(\exists \mathbf{b})\gamma$. We have to prove that in that case, if **b** is not free in δ, then $\vdash(\exists \mathbf{b})(M\delta \supset M(\delta \,.\, \gamma))$. Since $\vdash(\exists \mathbf{b})\gamma$, we have, by DR4 (p. 40), $\vdash M\delta \supset M(\delta \,.\, (\exists \mathbf{b})\gamma)$; hence by LPC5 (since **b** is not free in δ), $\vdash M\delta \supset M(\exists \mathbf{b})(\delta \,.\, \gamma)$; hence by the Barcan formula, $\vdash M\delta \supset (\exists \mathbf{b})M(\delta \,.\, \gamma)$; and hence finally by LPC6 (since **b** is not free in $M\delta$), $\vdash(\exists \mathbf{b})(M\delta \supset M(\delta \,.\, \gamma))$. **Q.E.D.**

Therefore Lemma 6 holds for all E_M-formulae.

LEMMA 7

If Λ is a consistent set of formulae, none of which contains any occurrence of **b**, and γ is an E_M-formula with respect to **b**, then $\Lambda \cup \{\gamma\}$ is consistent (i.e. γ can be consistently added to Λ).

The proof is as for Lemma 5 (p. 160) with obvious modifications and using Lemma 6 to obtain $\vdash(\exists \mathbf{b})\gamma$.

[119] The gist of our solution to the problem which faces us is this: if we give Γ_j the E_M'-property in a certain way to be described below, we can then add all the γ's (for $L\gamma \in \Gamma_i$) without fear of inconsistency. The E_M'-property is so defined that (i) it yields, in a maximal consistent set, the E-property, and (ii) in constructing a subordinate of any set which has E_M'-property we can always give the subordinate the E_M'-property too.

We now show how, beginning with any consistent wff, α, we can construct a system Γ, of maximal consistent sets (with respect to T + BF).

We assume that the wffs of T + BF have been arranged in some standard ordering. We make the same assumption about the E_M-forms.

For Γ_1 we begin with $\{\alpha\}$. We then add successively for each E_M-form some E_M-formula with respect to a variable not occurring earlier in the construction of the set. By Lemma 7 the set remains consistent. We then increase the set to a maximal consistent one in the standard way. The result is Γ_1, and clearly it has the E_M'-property, and hence the E-property.

We now show that given any maximal consistent set Γ_i with the E_M'-property, we can construct, for any wff $M\beta$ in Γ_i, a maximal consistent set Γ_j containing β and every wff γ such that $L\gamma \in \Gamma_i$, and itself having the E_M'-property.

We begin Γ_j with $\{\beta\}$. We next give the set the E_M'-property in the following way. We take the first E_M-form in the standard ordering. Let $\delta_{11}, \ldots, \delta_{1n}, \ldots$ be the E_M-formulae which have this E_M-form; i.e. $\delta_{11}, \ldots, \delta_{1n}, \ldots$ all have the first E_M-form, and differ only in that each is an E_M-formulae with respect to a different variable. The question is, which of these shall we put in Γ_j? We settle this by reference to Γ_i. Γ_i has the E_M'-property, and hence, by the definition of an E_M-formula, there will be in Γ_i not merely one of $\delta_{11}, \ldots, \delta_{1n}, \ldots$, but also some formula

$$(1)\ M\beta \supset M(\beta \,.\, \delta_{1*})$$

Where δ_{1*} is one of $\delta_{11}, \ldots, \delta_{1n}, \ldots$, and β is the formula with which we are starting Γ_j. And it is this δ_{1*} which we select to put in Γ_j. (Note that since (1) and $M\beta$ are in Γ_i, so is $M(\beta \,.\, \delta_{1*})$.)

We next take the second E_M-form, and from the E_M-formulae with this E_M-form, $\delta_{21}, \ldots, \delta_{2n}, \ldots$ we select for Γ_j some formula δ_{2*} such that

$$M(\beta \,.\, \delta_{1*}) \supset M(\beta \,.\, \delta_{1*} \,.\, \delta_{2*})$$

is in Γ_i. Since Γ_i has the E_M'-property, there must be some such formula in Γ_i. (Note also that since, as we have just shown, $M(\beta \,.\, \delta_{1*})$ is in Γ_i, so is $M(\beta \,.\, \delta_{1*} \,.\, \delta_{2*})$.)

In general, given that for the first m E_M-forms we have added the E_M-formulae $\delta_{1*}, \ldots, \delta_{m*}$, we then add a formula $\delta_{(m+1)*}$, of the $(m+1)$th E_M-form, which is such that

$$M(\beta . \delta_{1*} . \ldots . \delta_{m*}) \supset M(\beta . \delta_{1*} . \ldots . \delta_{m*} . \delta_{(m+1)*})$$

is in Γ_i. Since Γ_i has the E_M'-property, there will always be such a $\delta_{(m+1)*}$. (Also in general, $M(\beta . \delta_{1*} . \ldots . \delta_{m*} . \delta_{(m+1)*})$ will be in Γ_i. And by $\vdash M(p . q) \supset Mp$, where $\{\delta_1, \ldots, \delta_n\}$ is any finite subset of the E_M-formulae we have put into Γ_j by the process we have described, $M(\beta . \delta_1 . \ldots . \delta_n) \in \Gamma_i$.)

Let $\Gamma_{j'}$ be the set obtained by adding to $\{\beta\}$ an E_M-formula for each E_M-form, selected in the way just explained. Clearly $\Gamma_{j'}$ has the E_M'-property.

The next step in the construction of Γ_j is to add to $\Gamma_{j'}$ every wff, γ, such that $L\gamma \in \Gamma_i$. Let us call the resulting set, $\Gamma_{j''}$. We now show that $\Gamma_{j''}$ is consistent.

Consider any finite subset of $\Gamma_{j''}$,

$$(2) \quad \{\beta, \delta_1, \ldots, \delta_n, \gamma_1, \ldots, \gamma_m\}$$ [120]

where β is the initial formula in Γ_j, $n \geqslant 0$, $m \geqslant 0$, each of the δ's is an E_M-formula put into $\Gamma_{j'}$ in the process of giving it the E_M'-property, and each γ is such that $L\gamma \in \Gamma_i$. As we have observed above, $M(\beta . \delta_1 . \ldots . \delta_n)$ will be in Γ_i; and hence

$$(3) \quad \{M(\beta . \delta_1 . \ldots . \delta_n), L\gamma_1, \ldots, L\gamma_m\}$$

is a subset of Γ_i. But Γ_i is by hypothesis consistent; hence (3) is consistent, and therefore (by Lemma 4) (2) is consistent.

$\Gamma_{j''}$ has therefore no inconsistent finite subsets; i.e. $\Gamma_{j''}$ is consistent.

Having constructed $\Gamma_{j''}$ we finally complete the construction of Γ_j by increasing $\Gamma_{j''}$ to a maximal consistent set in the standard way.

The construction of Γ can now proceed as for propositional T. We form Γ_1 in the way described, beginning with the given consistent formula α. For every $\Gamma_i \in \Gamma$ and for every formula $M\beta$ in Γ_i there is to be in Γ a subordinate maximal consistent set Γ_j constructed in the way we have just explained. It is clear that Γ has the properties listed on p. 164.

[120] Of course β will not be in every subset of $\Gamma_{j''}$, but clearly if (2) is consistent, then (2) with β omitted is also consistent.

The rest of the completeness proof is now straightforward. We define the following T + BF-model ⟨W, R, D, V⟩, which will combine the features of the models we used earlier for T and for LPC. With each Γ_i in Γ we associate a world, w_i. Let W be the set of all such worlds. Let R be the relation such that $w_i R w_j$ iff Γ_j is either a subordinate of Γ_i or is Γ_i itself. Let D, as before, be the set of individual-variables, considered as objects. Again as before, we have V(**a**) = **a**, for every individual-variable **a**. For the value-assignment to an n-adic predicate variable, we have the following:

For any n-adic ϕ, V(ϕ) is the set of just those ordered $(n + 1)$-tuples $\{\langle x_1, \ldots, x_n, w_i\rangle, \ldots\}$ such that $\phi(x_1, \ldots, x_n) \in \Gamma_i$ (for every $\Gamma_i \in \Gamma$).

The value assigned by V to any wff can now be determined by the rules given on pp. 146f.

We can now state the theorem to which all this has been leading:

THEOREM 4

Given W, R, D *and* V *as defined above, for any wff,* β, *of* T + BF *and for any* $w_i \in$ W, V(β, w_i) = 1 *or* 0 *according as* $\beta \in \Gamma_i$ *or not.*

The proof is again by induction on the construction of a wff. For atomic formulae the theorem holds in virtue of the initial value-assignment, as we have pointed out above. For ~, ∨, L and quantification the proofs are as for PM, T or LPC as the case may be (with obvious modifications) and we shall not recapitulate them here.

The completeness of T + BF follows from the theorem in the same way as in the previous cases.

For the completeness of S4 + BF and LPC + S5 we make the same modifications as for the corresponding propositional systems (*vide* p. 158). I.e. for S4 + BF we let $w_i R w_j$ when Γ_j is a subordinate$_*$ of Γ_i; and for LPC + S5 we have in addition that whenever Γ_j is a subordinate of Γ_i, $w_j R w_i$ as well as $w_i R w_j$. The additions which need to be made to the completeness proof for T + BF are then the same as for the propositional systems.

CHAPTER TEN

Modality and Existence

Validity without the Barcan formula

The definition of validity which we gave for modal predicate calculi in Chapter 8 was one which made the Barcan formula come out valid, and in Chapter 9 we proved the completeness of systems containing the Barcan formula with respect to that definition. We remarked, however, that in the case of quantificational T and S4 (though not of S5) the Barcan formula is independent of the rest of the basis, and that we could therefore have two versions of these systems, one with the Barcan formula and one without it. The question therefore arises: can we give an account of validity which will fit the versions of these systems which do not contain the Barcan formula (viz. LPC + T and LPC + S4)?

This is not a question with a merely formal interest, for a number of objections have been brought against the validity of the formula from an intuitive point of view[121]. It is convenient here to consider the Barcan formula as the wff

$$(x)L\phi x \supset L(x)\phi x$$

Under the standard interpretation what this means is that if everything necessarily possesses a certain property ϕ, then it is necessarily the case that everything possesses that property. But now, it is sometimes argued, even if everything that actually exists is necessarily ϕ, this does not preclude the possibility that there might be (or might have been) some things which are not ϕ at all – and in that case it would not be a necessary truth that everything is ϕ.

This objection to the Barcan formula depends on the assumption that in various 'possible worlds', not merely might objects

[121] One of the earliest of these objections is found in Prior [1957], pp. 26–28 *et al.* There is also a discussion in Hintikka [1961]. For an objection of a somewhat different kind *vide* Myhill [1958], p. 80. For a defence of the formula *vide* Barcan Marcus [1962], pp. 88–90.

have different properties from those they have in the actual world, but there might even be objects which do not exist in the actual world at all. Now it is at least plausible to think of the semantics we have given for modal predicate calculi as implicitly denying this assumption, for in each model we have had a single domain of individuals, the same for each world. The validity of the Barcan formula is in fact connected with this feature of our semantics. And this suggests that we might obtain a semantics which does not bring the formula out as valid, by admitting models in which different domains are associated with different worlds. We shall now show how this can be done.

We define an LPC + T-model as an ordered quintuple $\langle W, R, D, Q, V\rangle$ where W, R and D are to be understood as before, V is a value-assignment of a kind to be specified in a moment, and Q is a function whose argument is a member of W and whose value is a subset of D. What this means is that Q assigns to each world w_i as its own peculiar domain a certain set selected from the total domain D. We write '$Q(w_i)$' for 'the domain assigned by Q to w_i', and normally abbreviate it to 'D_i'. We also make the requirement (the *inclusion requirement*) that whenever $w_i R w_j$, $D_i \subseteq D_j$; i.e. that where w_j is accessible to w_i, every member of w_i's domain is also a member of w_j's domain.

In an LPC + T-model, V differs from all the other kinds of value-assignments we have encountered, in that it sometimes assigns a value to a wff (in a given world) but sometimes does not assign it any value at all. When V assigns a value (1 or 0) to α in w_i, we shall say that V *defines* α (in w_i), or that $V(\alpha, w_i)$ is *defined*; when V assigns no value to α in w_i, we shall say that V leaves α *undefined* (in w_i), or that $V(\alpha, w_i)$ is *undefined*.

As before, V assigns to each individual-variable without exception some member of D. For an atomic formula, α, however, if the value assigned to any of the individual-variables in α is not in D_i, then $V(\alpha, w_i)$ is undefined; but if all these values are in D_i, then $V(\alpha, w_i) = 1$ or 0 by the same rules as before. Non-atomic formulae are evaluated or left undefined as explained below[122].

[122] This is one way of dealing with formulae containing individual symbols whose values do not exist (in a given world). *Vide* also p. 179. For the philosophical issues raised by identifying an object in one world with an object in another *vide* Chisholm [1967], Hintikka [1967], pp. 40–45 and Purtill [1968]; also footnote 151.

The conditions to be satisfied by V are:

1. For every individual-variable, **a**, there is some $u \in \mathrm{D}$ such that $\mathrm{V}(\mathbf{a}) = u$.

2. For every n-adic predicate variable, ϕ, $\mathrm{V}(\phi)$ is some set of ordered $(n+1)$-tuples, as described on p. 146.

3. If α is an atomic formula, $\phi(x_1, \ldots, x_n)$, then for any $w_i \in \mathrm{W}$, if each of $\mathrm{V}(x_1), \ldots, \mathrm{V}(x_n)$ is in D_i, $\mathrm{V}(\alpha, w_i) = 1$ or 0 according as $\langle \mathrm{V}(x_1), \ldots, \mathrm{V}(x_n), w_i \rangle \in \mathrm{V}(\phi)$ or not; otherwise $\mathrm{V}(\alpha, w_i)$ is undefined.

4. For any wff, α, and any $w_i \in \mathrm{W}$, $\mathrm{V}({\sim}\alpha, w_i) = 1$ iff $\mathrm{V}(\alpha, w_i) = 0$, and $\mathrm{V}({\sim}\alpha, w_i) = 0$ iff $\mathrm{V}(\alpha, w_i) = 1$. (Thus if $\mathrm{V}(\alpha, w_i)$ is undefined, so is $\mathrm{V}({\sim}\alpha, w_i)$.)

5. For any wffs, α and β, and any $w_i \in \mathrm{W}$, $\mathrm{V}((\alpha \vee \beta), w_i) = 1$ iff both $\mathrm{V}(\alpha, w_i)$ and $\mathrm{V}(\beta, w_i)$ are defined and either $\mathrm{V}(\alpha, w_i) = 1$ or $\mathrm{V}(\beta, w_i) = 1$; and $\mathrm{V}((\alpha \vee \beta), w_i) = 0$ iff both $\mathrm{V}(\alpha, w_i) = 0$ and $\mathrm{V}(\beta, w_i) = 0$. (Thus if either $\mathrm{V}(\alpha, w_i)$ or $\mathrm{V}(\beta, w_i)$ is undefined, $\mathrm{V}((\alpha \vee \beta), w_i)$ is also undefined.)

6. For any wff, α, any individual-variable, **a**, and any $w_i \in \mathrm{W}$, $\mathrm{V}((\mathbf{a})\alpha, w_i) = 1$ iff for every V' which assigns to **a** any member of D_i and is otherwise the same as V, $\mathrm{V}'(\alpha, w_i) = 1$; and $\mathrm{V}((\mathbf{a})\alpha, w_i) = 0$ iff there is some such V' for which $\mathrm{V}'(\alpha, w_i) = 0$. (Otherwise $\mathrm{V}((\mathbf{a})\alpha, w_i)$ is undefined.)

7. For any wff, α, and any $w_i \in \mathrm{W}$, $\mathrm{V}(L\alpha, w_i) = 1$ iff for every w_j such that $w_i\mathrm{R}w_j$, $\mathrm{V}(\alpha, w_j) = 1$; and $\mathrm{V}(L\alpha, w_i) = 0$ iff for every such w_j, $\mathrm{V}(\alpha, w_j)$ is defined and for some such w_j, $\mathrm{V}(\alpha, w_j) = 0$. (Thus $\mathrm{V}(L\alpha, w_i)$ is undefined iff for some such w_j, $\mathrm{V}(\alpha, w_j)$ is undefined.)

Note that, from the inclusion requirement, α will be defined in w_i iff it is defined in every w_j such that $w_i\mathrm{R}_*w_j$.

A wff, α, is LPC + T-valid iff for every LPC + T-model $\langle \mathrm{W}, \mathrm{R}, \mathrm{D}, \mathrm{Q}, \mathrm{V} \rangle$, and for every $w_i \in \mathrm{W}$, $\mathrm{V}(\alpha, w_i) = 1$ wherever $\mathrm{V}(\alpha, w_i)$ is defined (i.e. iff $\mathrm{V}(\alpha, w_i)$ is never 0).

Analogous definitions of LPC + S4- and LPC + S5-models are obtained by simply adding the familiar requirements that for LPC + S4 R be transitive and that for LPC + S5 it be symmetrical as well. The definitions of validity will then be as for LPC + T.

Every axiom of LPC + T (or LPC + S4 or LPC + S5, as the case may be) is valid in terms of the above definitions, and the transformation rules preserve validity. The proof of this is left to the reader[123].

Consider, however, the Barcan formula, in the form: $(x)L\phi x \supset L(x)\phi x$. Take a model $\langle W, R, D, Q, V\rangle$ in which the elements are defined as follows[124].

W has exactly two (distinct) members, w_1 and w_2. In addition to w_1Rw_1 and w_2Rw_2 we have w_1Rw_2 but not w_2Rw_1. D contains exactly two (distinct) members, u_1 and u_2. $Q(w_1)$ – i.e. D_1 – contains only u_1, and D_2 contains both u_1 and u_2. $V(x) = u_1$, while for any individual-variable, **a**, other than x, $V(\mathbf{a}) = u_2$. $V(\phi) = \{\langle u_1, w_1\rangle, \langle u_1, w_2\rangle\}$. We shall now evaluate the Barcan formula in w_1.

For atomic formulae we clearly have the following:

(1) $V(\phi x, w_1) = 1$;
(2) $V(\phi x, w_2) = 1$;
(3) $V(\phi y, w_1)$ is undefined;
(4) $V(\phi y, w_2) = 0$

By (1) and (2), $V(L\phi x, w_1) = 1$. Since u_1 is the only member of D_1, any V' which assigns to x any member of D_1 must assign the same value to x (viz. u_1) as V does. Hence for every such V', $V'(L\phi x, w_1) = V(L\phi x, w_1)$, which as we have seen is 1; and so $V((x)L\phi x, w_1) = 1$. Turning now to the consequent of the Barcan formula, we first evaluate $(x)\phi x$ in w_2. Let $V'(x)$ be u_2 (which is in D_2). Then V' assigns to x the same member of D_2 as V assigns to y, and so $V'(\phi x, w_2) = V(\phi y, w_2)$, which (4) gives as 0; hence $V((x)\phi x, w_2) = 0$. But since w_1Rw_2, we then have $V(L(x)\phi x, w_1) = 0$. Hence the Barcan formula has the value 0 in w_1, and is therefore invalid.

Now the model $\langle W, R, D, Q, V\rangle$ which we have described is both an LPC + T-model and an LPC + S4-model. So the Barcan formula is neither LPC + T-valid nor LPC + S4-valid. Moreover, we have here a proof of the independence of the Barcan formula in T + BF and S4 + BF: as we have remarked, all the theses derivable from the rest of the bases of those systems are valid in terms of our present definitions of validity, but we have just seen that the Barcan formula is not.

[123] Although straightforward, the proof depends on the inclusion requirement.

[124] This model is adapted from one used in Kripke [1963b], p. 87.

The model is not, however, an LPC + S5-model, for in it we have $w_1 R w_2$ but not $w_2 R w_1$, and so R is not symmetrical. In fact we can prove, in terms of our present semantics, that where R is an equivalence relation (i.e. reflexive, transitive and symmetrical), the Barcan formula is valid. We prove this by showing that, if R is any such relation, then if the value of the consequent of the Barcan formula in any world is 0, the value of the antecedent in that world is also 0. We begin by observing that since R is symmetrical, whenever we have $w_i R w_j$ we not merely have $D_i \subseteq D_j$ but $D_j \subseteq D_i$ as well – i.e., the domains for all worlds related by R are identical. Suppose now that for some $w_i \in W$, $V(L(x)\alpha, w_i) = 0$. Then for some w_j such that $w_i R w_j$, $V((x)\alpha, w_j) = 0$. Hence for some V′ which assigns to x some member of D_j, $V'(\alpha, w_j) = 0$. But (since R is symmetrical) D_j is identical with the domain of every $w_k \in W$ such that $w_i R w_k$ (including w_i itself). Hence since V′ defines α in w_j, it defines it in every such w_k; and therefore $V'(\alpha, w_i)$ is defined. But $V'(\alpha, w_j) = 0$; hence $V'(L\alpha, w_i) = 0$; and hence $V((x)L\alpha, w_i) = 0$.

(It is worth noting that since the proof we have just given depends on the symmetry, but not the transitivity, of R, the Barcan formula is also valid in LPC + B (*vide* p. 145) by our present semantics.)

Completeness without the Barcan formula

We shall now prove that LPC + T and LPC + S4 are complete with respect to these definitions of LPC + T and LPC + S4-validity. We shall again use a Henkin proof; i.e. we shall show, using the device of maximal consistent sets, that for every consistent formula a verifying model can be constructed. This time, however – for a reason to be given shortly – it will be unnecessary to introduce the E_M-forms and to give the maximal consistent sets the E_M'-property: the E′-property used for non-modal LPC (p. 159) will be sufficient, though of course the β in an E-formula may now contain modal operators. (In any case, in proving that each set could be consistently given the E_M'-property we used the Barcan formula (p. 166), which is not now available to us.) As before, we set out the proof for LPC + T, and then show the modifications needed for LPC + S4.

We show how, given a consistent wff, α, we can construct a certain system, Γ, of maximal consistent sets. As before, we begin Γ_1 with $\{\alpha\}$. Let the individual-variables occurring in α be $x_1, \ldots, x_n$. Let us now suppose that all the *other* individual-variables are arranged in an infinite series of infinite sets, each of which is to be associated, in a way to be explained in a moment, with one of the maximal consistent sets in Γ. For simplicity, let us write each individual-variable as x with a subscript and a superscript; we can then represent the sets of individual-variables as follows:

$$\mathbf{d_1} = \{x_1^1, \ldots, x_i^1, \ldots\}$$
$$\cdots\cdots\cdots\cdots\cdots\cdots$$
$$\mathbf{d_k} = \{x_1^k, \ldots, x_i^k, \ldots\}$$
$$\cdots\cdots\cdots\cdots\cdots\cdots$$

Let Λ_1 be the set of all those wffs of LPC + T all of whose individual-variables either occur in α or are members of $\mathbf{d_1}$. Having begun Γ_1 with $\{\alpha\}$ we then give the set the E′-property by adding, for each wff of the form $(\exists \mathbf{a})\beta$ in Λ_1, a wff $(\exists \mathbf{a})\beta \supset \beta[\mathbf{b}/\mathbf{a}]$, where the replacement-variable $\mathbf{b}$ is a fresh one in each case, drawn from $\mathbf{d_1}$. Finally we make the set maximal consistent for Λ_1: i.e. we add successively all those wffs in Λ_1 which we consistently can. It is clear that while Γ_1 contains no formulae with individual-variables other than those in α or in $\mathbf{d_1}$, yet it is maximal consistent with respect to a version of LPC + T in which the individual-variables are specified as those in α or in $\mathbf{d_1}$; and moreover that the relevant Lemmas in Chapter 9 hold for Γ_1 where their application is restricted to the wffs in Λ_1.

For each Γ_i in Γ and for each wff of the form $M\beta$ in Γ_i we form a subordinate, Γ_j, in the following way. Let Γ_j be the kth set being constructed in Γ. Let Λ_j be the set of all those wffs of LPC + T all of whose individual-variables either occur in α or are members of any of $\mathbf{d_1}$–$\mathbf{d_k}$ inclusive. (I.e. for each new maximal consistent set we construct we add a new set of individual-variables, and the wffs we now take into account are those whose variables are any of these new ones or any of the previously used ones.) We begin Γ_j with $\{\beta\}$. We then add every wff γ such that $L\gamma \in \Gamma_i$. By the same proof as for propositional T (p. 156), the set remains consistent. We then give the set the E′-property by adding for each wff of the form $(\exists \mathbf{a})\beta$ in Λ_j a wff $(\exists \mathbf{a})\beta \supset \beta[\mathbf{b}/\mathbf{a}]$

where the replacement-variable **b** is a fresh one in each case, and always drawn from the new set of variables $\mathbf{d_k}$. Since **b** will therefore not occur in the wff β with which the set was started or in any of the γ's, the set remains consistent for the reasons given in the case of non-modal LPC (pp. 160f). Finally we make the set maximal consistent for Λ_j.

Γ is the system of all maximal consistent sets constructed in the way described.

We now define the following LPC + T-model $\langle W, R, D, Q, V\rangle$. W and R are as in the previous chapter, each Γ_i being associated with some $w_i \in W$ in the familiar way. Let D be the set of all individual-variables (considered as objects). Let each $Q(w_i)$ – i.e. each D_i – be the set of individual-variables in any wff in Γ_i: i.e. the domain for each world, w_i, is the set of individual-variables occurring in any wff in Γ_i. (Since if Γ_j is subordinate to Γ_i then Γ_j is constructed after Γ_i, the requirement that if $w_i R w_j$, $D_i \subseteq D_j$ is automatically complied with.) V is defined as follows. For every individual-variable, **a**, $V(\mathbf{a}) = \mathbf{a}$, as before. For every n-adic variable ϕ and every $\Gamma_i \in \Gamma$, $V(\phi)$ is again defined as before (p. 169), but with the result this time that $V(\phi(x_1, \ldots, x_n), w_i)$ has the value 1 iff $\phi(x_1, \ldots, x_n) \in \Gamma_i$, has the value 0 iff each of $x_1, \ldots, x_n$ occurs in some wff in Γ_i but $\phi(x_1, \ldots, x_n) \notin \Gamma_i$, and is undefined iff at least one of $x_1, \ldots, x_n$ does not occur in any wff in Γ_i at all.

We now have:

THEOREM 5

Given W, R, D, Q *and* V *as defined above, for any wff*, β, *of* LPC + T, *and for any* $w_i \in W$, $V(\beta, w_i) = 1$ *iff* $\beta \in \Gamma_i$.

The reader who has followed the proofs of the theorems in Chapter 9 should be able to prove this one for himself. (The one new feature of the present case which might be thought to cause difficulty is that each Γ_i is maximal consistent not with respect to all the wffs of LPC + T but only with respect to those in Λ_i. As a result there will be wffs which are not in Γ_i but are consistent with it, and for such wffs Lemma 2 (p. 153), for example, will not hold. Nevertheless from the construction of Γ and the conditions for V, all such wffs will be undefined in w_i, and so their absence from Γ_i will not prevent the establishment of the theorem.)

Since each subordinate$_*$ of Γ_i contains all the variables in Γ_i, the proof can be extended to S4 in the way noted on p. 169.

Interpretation of modal LPC-models

From an intuitive point of view the models we have been considering for modal LPC, like those in Part I for modal propositional logic, can be thought of as developing the idea of necessity as truth not merely in the actual world but in every possible world as well. Since we have wanted to consider various restrictions on what is to count as a 'possible' world, we have usually spoken instead of truth in every world 'accessible to' a given world; though, as we remarked earlier, the word 'accessible' is deliberately open to a variety of intuitive or imaginative interpretations. Now it is natural to think of a 'world' as consisting of a set of objects with various properties and standing in various relations to one another (we shall ignore the metaphysical problems to which this notion gives rise), and if we were to ask ourselves what kinds of worlds or states of affairs other than the actual one we were prepared to envisage or count as possible, there are several answers we might give.

(a) One answer might be that we could envisage only certain worlds containing just the same objects as the actual world does, though with new properties and standing in different relations. We could think of the semantics of Chapter 9 as expressing this kind of answer, since in each model there is a single invariant domain for all worlds. From the point of view of such models, when we say that a proposition is necessary we can take this to mean that it not merely is true as things are, but would still be true even if things were to alter their properties or relations; and as we have seen, the Barcan formula then comes out valid for all the three systems we considered.

(b) Another answer, however, might be that we were also prepared to envisage worlds in which not only were the properties and relations of actual objects different but some new objects had been added as well. Thus we might perhaps envisage a world with all the present objects in it and, say, Pegasus too. The models so far discussed in the present chapter could be thought of as based on this answer, since in them we allow each world accessible to a given world to have a more extensive domain. In this case, when we say that a proposition is necessary we mean that it would continue to be true even if not merely things were to change their properties or relations but other things were to come into existence as well; and as we have seen, the Barcan formula now turns out to be invalid, at least as far as T and S4 are concerned.

(c) A more liberal answer still would be that we should count as possible (relative to this world) not merely worlds of the kinds we have already indicated but also certain worlds in which some of the present objects had been removed (whether or not some had been added as well). Semantically this would mean that we should drop the requirement that whenever $w_i R w_j$, $D_i \subseteq D_j$ – i.e. that we should put no restrictions on the way in which one world's domain differs from another's. We shall look into some of the consequences of doing this shortly.

Now if we allow the domains to vary, either in the restricted way permitted by the second answer or in the unrestricted way permitted by the third, the following question arises. What are we to say is the truth-value in a given world of a proposition about some object which does not exist in the domain of that world? What, for example, are we to say is the truth-value in the actual world of 'Pegasus is piebald' or 'Mr Pickwick lived in London'? And here we seem to have a choice. (i) We might say that in the world in question such a proposition has no truth-value at all; and the semantics we have given earlier in this chapter can be thought of as expressing this answer. (ii) We might say that even in the world in question such a proposition might be true or might be false but does have *some* truth-value – i.e. that the mere absence of the object from the domain does not debar it from having a truth-value and does not automatically settle its truth-value, any more than the mere presence of the object in the domain settles its truth-value. We shall now give a brief account of a semantics suggested by Kripke, which is based on this second answer, as well as on the third answer, (c), to our previous question.

Kripke's semantics for modal LPC

Kripke[125] has given a semantics for modal LPC which differs from the one we have presented in this chapter in the following two ways:

1. It does not contain the requirement that whenever $w_i R w_j$, $D_i \subseteq D_j$. In other words, we now simply say that different worlds

125 Kripke [1963b].

in W may have the same domain or may have different ones, and leave it at that.

2. In every model, V assigns a value (1 or 0) to every atomic formula in every $w_i \in$ W. V(ϕ) is defined, as before, as a set of ordered $(n+1)$-tuples, each of the form $\langle u_1, \ldots, u_n, w_i \rangle$, where $u_1, \ldots, u_n$ are all in D, though they need not all be in D_i. But this time, if α is an atomic formula, the rule for evaluating V(α, w_i) is simply as in Chapter 8 (p. 146). Thus, for example, V(ϕy, w_i) = 1 or 0 according as $\langle$V(y), $w_i\rangle \in$ V(ϕ) or not, irrespective of whether or not V(y) $\in D_i$, and is never undefined.

In other respects Kripke's semantics are the same as ours given earlier in this chapter, except that the clauses stating the conditions under which formulae are undefined will now become redundant, since if every atomic formula is defined the rules ensure that every formula without exception is defined.

It is worth noting that the rule for evaluating quantified formulae (in both types of semantics) gives V((**a**)α, w_i) = 1 iff for every V′ *which assigns to* **a** *some member of* D_i and is otherwise the same as V, V′(α, w_i) = 1. That is, intuitively speaking, we are interpreting '"everything is ϕ" is true in w_i' as meaning that everything *in* D_i is ϕ. Now it may of course be the case that everything in D_i is ϕ but that some object, u_j, not in D_i, is not ϕ; and if that is so then we shall have V((x)ϕx, w_i) = 1 but (by Kripke's semantics) V(ϕy, w_i) = 0 if u_j is assigned to y. Since this is a perfectly possible assignment, we have the result that

$$(1)\ (x)\phi x \supset \phi y$$

is not valid. But (1) is a straightforward instance of the axiom-schema $\forall$1; hence Kripke's semantics invalidates $\forall$1[126].

[126] One can of course have a logic without $\forall$1. If we add an 'existence' predicate E (cf. Rescher [1959]) we can let $\langle u_i, w_j\rangle \in$ V(E) iff $u_i \in D_j$. (I.e. V(E**a**, w_j) = 1 iff V(**a**) $\in D_j$). In this case $\forall$1 would be replaced by ((**a**)α . E**b**) $\supset \beta$ where β is α with free **b** replacing free **a**. Sometimes E is defined in terms of identity; *vide* Hintikka [1963], pp. 70–71 and Kripke [1963b], p. 90. Systems of logic which deviate in this respect from standard LPC have been called 'free' logics by Lambert ([1963] and elsewhere). For a completeness proof for free logic *vide* van Fraassen [1966]. Although free logic need not be tied to modal logic the non-existent seems to raise more questions in contexts where what might exist but doesn't is being considered.

Kripke's way of dealing with this situation can be expressed as follows. Although (1) is invalid by his semantics,

$$(2)\quad (y)((x)\phi x \supset \phi y)$$

is valid. We shall call (2) the *universal closure* of (1). In general, if α is a wff whose free variables are $x_1, \ldots, x_n$, we shall say that

$$(x_1)\ldots(x_n)\alpha$$

– i.e. the formula obtained by binding each free variable in α by a universal quantifier at the beginning – is the universal closure of α. (We shall call a formula with no free variable in it a *closed* formula; and it is convenient to speak of a closed formula as the universal closure of itself.) Now Kripke's semantics brings out as valid, not every instance of ∀1, but the universal closure of every instance of ∀1; and the same holds for every theorem derived by ∀2. Hence Kripke is led to use an axiomatization of non-modal LPC in which the axiom-schemata yield only closed formulae as theorems. None of the formulae with free variables which are theorems in the version of LPC presented in Chapter 8, therefore, are theorems in Kripke's version; in every case, however, their universal closures are, and all these closed formulae are valid in terms of his semantics.

Now it can be plausibly argued that, at least as far as non-modal LPC is concerned, Kripke's style of axiomatization is as satisfactory as the more usual kind we presented earlier. For by the standard accounts of validity in LPC, a formula with free variables is valid iff its universal closure is valid; and therefore, it may well be held, we really lose nothing by having only closed formulae as theorems[127]. At any rate, it is to a basis of this kind for LPC that Kripke subjoins the modal elements, viz. A5, A6 and N (together with A7 for S4 and A8 for S5); and by his semantics these axiom-schemata come out valid too, and the rules are validity-preserving.

The interesting thing from our present point of view is that by Kripke's semantics the Barcan formula turns out to be

[127] We might even say that when we assert a formula with free variables as a theorem, what we are implicitly claiming is that it is true for all values of its free variables, and that asserting its universal closure simply makes this implicit claim explicit. (Cf. Kripke, op. cit., p. 89, esp. note 1.)

invalid even in quantificational S5, and hence is not a theorem when the system is axiomatized in the way we have just described. Its invalidity can be shown by the following model[128]. W contains exactly two distinct worlds, w_1 and w_2. We have $w_1 \mathrm{R} w_1$, $w_2 \mathrm{R} w_2$, $w_1 \mathrm{R} w_2$ and $w_2 \mathrm{R} w_1$ (and hence the model is an LPC + S5 one). $\mathrm{D} = \{u_1, u_2\}$ $(u_1 \neq u_2)$. $\mathrm{D}_1 = \{u_1\}$; $\mathrm{D}_2 = \{u_1, u_2\}$. $\mathrm{V}(x) = u_1$. $\mathrm{V}(\phi) = \{\langle u_1, w_1\rangle, \langle u_1, w_2\rangle\}$. Let us now evaluate in w_1 the Barcan formula in the form

$$(x)L\phi x \supset L(x)\phi x$$

Clearly by the definitions of D_1, D_2 and V we have:

$$(1)\ \ \mathrm{V}(\phi x, w_1) = 1$$

and

$$(2)\ \ \mathrm{V}(\phi x, w_2) = 1$$

and hence

$$(3)\ \ \mathrm{V}(L\phi x, w_1) = 1$$

Since u_1 is the only member of D_1, we therefore have $\mathrm{V}'(L\phi x, w_1) = 1$ for every V' which assigns to x some member of D_1; and hence

$$(4)\ \ \mathrm{V}((x)L\phi x, w_1) = 1$$

Since, however, $u_2 \in \mathrm{D}_2$ but $\langle u_2, w_2\rangle \notin \mathrm{V}(\phi)$, there will be some V' which assigns to x some member of D_2 (viz. u_2) and for which $\mathrm{V}'(\phi x, w_2) = 0$; hence

$$(5)\ \ \mathrm{V}((x)\phi x, w_2) = 0$$

and so

$$(6)\ \ \mathrm{V}(L(x)\phi x, w_1) = 0$$

From (4) and (6) we obtain

$$(7)\ \ \mathrm{V}(((x)L\phi x \supset L(x)\phi x), w_1) = 0$$

and this shows the Barcan formula to be invalid even in LPC + S5.

(Note, however, that the model we have just been discussing does not comply with all the conditions we laid down earlier; for we have $w_2 \mathrm{R} w_1$ but not $\mathrm{D}_2 \subseteq \mathrm{D}_1$.)

Another result which follows from Kripke's semantics is that the converse of the Barcan formula, which by our earlier semantics was valid even in T, is now invalid even in S5. To prove this, consider a model which is identical with the previous one except

[128] *Vide* Kripke [1963b], p. 87.

that (a) $D_1 = \{u_1, u_2\}$ and $D_2 = \{u_1\}$, and (b) $V(\phi) = \{\langle u_1, w_1\rangle, \langle u_2, w_1\rangle, \langle u_1, w_2\rangle\}$. In this case we have $V'(\phi x, w_1) = 1$ for every V' assigning to x some member of D_1, and hence $V((x)\phi x, w_1) = 1$. Similarly we have $V((x)\phi x, w_2) = 1$; and therefore:

$$(8)\ V(L(x)\phi x, w_1) = 1$$

But if V' assigns u_2 to x, we have $V'(\phi x, w_2) = 0$ (since $\langle u_2, w_2\rangle \notin V(\phi)$), and hence $V'(L\phi x, w_1) = 0$. Therefore, since V' assigns to x some member of D_1, we have:

$$(9)\ V((x)L\phi x, w_1) = 0$$

and (8) and (9) then give

$$(10)\ V((L(x)\phi x \supset (x)L\phi x), w_1) = 0$$

which shows the converse of the Barcan formula to be invalid even in S5 and *a fortiori* in T and S4.

Since, as we remarked earlier, all the theses of Kripke's axiomatic version of modal LPC are valid by his semantics, we therefore have the result that neither the Barcan formula nor its converse is a theorem in his systems.

There is an important cautionary moral to be drawn from all this for axiomatic theory in general. As we have seen, our bases for modal LPC and Kripke's are identical in their modal parts and differ only in the (non-modal) LPC basis to which these modal elements are subjoined. Moreover, as far as non-modal LPC is concerned, our system and his differ only in what is in a sense the trivial respect that where a formula containing free variables is a theorem for us, then although it is not a theorem for him, its universal closure is; thus the *closed* theses are precisely the same in both systems. Now we might well expect that when we add identical modal elements to these bases, this would continue to be the only difference between the two systems. But we have just seen that this is not so; for the Barcan formula and its converse are both closed formulae. It is in fact the absence from Kripke's system of certain *open* non-modal formulae which prevents the derivation of these *closed* modal formulae[129].

[129] If the reader will go through the proof of the Barcan formula in LPC + S5 given on p. 145, using only the universal closures of theses, he will see that the proof cannot then be carried through.

In Part III we shall come across some further examples of the pitfalls that can be concealed behind the innocent-sounding phrase, 'subjoin to some basis for system S . . .'

Modality *de dicto* and modality *de re*

In connection with formulae which contain both quantifiers and modal operators a number of logicians have distinguished between what they have called modality *de dicto* and modality *de re*[130]. To illustrate this distinction, consider the Barcan formula once more. Its consequent, viz.

$$(1)\ L(x)\phi x$$

says that it is a necessary truth that everything is ϕ; its antecedent, viz.

$$(2)\ (x)L\phi x$$

says that everything is necessarily ϕ. (1) is said to express a modality *de dicto* and (2) a modality *de re*. These terms are often explained by saying that in a modality *de dicto* necessity (or possibility) is attributed to a proposition (or *dictum*), but that in a modality *de re* it is attributed to the possession of a property by a thing (*res*); thus in asserting a modality *de dicto* we are saying that a certain proposition is bound to be (or may be) true, while in asserting a modality *de re* we are saying that a certain object is bound to have (or may have) a certain property. But although this explanation may accord well with the literal meanings of the Latin phrases, it does not seem to give a satisfactory account of the difference in meaning between (1) and (2); for even in (2) the argument of L is a propositional expression (ϕx), as we can perhaps see more clearly if we translate (2) at length as 'For every value of x, it is a necessary truth that x is ϕ'[131]. From a formal point of view the feature of (2) which has

[130] *Vide*, e.g., Von Wright [1951], pp. 25–28; Prior [1955], pp. 209–215. The expressions '*de dicto*' and '*de re*' are derived from mediaeval logic, but we shall not enquire here into the difference between their mediaeval and modern uses. (Cf. Kneale [1962a].)

[131] The explanation we have mentioned strongly suggests, in fact, a distinction which our symbolism is not able to express, though it might be modified to do so, viz. the distinction between taking (i) a *proposition* and (ii) merely a *predicate* as the argument of L or M. Introducing extra brackets to mark this distinction, we might take $L(\phi x)$ to mean 'it is a necessary truth that x is ϕ', and $(L\phi)x$ to mean 'x has the property of being necessarily-ϕ'. Whether a satisfactory sense can be given to this

led to its being called a modality *de re* is that the expression within the scope of the modal operator contains a free individual-variable[132]. For our present purposes, therefore, we shall define modality *de dicto* and *de re* as follows:

A wff, α, containing a modal operator (L or M) will be said to express a modality *de re* iff the scope of some modal operator in it contains some free occurrence of an individual-variable; otherwise α will be said to express a modality *de dicto*.

Examples of formulae expressing modalities *de re* would be $(\exists x)L\phi x$, $(x)M\phi x$, $(x)L(y)\phi xy$ – as well, of course, as (2) above.

It has been suggested, e.g. by Von Wright[133], that in a satisfactory modal predicate logic all modalities *de re* would be eliminable in favour of modalities *de dicto*. What this suggestion amounts to is that for every wff, α, which contains a modality *de re*, we should be able to construct a wff, α', which contains no modality *de re* and which is provably equivalent to α, i.e. which is such that $\vdash(\alpha \equiv \alpha')$. Now it is almost certain that even in the most promising of the systems we have examined (viz. LPC + S5 as axiomatized in Chapter 8), this cannot be done – though in fact no one seems actually to have demonstrated this impossibility. And Von Wright does not claim that it can be done in any of these systems. What he does suggest is that it could be done if we were to adopt as a thesis what he calls the Principle of Predication[134]. According to this principle, all properties can be divided into two types: (a) those whose belonging to an object is always either necessary or impossible, and which we may call *formal* properties; and (b) those whose belonging to an object is always contingent, never a matter of necessity or impossibility, and which we may call *material* properties. If ϕ is a formal property we therefore have: $(x)(L\phi x \vee L{\sim}\phi x)$; and if

distinction is a difficult question; but if it can, it would be an interesting task to devise axiomatic bases for systems in which the distinction could be expressed (*vide infra* footnote 151).

132 Of course x is not free in (2) as a whole; but it is free in ϕx, which constitutes the scope of L. In (1), by contrast, the scope of L is $(x)\phi x$, in which x is not free.

133 Von Wright [1951], pp. 26–28.

134 Von Wright does not perhaps make this suggestion quite explicitly, but this is what we take to be the substance of his remarks; and even if we have misinterpreted him, the suggestion is worth discussing for its own sake.

ϕ is a material property we have: $(x)(M\phi x \,.\, M{\sim}\phi x)$. So we can formulate the Principle of Predication as follows[135]:

$$(x)(L\phi x \vee L{\sim}\phi x) \vee (x)(M\phi x \,.\, M{\sim}\phi x)$$

We can generalize this into a schema to suit the style of axiomatization we have been using:

Pr $(\mathbf{a})(L\alpha \vee L{\sim}\alpha) \vee (\mathbf{a})(M\alpha \,.\, M{\sim}\alpha)$

where **a** is any individual-variable and α is any wff.

As we shall show later on, the Principle of Predication is not a thesis even of LPC + S5. It could, however, be consistently added to LPC + S5 as an extra axiom, to give a system which we shall call LPC + S5 + Pr, and we can now raise the question whether in this system all modalities *de re* are eliminable.

Von Wright does not in fact prove that they are, in the sense we have defined. What he does prove is a weaker, though related result which, adapting his argument slightly, we can express by saying that in LPC + S5 + Pr we can derive:

$$(3)\quad (L\alpha = \alpha) \vee (L\alpha = (\beta \,.\, {\sim}\beta))$$

where α and β are any wffs – and in particular we can let β be some arbitrary wff containing no modality *de re*.

In outline the proof of (3) is this. Even without Pr we can easily derive:

$$(4)\quad (\mathbf{a})(L\alpha \vee L{\sim}\alpha) \supset (L\alpha = \alpha)$$

and

$$(5)\quad (\mathbf{a})(M\alpha \,.\, M{\sim}\alpha) \supset (L\alpha = (\beta \,.\, {\sim}\beta))$$

from which, by PC, we obtain:

$$(6)\quad ((\mathbf{a})(L\alpha \vee L{\sim}\alpha) \vee (\mathbf{a})(M\alpha \,.\, M{\sim}\alpha)) \supset ((L\alpha = \alpha) \vee (L\alpha = (\beta \,.\, {\sim}\beta)))$$

from which in turn, if we do have Pr as an axiom, we obtain (3) by MP.

Now if α contains a free individual-variable, then the left-hand side of each equivalence in (3) contains a modality *de re* and in the right-hand side that modality is absent; nevertheless (3) does

[135] Von Wright does not formulate the Principle of Predication in symbolic terms. For a formulation only trivially different from ours *vide* Prior [1955], p. 211.

not enable us automatically to eliminate modalities *de re*. (3) should not be confused with:

$$(7)\ L\alpha = (\alpha \vee (\beta\,.\sim\beta))$$

which is not a thesis at all. (If we suppose (7) to be derivable from (3), this is probably because we mistakenly imagine that $((p = q) \vee (p = r)) \supset (p = (q \vee r))$ is a thesis of S5.) If (7) *were* a thesis we could of course use it to eliminate modalities *de re* very simply, by replacing $L\alpha$ wherever necessary by $\alpha \vee (\beta\,.\sim\beta)$ – or more simply still by its obvious equivalent, α itself.

Nor should (3) be confused with the metalogical statement:

$$(8)\ \text{Either} \vdash(L\alpha = \alpha) \text{ or } \vdash(L\alpha = (\beta\,.\sim\beta))$$

(That the step from (3) to (8) is illegitimate can easily be seen from the fact that in PC we have $\vdash(p \vee \sim p)$ but not: either $\vdash p$ or $\vdash\sim p$.) Even if we had (8), however, it would not follow that all modalities *de re* could be eliminated; for all that (8) assures us is that for every modality *de re* one or other of two equivalences is a thesis. Each of these, if given as a thesis, would indeed enable the modality in question to be eliminated, but we appear to have no effective method of telling from the form of the formula alone *which* one of the equivalences is a thesis, and it is not at all clear how we could have.

The position in fact seems to be that no one has demonstrated that the addition to LPC + S5 of the Principle of Predication[136] would enable all modalities *de re* to be eliminated. On the other hand no one seems to have proved that it would not. It is true that even with the help of Pr we cannot derive, for example,

$$(9)\ (\exists x)L\phi x \equiv L(\exists x)\phi x$$

and we have already commented on the invalidity of this formula[137]. But this does not by itself prove the non-eliminability of modalities *de re*; for there might be some formula other than $L(\exists x)\phi x$ (perhaps a much more complex one), containing no modality *de re*, which is provably equivalent to

136 Or, for that matter, of the related though distinct (8).

137 *Vide supra*, p. 144. Cf. Prior [1955], pp. 211–214. Prior appears to think that the invalidity of (9) automatically proves the non-eliminability of modalities *de re* (if he does not quite explicitly assert this, he at least comes *very* near to doing so).

$(\exists x)L\phi x$; and for all we have said, there might be an effective method of constructing such a formula for every modality *de re*. Our conjecture, however, for what it is worth, is that there is no such effective method, even in LPC + S5 + Pr.

We remarked above that the Principle of Predication is not a thesis of LPC + S5. We can show this by the following LPC + S5-model, in which Pr has the value 0 in w_1: $W = \{w_1, w_2\}$; $D = \{u_1, u_2\}$; $V(x) = u_1$; $V(\phi) = \{\langle u_1, w_1\rangle, \langle u_1, w_2\rangle, \langle u_2, w_1\rangle\}$[138]. Let us suppose that $V'(x) = u_2$, but that V′ is otherwise the same as V.

We consider Pr in the slightly more convenient form

$$(x)(L\phi x \vee L{\sim}\phi x) \vee (x)(M\phi x \,.\, {\sim}L\phi x)$$

Now from the model and the definition of V′ we clearly have:

$$\text{(i)}\ V(\phi x, w_1) = 1$$
$$\text{(ii)}\ V(\phi x, w_2) = 1$$
$$\text{(iii)}\ V'(\phi x, w_1) = 1$$
$$\text{(iv)}\ V'(\phi x, w_2) = 0$$

By (iii) and (iv) we have $V'(L\phi x, w_1) = 0$ and $V'(L{\sim}\phi x, w_1) = 0$, hence $V'((L\phi x \vee L{\sim}\phi x), w_1) = 0$, and hence in turn $V((x)(L\phi x \vee L{\sim}\phi x), w_1) = 0$; by (i) and (ii) we have $V(L\phi x, w_1) = 1$, hence $V({\sim}L\phi x, w_1) = 0$, hence $V((M\phi x \,.\, {\sim}L\phi x), w_1) = 0$; and this gives $V((x)(M\phi x \,.\, {\sim}L\phi x), w_1) = 0$. These two results then clearly give $V(\mathbf{Pr}, w_1) = 0$.

The question now arises: can we give a semantics for LPC + S5 + Pr? To do so we should have to modify our account of LPC + S5-models and LPC + S5-validity so that not only all theses of LPC + S5 but Pr as well would come out valid. What we need to change is the rule for assigning a value to a predicate variable – everything else can remain as set down for LPC + S5 on pp. 146f. If ϕ is *monadic*, we have, for an LPC + S5 + Pr-model, to restrict $V(\phi)$ to a set of ordered pairs, each of the form $\langle u_i, w_j\rangle$, satisfying *one or other* of the following conditions: (i) every member of D is paired either with every member of W or with none; (ii) every member of D is both paired with some member of W and not paired with some other member of W. (Thus, to

[138] Since we are dealing with S5, we do not need to take account of a relation R.

take a simple case, if $W = \{w_1, w_2\}$ and $D = \{u_1, u_2\}$, we could have $V(\phi) = \{\langle u_1, w_1\rangle, \langle u_1, w_2\rangle\}$, satisfying condition (i) since u_1 is paired with each of w_1 and w_2 and u_2 is paired with neither; or $V(\phi) = \{\langle u_1, w_1\rangle, \langle u_2, w_2\rangle\}$, satisfying condition (ii) since u_1 is paired with w_1 but not w_2 and u_2 is paired with w_2 but not w_1; but not $V(\phi) = \{\langle u_1, w_1\rangle, \langle u_1, w_2\rangle, \langle u_2, w_1\rangle\}$ – which is incidentally the assignment we used above in falsifying Pr in LPC + S5 – since this would satisfy neither condition.) If we now consider Pr in its simplest form, viz.

$$(x)(L\phi x \vee L{\sim}\phi x) \vee (x)(M\phi x \,.\, M{\sim}\phi x)$$

we shall find that every value-assignment to ϕ for which condition (i) holds yields a model satisfying the first disjunct, and every value-assignment to ϕ for which condition (ii) holds yields a model satisfying the second disjunct. In the first case the model represents ϕ as a formal predicate and in the second as a material one. Analogous, though more complicated, restrictions would have to be made on the assignments to predicate variables of higher degree. We should in this way obtain a semantics which would validate Pr in its general form. Moreover, since LPC + S5 + Pr-models are clearly a subset of LPC + S5-models, every thesis of LPC + S5 will remain valid. It would also be possible to use a Henkin proof to show the completeness of LPC + S5 + Pr.

CHAPTER ELEVEN

Identity and Description in Modal LPC

LPC with identity

A non-modal LPC such as we have described is often augmented by the introduction of a primitive dyadic predicate constant to represent identity. For this predicate we use the symbol =, which we write between its arguments[139]. (Thus we have $x = y$ as a possible replacement for ϕxy.) The intended interpretation is that $x = y$ is to mean that x is the same individual as y; the most convenient reading is, 'x is identical with y'. We usually abbreviate $\sim(\mathbf{a} = \mathbf{b})$ to $\mathbf{a} \neq \mathbf{b}$.

We add to the basis for LPC the following two axiom-schemata, where **a** and **b** are any individual-variables:

I1 $\mathbf{a} = \mathbf{a}$

I2 $(\mathbf{a} = \mathbf{b}) \supset (\alpha \supset \beta)$, where α and β differ only in that in one or more places where α has free **a**, β has free **b**.

Modal LPC with identity

We can make the same additions to any of the modal predicate

[139] In LPC = is primitive. In a second-order predicate calculus (i.e. one in which we can quantify over predicate variables as well as over individual-variables) it could be introduced by the definition:

$$(\mathbf{a} = \mathbf{b}) =_{\mathrm{Df}} (\phi)(\phi\mathbf{a} \supset \phi\mathbf{b})$$

Identity is frequently referred to as *Equality*, especially for systems of LPC whose intended interpretation is a mathematical one. The use of the same symbol for identity and for strict equivalence should not give rise to any confusion, since the arguments of the identity sign are always individual-variables and those of the strict equivalence sign are always wffs. (Our frequent use of = as a metalogical symbol, as when we write '$V(\phi x) = 1$' to mean that the value assigned to ϕx by V is 1, should not lead to any confusion either.)

systems we have considered[140]. We shall call T + BF, S4 + BF and LPC + S5 with the addition of I1 and I2: T + I, S4 + I and S5 + I respectively. In the systems so obtained, α and β in I2 may of course be wffs containing modal operators.

Among the theorems we can now derive are two which under the intended interpretation are apt to appear intuitively unacceptable. The first of these is derivable even in T + I:

LI $(x = y) \supset L(x = y)$[141]

PROOF

I2:	(1) $(x = y) \supset (L(x = x) \supset L(x = y))$
(1) × Perm:	(2) $L(x = x) \supset ((x = y) \supset L(x = y))$
I1 × N:	(3) $L(x = x)$
(3), (2) × MP:	(4) $(x = y) \supset L(x = y)$ **Q.E.D.**

(In this proof (1) is obtained from I2 by taking α as $L(x = x)$ and replacing the (free) second occurrence of x by free y.)

The second theorem is closely related to LI but is derivable only in S5 + I, not in the weaker systems.

LNI $(x \neq y) \supset L(x \neq y)$[142]

The proof is from LI by Transp, LMI and DR5.

What LI means is that whenever x and y are the same object it is a necessary truth that they are, or that every true statement of identity is necessarily true, or that there are no true contingent statements of identity. Now it seems easy to think of counter-examples to this. E.g., the sentence:

(1) The man who lives next door is the mayor

[140] We shall only discuss systems which contain the Barcan formula though many of our results will probably apply to the corresponding systems without it.

[141] Cf. Prior [1955], p. 205 and Barcan [1947]. Barcan derives the theorem in quantificational S2 (for S2 *vide infra*, Chapter 12): she also derives the stronger $(x = y) \prec L(x = y)$ in S4 (it can of course be derived by N in T, but this is not one of the systems she discusses). We could also prove LI as the schema $(\mathbf{a} = \mathbf{b}) \supset L(\mathbf{a} = \mathbf{b})$, where **a** and **b** are any individual-variables.

[142] We can of course also prove the corresponding schema

$$(\mathbf{a} \neq \mathbf{b}) \supset L(\mathbf{a} \neq \mathbf{b})$$

seems to assert an identity between the man who lives next door and the mayor; and if so we could re-write (1), semi-formally, as:

The man who lives next door = the mayor

Yet surely, it may be said, this is contingent, for it is logically possible that the man who lives next door might not have been the mayor. Or, to use a classic example[143], although the morning star is in fact the same body as the evening star, this is a contingent truth of astronomy and not a necessary truth of logic. So if we are to regard LI as valid we shall have to show that cases like these are not genuine counter-examples.

One way of doing this would be to construe (1) in such a way that it did express a necessary truth. Now it is contingent that the man who is in fact the man who lives next door is the man who lives next door, for he might have lived somewhere else; that is, *living next door* is a property which belongs contingently, not necessarily, to the man to whom it does belong. And similarly, it is contingent that the man who is in fact the mayor is the mayor; for someone else might have been elected instead. But if we understand (1) to mean that the object which (as a matter of contingent fact) possesses the property of being the man who lives next door is identical with the object which (as a matter of contingent fact) possesses the property of being the mayor, then we are understanding it to assert that a certain object (variously described) is identical with itself, and this we need have no qualms about regarding as a necessary truth. This would give us a way of construing identity statements which makes LI perfectly acceptable: for whenever $x = y$ is true we can take it as expressing the necessary truth that a certain object is identical with itself.

143 Although this example dates from Frege [1892] the difficulties for modal predicate logic to which it draws attention were first raised by Quine [1947] and have since been forcefully presented by him in [1953] [1960] and [1966]. It is not our purpose to enter into a discussion of the philosophical problems with which this whole area bristles, and we shall confine any philosophical remarks we do make to those which bear directly on the interpretation of modal systems. Those interested in the wider issues, which we do not in any way suggest are unimportant, may be referred to the following, in addition to the texts by Quine: Donnellan [1966], Geach [1963], Hintikka [1961], [1963], Kanger [1957b], Kenny [1963], Linsky [1966], Prior [1963b], [1963c], Rundle [1965], Slater [1963]. (This list is not intended to be complete or definitive.)

It is also at least arguable that intuitively LI and LNI stand or fall together, and that if a satisfactory modal system is to contain LI it should contain LNI as well. As we have seen, both are in S5 + I, but only LI is in T + I and S4 + I. We could, however, consistently add LNI as an extra axiom to either of these weaker systems without strengthening it to S5 + I. (We should of course add it in the form of the schema

$$(\mathbf{a} \neq \mathbf{b}) \supset L(\mathbf{a} \neq \mathbf{b})$$

to which we shall also refer as LNI.) We can thus have two versions of T + I, one obtained by adding simply I1 and I2, and the other by adding LNI as well, to the basis for T + BF; and the same holds for S4 + I. We shall call these latter versions, T + LNI and S4 + LNI, reserving the names T + I and S4 + I to the systems without LNI.

We have here a situation parallel to the one we met in connection with the Barcan formula (p. 144). The converse of that formula, it will be remembered, was, like LI, in all three systems as originally formulated; but the Barcan formula itself, like LNI, was only in LPC + S5, though it could be added to either of the weaker systems. The parallel in fact goes further than this: we can also modify the systems with identity so that none of them contain either LNI or LI, just as Kripke's axiomatizations of modal predicate calculi yield neither the Barcan formula nor its converse (pp. 180–182).

Semantics for modal LPC with identity

We begin with the systems containing LNI (as an axiom in T + LNI and S4 + LNI, as a theorem in S5 + I). We can define validity for these systems by simply adding to the accounts of the models on pp. 146f the rule:

[V=]. For any individual-variables, **a** and **b**, and for any $w_i \in$ W, V((**a** = **b**), w_i) = 1 or 0 according as V(**a**) = V(**b**) or not.

(By this rule, $(x = y)$ is to count as true when and only when x and y are assigned the same object; this is clearly in accordance with the interpretation of $(x = y)$ to mean 'x is the same object as y'. In effect, since = is a dyadic predicate, we are defining V(=) as the set of triples $\langle u, u, w_i \rangle$ for every $u \in$ D and $w_i \in$ W.)

The definition of validity is then as before.

Clearly the mere addition of the rule for identity cannot affect the validity of formulae not containing the identity operator. Moreover, wherever we have $V((\mathbf{a} = \mathbf{b}), w_i) = 1$ we have $V(\mathbf{a}) = V(\mathbf{b})$ and hence $V((\mathbf{a} = \mathbf{b}), w_j) = 1$ for every $w_j \in W$, i.e. $V(L(\mathbf{a} = \mathbf{b}), w_i) = 1$. Similarly, whenever we have $V((\mathbf{a} = \mathbf{b}), w_i) = 0$ (i.e. $V((\mathbf{a} \neq \mathbf{b}), w_i) = 1$), we have $V((\mathbf{a} = \mathbf{b})\ w_j) = 0$ (i.e. $V((\mathbf{a} \neq \mathbf{b}), w_j) = 1$) for every $w_j \in W$, and hence $V(L(\mathbf{a} \neq \mathbf{b}), w_i) = 1$. So this semantics validates LI and LNI in all systems. The validity of I1 is obvious, since $V(\mathbf{a}) = V(\mathbf{a})$. I2 is likewise valid, since if $V((\mathbf{a} = \mathbf{b}), w_i) = 1$, then α and β differ only in respect of some free variable to which the same value is assigned; hence if $V(\alpha, w_i) = 1$, $V(\beta, w_i) = 1$. Therefore all the theses of T + LNI, S4 + LNI and S5 + I are valid by our present semantics.

To prove completeness for these systems we again construct a system of maximal consistent sets, Γ, as described on pp. 164–168, making each Γ_i of course consistent relative to the augmented systems with the axioms for identity. In making a value-assignment which will yield the result that $V(\alpha, w_i) = 1$ iff $\alpha \in \Gamma_i$, however, we run into a complication. Our earlier plan was to assign to each individual-variable, **a**, itself (considered as an object). But this plan will not work here, for since x and y, for example, are distinct variables, they would be assigned distinct values, and we should therefore always have $V((x = y), w_i) = 0$, regardless of whether $(x = y) \in \Gamma_i$ or not. One way of coping with this difficulty is the following. We still assign to each individual-variable some individual-variable (considered as an object), but this time not necessarily itself. Instead we make the assignment in this way. Let us assume that all the individual-variables are arranged in some determinate order. Let **b** be any of these individual-variables. If there is any variable, **a**, occurring *earlier* in the ordering, such that $(\mathbf{a} = \mathbf{b}) \in \Gamma_1$, then we assign to **b** not itself but the variable previously assigned to **a** – i.e. we let $V(\mathbf{b}) = V(\mathbf{a})$. If, however, there is no such earlier variable, then we simply let $V(\mathbf{b})$ be **b** itself as before. This assignment will ensure that whenever any wff of the form $(\mathbf{a} = \mathbf{b}) \in \Gamma_1$, then $V(\mathbf{a})$ is the same as $V(\mathbf{b})$, and hence $V((\mathbf{a} = \mathbf{b}), w_1) = 1$. It will also ensure the converse, viz. that whenever $V((\mathbf{a} = \mathbf{b}), w_1) = 1$, $(\mathbf{a} = \mathbf{b}) \in \Gamma_1$, for the following reason. If $V((\mathbf{a} = \mathbf{b}), w_1) = 1$, then by [V=], $V(\mathbf{a})$ is the same as $V(\mathbf{b})$; but by the nature of the assignment this can only be because *either:*

(i) V(**a**) and V(**b**) are each some individual-variable, **c**, distinct from **a** and **b** and such that $(\mathbf{c} = \mathbf{a}) \in \Gamma_1$ and $(\mathbf{c} = \mathbf{b}) \in \Gamma_1$; in which case since $\vdash(\mathbf{c} = \mathbf{a}) \supset ((\mathbf{c} = \mathbf{b}) \supset (\mathbf{a} = \mathbf{b}))$, $(\mathbf{a} = \mathbf{b}) \in \Gamma_1$; *or:*

(ii) **a** and **b** are distinct variables and are both assigned **a** or are both assigned **b**; now this presupposes that either $(\mathbf{a} = \mathbf{b}) \in \Gamma_1$ or $(\mathbf{b} = \mathbf{a}) \in \Gamma_1$, and hence, since $\vdash(\mathbf{b} = \mathbf{a}) \supset (\mathbf{a} = \mathbf{b})$, $(\mathbf{a} = \mathbf{b}) \in \Gamma_1$; *or:*

(iii) **a** and **b** are the same variable (say **a**), and V(**a**) is **a** itself; in which case, by I1, $(\mathbf{a} = \mathbf{b}) \in \Gamma_1$ once more.
Hence in every case, if $V((\mathbf{a} = \mathbf{b}), w_1) = 1$, $(\mathbf{a} = \mathbf{b}) \in \Gamma_1$[144].

Note at this point that if $(\mathbf{a} = \mathbf{b})$ is in any Γ_i in Γ it is also in every other maximal consistent set in Γ, for (i) if $(\mathbf{a} = \mathbf{b}) \in \Gamma_i$ then by LI, $L(\mathbf{a} = \mathbf{b}) \in \Gamma_i$, and so if Γ_j is subordinate to Γ_i, $(\mathbf{a} = \mathbf{b}) \in \Gamma_j$; and (ii) if $(\mathbf{a} = \mathbf{b}) \notin \Gamma_i$, then $(\mathbf{a} \neq \mathbf{b}) \in \Gamma_i$, hence by LNI, $L(\mathbf{a} \neq \mathbf{b}) \in \Gamma_i$, hence $(\mathbf{a} \neq \mathbf{b}) \in \Gamma_j$, and so $(\mathbf{a} = \mathbf{b}) \notin \Gamma_j$. Thus although the assignment is defined for Γ_1 it will hold for every Γ_i; i.e., for every $\Gamma_i \in \Gamma$, $(\mathbf{a} = \mathbf{b}) \in \Gamma_i$ iff $V(\mathbf{a}) = V(\mathbf{b})$, and thus $(\mathbf{a} = \mathbf{b}) \in \Gamma_i$ iff $V((\mathbf{a} = \mathbf{b}), w_i) = 1$.

We have however now to show that with this revised assignment to individual variables, $V(\alpha, w_i) = 1$ iff $\alpha \in \Gamma_i$, not merely when α is of the form $(\mathbf{a} = \mathbf{b})$, but when α is any wff whatever; i.e. that Theorem 4 (p. 169) still holds. Suppose that α is an atomic formula $\phi(x_1, \ldots, x_n)$. By the nature of the assignment we shall have $V(x_1) = y_1, \ldots, V(x_n) = y_n$, where $y_1, \ldots, y_n$ are members of D (individual-variables) such that $(x_1 = y_1), \ldots, (x_n = y_n) \in \Gamma_i$[145]. Now $V(\phi(x_1, \ldots, x_n), w_i) = 1$ iff $\phi(y_1, \ldots, y_n) \in \Gamma_i$; but from I2 we can easily derive $((x_1 = y_1) \cdot \ldots \cdot (x_n = y_n)) \supset (\phi(x_1, \ldots, x_n) \equiv \phi(y_1, \ldots, y_n))$; hence $V(\phi(x_1, \ldots, x_n), w_i) = 1$ iff $\phi(x_1, \ldots, x_n) \in \Gamma_i$. So Theorem 4 still holds for atomic formulae; and since we have already shown that it holds for wffs

144 This method is sometimes called *Lindenbaum's* method and is one of very general applicability. Essentially it shows that whenever in an axiomatic system we have a number of symbols whose values are indistinguishable in the system then we can always construct a model in which they are assigned the same value, by taking some value to 'represent' all of them. What is often used as this value is the *class* of all the indistinguishable elements (called the *equivalence class* of these elements); but where, as in our case, the elements are denumerable we can more conveniently take the first in the standard ordering to represent them all.

145 Note that if $V(x_k) = x_k$ we have, from I1, that $(x_k = x_k) \in \Gamma_i$.

of the form (**a** = **b**), the original induction will show that it holds for all wffs.

Contingent identity systems

We have seen that, when looked at in a certain light, LI can seem intuitively unacceptable; and the same applies to LNI. For this reason some logicians have proposed systems of modal LPC in which these theorems cannot be derived. We can call such systems, *contingent identity systems*. One suggestion has been to introduce a restriction into I2, so that it will now run:

I2′ (CI). $(\mathbf{a} = \mathbf{b}) \supset (\alpha \supset \beta)$, where α and β differ only in that in one or more places where α has free **a**, *not occurring within the scope of a modal operator*, β has free **b**[146].

Following this suggestion, we can form the systems T + CI, S4 + CI, S5 + CI, containing I1 and I2′ (but not of course I2 or LNI).

The replacement of I2 by I2′ leaves the systems unchanged as far as non-modal theses are concerned; but it immediately blocks the proofs we gave earlier of LI and LNI, and in fact we shall show later on that these cannot now be proved, even in S5 + CI.

In constructing a semantics for these systems we shall want to keep the rule that *within any given world* a wff of the form (**a** = **b**) is to count as true iff **a** and **b** are assigned the same value, since the validity of non-modal theses containing the identity sign depends on this. But we can avoid having LI as valid – we can, that is, admit the possibility of having $V((x = y), w_i) = 1$ but $V((x = y), w_j) = 0$, where $w_i R w_j$ – by requiring V to make, not an absolute assignment to a variable **a**, but an assignment to **a** *within a world* w_i[147]; for then we can have a single object assigned to x and y in w_i, and hence $V((x = y), w_i) = 1$, but distinct objects assigned to x and y in w_j, and hence $V((x = y), w_j) = 0$. We therefore substitute for the condition to be satisfied by a value-assignment to individual-variables in the account of an LPC-model on p. 146, the following:

For any individual-variable, **a**, and for any $w_i \in W$, $V(\mathbf{a}, w_i)$ is some member of D.

146 *Vide* Kanger [1957b], Hintikka [1961], [1963], Cohen [1960].

147 The semantics of Kanger [1957b], pp. 2–3, does in effect make this assumption.

This will require a consequential re-phrasing of the rule for atomic formulae, which will now run:

For any n-adic predicate variable, ϕ, and for any $w_i \in W$, $V(\phi(x_1, \ldots, x_n), w_i) = 1$ or 0 according as $\langle V(x_1, w_i), \ldots, V(x_n, w_i), w_i \rangle \in V(\phi)$ or not.

For identity we replace the rule [V=] by:

[V=′]. For any individual-variables, **a** and **b**, and for any $w_i \in W$, $V((\mathbf{a} = \mathbf{b}), w_i) = 1$ or 0 according as $V(\mathbf{a}, w_i) = V(\mathbf{b}, w_i)$ or not.

This change in making assignments to the individual-variables, however, causes a difficulty in interpreting [V∀] (p. 147). [V∀] states that $V((\mathbf{a})\alpha, w_i) = 1$ iff for every V′ differing from V *only in assignment to* **a**, $V'(\alpha, w_i) = 1$.

The most obvious way of interpreting this rule here would seem to be as:

[V∀′]. For any wff, α, any individual-variable, **a**, and any $w_i \in W$, $V((\mathbf{a})\alpha, w_i) = 1$ iff for every V′ differing from V only in assignment to **a** in any world or worlds, $V'(\alpha, w_i) = 1$[148].

However, it is easy to show that the semantics so obtained would make

$$(1)\quad L(\exists x)\phi x \supset (\exists x)L\phi x$$

valid. For suppose that in some $w_i \in W$ in a CI-model, $V(L(\exists x)\phi x, w_i) = 1$. Then $V((\exists x)\phi x, w_j) = 1$ for every w_j such that $w_i R w_j$. Hence for every such w_j there is some assignment, say V_j', differing from V only in assignment to x, and such that $V_j'(\phi x, w_j) = 1$ (i.e., $\langle V_j'(x, w_j), w_j \rangle \in V(\phi)$). This means that we can define an assignment, V′, throughout the whole model, simply by letting $V'(x, w_j)$ be that member of D which V_j' assigns to x in w_j in each case, and otherwise letting V′ be the same as V. Clearly $V'(\phi x, w_j) = 1$ for every w_j such that $w_i R w_j$, and hence $V'(L\phi x, w_i) = 1$. But V′ differs from V only in assignment to x, and therefore $V((\exists x)L\phi x, w_i) = 1$. Thus for any $w_i \in W$ in any CI-model, we should have $V(L(\exists x)\phi x \supset (\exists x)L\phi x, w_i) = 1$, and therefore the formula would be valid.

148 Cf. Kanger, loc. cit. An alternative would be to let $V((\mathbf{a})\alpha, w_i)$ be 1 iff for every V′ differing from V only in assignment to **a** *in* w_i, $V'(\alpha, w_i) = 1$; (thus in forming any V′ we should have to keep $V'(\mathbf{a})$ the same as $V(\mathbf{a})$ in every world other than w_i). This would give us a semantics for the quantifier (Ux) of Kanger [1957c]. As Kanger himself observes there, however, such a semantics would fail to validate $(x)L\phi x \supset L\phi y$.

Now this formula (1) is one which we have encountered before (p. 144), and we remarked then that it is not intuitively plausible. To adapt an example given by Quine[149], in certain games it is necessary that some player will win, but there is no individual player who is bound to win. There is, however, one way in which we *could* make (1) sound plausible, and that is by thinking of an expression such as 'the winner' as in a sense standing for a single 'object', though one which in a more usual sense of 'object' may be one object in a certain situation but a different one in another. For in that case, if it is necessary that someone will win then there *is* someone, viz. *the winner*, who is bound to win. Now we do often use phrases of the form 'the so-and-so' in such a way as this. Consider, for example, the expression 'the top card in the pack', as it occurs in the rules of a card game. The rules may, without ambiguity, specify that at a certain point in the play the top card is to be dealt to a certain player; yet on one occasion the top card may be the Ace of Spades and on another it may be the Queen of Hearts. Thus the phrase 'the top card in the pack' does not designate any particular card (individual piece of pasteboard), except in the context of a particular state of the pack; yet we can in one sense think of it as standing for a single object, contrasted with *the bottom card in the pack* and so forth.

Such 'objects' are often called *intensional objects*[150], and the rules we have been considering would seem to provide a semantics for a logic in which (unlike the other systems we have considered) the individual-variables range over intensional objects. We shall, however, not attempt to construct such a logic here, though as the above discussion has indicated,

$$(1)\quad L(\exists x)\phi x \supset (\exists x)L\phi x$$

would be valid in it. (Where ϕx means 'x is at the top of the pack', and x ranges over intensional objects, then (1) will be true because if it must be the case that there is a card at the top of the pack, then although no individual piece of pasteboard is

149 Quine [1953], p. 148, where however the example is used to illustrate a somewhat different point.

150 Intensional objects would be like the *individual concepts* of Carnap [1947], p. 47, and Frege [1892]. For a formalization of Frege's intensional logic *vide* Church [1951].

bound to be at the top of the pack, yet there *is* something, viz. *the top card*, which is bound to be at the top of the pack. This parallels what we said about the game in which it must be the case that someone will win.)

However, (1) is not in fact a theorem of any of the contingent identity systems, since these are obtained by weakening the LNI systems and (1) is not a theorem of any of *them*. And this shows that what seemed to be the most natural semantics for the CI systems turns out not to characterize them after all.

The fault seems to be that the semantics enables us to make an 'object' out of any string of members of D whatever. E.g., suppose there are two worlds, w_1 and w_2, then where u_1 and u_2 are members of D we seem to be entitled to make up the 'object' which is u_1 in w_1 and u_2 in w_2. (Roughly, what (1) says is that any string of members of D each of which is ϕ in some world accessible to w_i entitles us to assume the existence of an 'object' which is ϕ in all of them.)

When we look at the matter in this way we might think of the LNI systems as requiring that the only strings of members of D which count as objects are strings consisting of the same member of D in each world (i.e., the only objects recognized in these systems are the straightforward members of D themselves). It seems therefore that an adequate semantics for the CI systems would have to require neither that only strings consisting solely of a single member of D should count as objects, nor that any string whatever of members of D should count as an object.

One way in which we might achieve such a semantics is to let the set of strings which are to count as objects be determined by each model; i.e., to let the model specify what assignments to individual-variables are to be permissible. Formally we could do this by having the model include a set, Θ, of value-assignments, taking one of them, V_1, as the designated assignment and letting a wff, α, be valid iff $V_1(\alpha, w_i) = 1$ for every $w_i \in W$ in every such model.

A T + CI (S4 + CI, S5 + CI)-model can then be defined as an ordered quintuple $\langle W, R, D, V_1, \Theta \rangle$, where W, R and D are as before, $V_1 \in \Theta$, and Θ is a set of value-assignments such that for every $V \in \Theta$ and for any individual-variables **a** and **b** there is some $V' \in \Theta$ which is the same as V except that $V(\mathbf{a}, w_i) = V'(\mathbf{b}, w_i)$ for every $w_i \in W$.

[V~], [V∨] and [VL] will all be as before (but generalized for every $V \in \Theta$), but [V∀] will be replaced by:

[V∀″]. For any wff, α, any individual-variable, **a**, and any $w_i \in W$, $V((\mathbf{a})\alpha, w_i) = 1$ iff for every $V' \in \Theta$ differing from V only in assignment to **a**, $V'(\alpha, w_i) = 1$.

The models for the LNI systems now become equivalent to that subset of CI-models in which Θ contains precisely those assignments which give to any individual-variable the same value in every world in the model (and where V_1 is the V of the LNI-models). From this it follows that formula (1), since it is not valid in any of the LNI systems, is not valid by these semantics either.

The PC, modal and quantificational axioms of T + CI, S4 + CI and S5 + CI are all valid by the semantics we have now described. I1 is likewise valid, since every $V \in \Theta$ assigns a unique value to any individual-variable in a given world, and hence $V((\mathbf{a} = \mathbf{a}), w_i) = 1$ for every $w_i \in W$. And I2′ is valid, for any modal part of α will be repeated exactly in β, and, given the values (in w_i) of these modal sub-formulae, the values of α and β in w_i will depend solely on the value-assignments made in w_i itself to the variables; but α differs from β only in that it has free **b** where α has free **a**, and by hypothesis **a** and **b** have the same value in w_i.

The unrestricted axiom-schema I2, however, is no longer valid. For while

$$(2)\ (x = y) \supset (\phi x \supset \phi y)$$

(a simple instance of I2′) is valid,

$$(3)\ (x = y) \supset (L\phi x \supset L\phi y)$$

which is an instance of I2 but not of I2′, is falsified in the following model: $W = \{w_1, w_2\}$; $D = \{u_1, u_2\}$; $V(x, w_1) = u_1$, $V(y, w_1) = u_1$, $V(x, w_2) = u_1$, $V(y, w_2) = u_2$; $V(\phi) = \{\langle u_1, w_1\rangle, \langle u_1, w_2\rangle\}$. Here x and y are assigned the same value (u_1) in w_1, so that we have $V((x = y), w_1) = 1$; but from the assignments to x, y and ϕ we have $V(\phi x, w_1) = 1$, $V(\phi x, w_2) = 1$ and $V(\phi y, w_2) = 0$, and hence $V(L\phi x, w_1) = 1$ but $V(L\phi y, w_1) = 0$, so that $V((3), w_1) = 0$[151]. The same model – in fact any model in which x and y are assigned

[151] The fact that (2) is a theorem of the contingent identity systems but (3) is not brings out sharply another awkward feature of these systems, viz. that the rule of uniform substitution for predicate variables does not hold in them; for if it did we could replace ϕ by $L\phi$ in (2) and thus obtain

the same value in w_1 but distinct values in w_2 – will falsify LI, for in such a model we shall have $V((x = y), w_1) = 1$ but $V((x = y), w_2) = 0$ and therefore $V(L(x = y), w_1) = 0$. And LNI is falsified in an analogous way if x and y have the same value

---

(3) as a theorem. It thus seems that acceptance of these systems as they stand involves a refusal to treat an expression of the form $L\phi$ as designating a genuine property. This might be unworrying if there were no intelligible way of construing $L\phi$ as a property; but the semantics we have been using for the CI systems in fact gives us a perfectly clear way of defining $L\phi$ as a property by the same technique as we have used for ϕ itself, viz. by correlating it in a model with a set of ordered pairs. For we can say that $V(L\phi)$ is the set of pairs $\langle u_k, w_i \rangle$ such that for every w_j such that $w_i R w_j$, $\langle u_k, w_j \rangle \in V(\phi)$; in other words, a member of D is in $V(L\phi)$ for a given world iff it is in $V(\phi)$ for every world accessible to that world. $L\phi x$ will then be true in w_i iff the value assigned to x *in* w_i is ϕ in every w_j accessible to w_i (irrespective of whether the value assigned to x in w_j is ϕ in w_j or not). All this clearly defines $L\phi$ as a property in the relevant sense, and gives us moreover one very natural way of construing 'x is necessarily ϕ'; so the non-substitutability of $L\phi$ for ϕ in (2) involves a severe restriction on the range of properties which can be values of the predicate variables in the CI systems.

It appears, in fact, that the assignment of values to individual-variables world by world instead of absolutely, which is an essential feature of the semantics for the CI systems, leaves open two quite distinct ways of evaluating formulae containing $L\phi$. We can illustrate these by $L\phi y$ in the model cited in the text. By the rules for the CI systems we have, as we noted, $V(L\phi y, w_1) = 0$, since although the object assigned to y in w_1 (viz. u_1) is ϕ in w_1, the object assigned to y in w_2 (viz. u_2) is not ϕ in w_2. But by the revised rule we have just given we should have $V(L\phi y, w_1) = 1$, since the object assigned to y in w_1 (u_1) *is* ϕ in w_2 as well as in w_1. This difference in the ways of interpreting $L\phi x$ seems to be analogous to, if not the same as, the distinction between *de dicto* and *de re* modalities in the sense referred to in footnote 131 on p. 183. Following the suggestion we made there, we could express $L\phi x$, when construed in the first way, as $L(\phi x)$, and when construed in the second, as $(L\phi)x$. We should then find that

$$(4)\ (x = y) \supset ((L\phi)x \supset (L\phi)y)$$

is valid, though

$$(5)\ (x = y) \supset (L(\phi x) \supset L(\phi y))$$

is not. The CI systems, in fact, by making (3) invalid, seem to be insisting that we interpret it in the sense of (5) rather than (4). This is another way of bringing out the fact that they do not allow us to treat $L\phi$ as a genuine property, though what we have said suggests that by suitable modifications they could be made to do so.

If one were to hold that it never makes sense to identify an object in one world with an object in another (*vide* footnote 122), one would presumably wish to have a CI system in which no two domains overlapped.

in w_2 but distinct values in w_1. These results for formulae with identity are the same whether we are dealing with T + CI, S4 + CI or S5 + CI.

Semantics for T + I and S4 + I

By making a single change in the semantics we have just been considering for the CI systems we can define validity for the systems which we mentioned at the beginning of this chapter and then set aside, viz. T + I and S4 + I, in which I1 and I2 but not LNI are theses. (LNI is of course derivable as a theorem in S5 + I). The change in question is a restriction in the rule for value-assignments to individual-variables. We still allow a $V \in \Theta$ to assign different values to an individual-variable in different worlds, but we now insist that wherever two individual-variables have the same value in a given world, say w_i, they must also have the same value (though not necessarily *that* one) in every world accessible to w_i. The rule will therefore now run:

For any individual-variable, **a**, any $V \in \Theta$ and for any $w_i \in W$, $V(\mathbf{a}, w_i)$ is some member of D; provided that (for any individual-variable, **b**) if $V(\mathbf{a}, w_i) = V(\mathbf{b}, w_i)$ then for any $w_j \in W$ such that $w_i R w_j$, $V(\mathbf{a}, w_j) = V(\mathbf{b}, w_j)$.

Everything else is to remain as in the account of models and validity which we have just given for the CI systems.

It is clear that T + I (S4 + I)-models on the present account are a subset of the T + CI (S4 + CI)-models, and that therefore everything that is valid in the systems of contingent identity remains valid here. But LI now becomes valid too, even in T + I: for given that $V((\mathbf{a} = \mathbf{b}), w_i) = 1$, **a** and **b** have the same value not only in w_i but also in every world accessible to it, and hence $V(L(\mathbf{a} = \mathbf{b}), w_i) = 1$. Similarly I2 is now valid without restriction: for the values of α and β in w_i depend solely on the value-assignments to variables in w_i and in worlds accessible$_*$ to it; and therefore, since α and β differ only in that β has free **b** where α has free **a**, wherever we have $V((\mathbf{a} = \mathbf{b}), w_i) = 1$ and $V(\alpha, w_i) = 1$ we also have $V(\beta, w_i) = 1$.

When, however, **a** and **b** have distinct values in w_i, nothing is prescribed about whether they have the same or distinct values in w_j. So we can have $V((\mathbf{a} \neq \mathbf{b}), w_i) = 1$ but $V((\mathbf{a} \neq \mathbf{b}), w_j) = 0$ and hence $V(L(\mathbf{a} \neq \mathbf{b}), w_i) = 0$; therefore LNI is invalid in T + I (and likewise in S4 + I). On the other hand, when we add the S5 condition that whenever $w_i Rw_j$, $w_j Rw_i$, the result is that variables assigned the same value in either world must be assigned the same value in the other, and therefore variables assigned distinct values in either world must be assigned distinct values in the other. So in S5 + I, if we have $V((\mathbf{a} \neq \mathbf{b}), w_i) = 1$ we also have $V((\mathbf{a} \neq \mathbf{b}), w_j) = 1$ whenever $w_i Rw_j$, and hence $V(L(\mathbf{a} \neq \mathbf{b}), w_i) = 1$; and therefore LNI is valid.

Descriptions

We saw earlier that a perplexity can arise about the propositions expressed by such English sentences as:

(1) The man who lives next door is the mayor

or (2) The morning star is the evening star.

Expressions such as 'the morning star' or 'the man who lives next door' are often called *definite descriptions*; they have their typical uses when there is some property which belongs to one and only one object, for then we frequently want to speak about that object as *the thing which possesses that property*. If ϕ is any such property, we want to be able to speak of *the thing which is* ϕ (or more briefly, *the* ϕ), and to form such propositions as 'x is the ϕ', 'only the ϕ is ψ', 'the ϕ is the ψ', and so on. In order to be able to represent such propositions we shall use the expression $(\iota x)\phi x$ to mean 'the x such that ϕx' (i.e. 'the (thing which is) ϕ')[152]. We call (ιx) a *description operator*. The general form of which $(\iota x)\phi x$ is an instance is $(\iota \mathbf{a})\alpha$, where $\mathbf{a}$ is any individual-variable and α is any wff. (The interesting cases will of course be those in which α contains some free occurrence of $\mathbf{a}$, but it is simpler to allow α to be any wff at all.) A description operator is analogous to a quantifier in that every free occurrence of $\mathbf{a}$ in α counts as bound in $(\iota \mathbf{a})\alpha$[153], though what it forms out of α is not a new wff (propositional expression) but a certain kind of individual-designating expression, called a *term*. Now we want expressions of the form $(\iota \mathbf{a})\alpha$ (terms) to function in the formal

[152] The symbol ι is introduced with this use in Whitehead and Russell [1910], p. 30, though it is not taken there as a primitive symbol (*vide infra*, pp. 206–209).

[153] Thus, just as $(x)\phi x$ and $(y)\phi y$ each mean simply that everything is ϕ, so $(\iota x)\phi x$ and $(\iota y)\phi y$ each mean simply 'the ϕ', and hence they too can be regarded as bound alphabetic variants of each other.

systems as other individual-symbols do, so we extend the formation rules to allow terms to replace individual-variables as the arguments of predicates. Thus we shall have wffs such as $\psi(\iota x)\phi x$ ('the ϕ is ψ'), $x = (\iota y)\phi y$ ('x is the ϕ') and $(\iota x)\phi x = (\iota y)\psi y$ ('the ϕ is the ψ').

We also enlarge $\forall 1$ so that it also gives $((\mathbf{a})\alpha \supset \beta)$ whenever β differs from α only in having any term in place of a free occurrence of $\mathbf{a}$ in α (provided that no variable free in the term is bound in β). Thus a simple instance of $\forall 1$ would be: $(x)\phi x \supset \phi(\iota x)\psi x$ – 'if everything is ϕ, the ψ is ϕ'.

We finally add one further axiom-schema. In order to state this succinctly we introduce the notation $(\exists^1 x)\phi x$ to mean 'exactly one thing is ϕ' – or more generally $(\exists^1 \mathbf{a})\alpha$ to mean 'there is exactly one value of $\mathbf{a}$ which makes α true'. We can define $(\exists^1 \mathbf{a})$ as follows:

$$(\exists^1 \mathbf{a})\alpha =_{\text{Df}} (\exists \mathbf{b})(\mathbf{a})(\alpha \equiv (\mathbf{a} = \mathbf{b}))^{154}$$

where $\mathbf{b}$ is an individual-variable which does not occur free in α. (The sense of the definiens is: 'there is an object, $\mathbf{b}$, such that α is true when $\mathbf{a}$ is $\mathbf{b}$ but not otherwise'.) We now state the new axiom-schema:

D1 $(\exists^1 \mathbf{a})\alpha \supset \beta$, where β differs from α only in having $(\iota \mathbf{a})\alpha$ wherever α has free $\mathbf{a}$[155].

The sense of D1 can be seen by considering a simple instance, viz. $(\exists^1 x)\phi x \supset \phi(\iota x)\phi x$, which may be rendered as 'if there is exactly one thing that is ϕ, then it is the ϕ which is ϕ'.

These additions prove to be highly convenient in non-modal LPC: they give us a suitable notation for definite descriptions, and theorems which contain description operators. At the same time it can be proved that they do not give us any new theorems

154 Cf. Whitehead and Russell [1910], p. 30. (Russell uses $E!(\iota x)(\phi x)$ in place of $(\exists^1 x)\phi x$.) For alternative definitions *vide* Mendelson [1964], p. 79 and Kleene [1952], p. 199.

155 Provided that no variable free in the term is bound in β. Cf. Rosser [1953], p. 185. Since we are only interested in the peculiarly modal problems about descriptions and are assuming the Barcan formula we do not intend to discuss questions connected with non-referring descriptions. Formally we shall let $(\iota x)\phi x$ be the ϕ if there is exactly one thing which is ϕ (i.e. if the antecedent of D1 is satisfied); otherwise we shall let $(\iota x)\phi x$ be any arbitrary member of the domain of individuals.

which do *not* contain description operators[156]. Our present concern, however, is with the consequences of making the same additions to the modal systems, and here we run into a difficulty. Consider the following derivation:

D1:	(1) $(\exists^1 x)\phi x \supset \phi(\iota x)\phi x$
(1) × DR1:	(2) $L(\exists^1 x)\phi x \supset L\phi(\iota x)\phi x$
∀1:	(3) $(x){\sim}L\phi x \supset {\sim}L\phi(\iota x)\phi x$
(3) × Transp, × Def ∃:	(4) $L\phi(\iota x)\phi x \supset (\exists x)L\phi x$
(2), (4) × Syll:	(5) $L(\exists^1 x)\phi x \supset (\exists x)L\phi x$

This last formula, (5), is only a slightly weaker version of formula (1) on p. 196 (viz. $L(\exists x)\phi x \supset (\exists x)L\phi x$), and intuitively both would seem to stand or fall together; in particular they would seem to *fall* in systems which do not allow intensional objects.

Now to investigate whether formulae containing description operators, such as those that are derivable by D1 or related schemata, are valid in modal systems we need a rule for evaluating $(\iota x)\phi x$ in a semantical model. In non-modal LPC it is easy to say: if there is exactly one member of D which is ϕ, let $V((\iota x)\phi x)$ be that member of D; and if there is no such member of D let $V((\iota x)\phi x)$ be some arbitrary object. This rule will obviously bring out D1 as valid[157]. And in modal systems which are prepared to admit intensional objects we could, analogously, let $V((\iota x)\phi x, w_i)$ be the member of D which is ϕ in w_i if there is exactly one such and some arbitrary member of D if there is not. But for the other modal systems it is not at all clear what the corresponding rule should be. For even if in each world (in a

[156] Indeed a stronger result can be proved. For it can be shown that there is an effective method of transforming any wff, α, containing a description operator, into a wff, α', which does not and which is such that α is a thesis of the extended system iff α' is a thesis of the system without D1 or the extension to ∀1. In this sense we can say that D1 is non-independent in non-modal LPC. *Vide* Mendelson [1964], pp. 83–85, and Kleene [1952], pp. 405–413; and cf. *infra*, pp. 206–209.

[157] From a result of Smullyan [1948] it follows that if we do not allow terms to occur within the scope of modal operators we can have D1 in the modal systems too without running into the trouble we have been discussing; but this restriction is a very drastic one – we might almost as well dispense with terms in modal logic altogether.

given model) there is exactly one thing which is ϕ, this may not be the same thing in all worlds; and in such a case there will be no member of D which we can fix on as clearly the value to be assigned to $(\imath x)\phi x$. The one kind of model in which it *is* clear what we should do is one in which one and the same member of D is the only thing which is ϕ in each of the worlds; i.e. in which there is some $u_k \in \mathrm{D}$ such that for every $w_i \in \mathrm{W}$, u_k is ϕ in w_i and nothing else is; for then it is obvious that this u_k is 'the ϕ' in that model and should be assigned to $(\imath x)\phi x$. This suggests one rule we might adopt for value-assignments to terms, viz. that if in a given model there is some $u_k \in \mathrm{D}$ which satisfies the condition we have just stated, then $\mathrm{V}((\imath x)\phi x) = u_k$; and if there is no such u_k then, as in the non-modal case, we can let $\mathrm{V}((\imath x)\phi x)$ be some arbitrary member of D. We can call this rule, [V$\imath$].

With this rule, D1 will be invalid even in S5 + I. The simplest instance of D1 is formula (1) in the derivation above, viz.

$$(1)\ (\exists^1 x)\phi x \supset \phi(\imath x)\phi x$$

and this is falsified in a model in which $\mathrm{W} = \{w_1, w_2\}$, $\mathrm{D} = \{u_1, u_2\}$ and $\mathrm{V}(\phi) = \{\langle u_1, w_1\rangle, \langle u_2, w_2\rangle\}$. For in each world in such a model $(\exists^1 x)\phi x$ will be true, yet whichever member of D we choose as the value of $(\imath x)\phi x$, there will be one world in which $\phi(\imath x)\phi x$ is false.

Moreover the same model will falsify (5). For since $(\exists^1 x)\phi x$ is true in each world, we have $\mathrm{V}(L(\exists^1 x)\phi x, w_1) = 1$; but $(\exists x)L\phi x$ will be true in w_1 only if there is some member of D which is ϕ in each world, and since there is no such member of D,

$$\mathrm{V}((\exists x)L\phi x, w_1) = 0$$

The condition under which [V$\imath$] gives a non-arbitrary assignment to $(\imath x)\phi x$ is, however, precisely the condition under which *in an S5 + I-model* (where every world is accessible to every other) we have $\mathrm{V}((\exists^1 x)L\phi x, w_i) = 1$ for every $w_i \in \mathrm{W}$. So although (1) is invalid, the weaker formula:

$$(6)\ (\exists^1 x)L\phi x \supset \phi(\imath x)\phi x$$

and in general

D2 $(\exists^1 \mathbf{a})L\alpha \supset \beta$

(where α and β are related as they are in D1) will be valid in

S5 + I[158]. We could thus add D2 as an axiom-schema in place of D1 to S5 + I without fearing the unwelcome consequences, such as (5), that D1 brought[159].

We could also consider adding D2 as an axiom-schema to quantificational T and S4, but there is then a problem about what the semantics for the systems so obtained should be. To get such a semantics we should certainly have to modify [V℩], since this rule will make (6), and therefore D2, invalid in these systems. For suppose that u_k is the one thing which is ϕ in every world accessible to w_i, but not in some world inaccessible to w_i. Then we shall have $V((\exists^1 x)L\phi x, w_i) = 1$; but if we use [V℩], $(\iota x)\phi x$ will be assigned an arbitrary member of D which may not be ϕ in w_i at all, and so we may have $V(\phi(\iota x)\phi x, w_i) = 0$. Now [V℩] seems clearly to be the appropriate semantic rule for the systems which contain LNI, so what this shows is that we could not add D2 to T + LNI or S4 + LNI if we wish, as we obviously do, to have all theorems valid by the relevant semantics. It might prove possible, however, to add D2 to T + I and S4 + I, or to T + CI and S4 + CI, since in these systems individual-symbols can have different values in different worlds, and so some modification of [V℩] would seem to be called for in any case. But the general complexity of the semantics for these systems and the particular complexity of the problem of what the required revision of [V℩] would be make it unclear just what results would be obtained.

Introducing descriptions by definition

A quite different method of introducing terms into modal (or for that matter, non-modal) systems would be to regard formulae containing terms as abbreviations of formulae in which terms do not occur at all. If we could do this, we should not require a separate semantical rule for terms, such as [V℩], but could always work out the value of such formulae by the other rules. (The eliminability of terms would of course then be automatically guaranteed.) To carry out this programme it would be

158 In view of the thesishood in S5 + I of $(\exists^1 \mathbf{a})L\alpha \supset L(\exists^1 \mathbf{a})L\alpha$ we may easily obtain from D2 (by DR1) the stronger $(\exists^1 \mathbf{a})L\alpha \supset L\beta$; and more generally we can derive D1 itself ($(\exists^1 \mathbf{a})\alpha \supset \beta$) wherever α is fully modalized.

159 And here D2 is non-independent in S5 + I in the sense of footnote 156.

sufficient to give definitions of (a) atomic formulae in which some individual-variable is replaced by a term, and (b) identity formulae in which one argument of the identity sign is a term.

We content ourselves with a few indications of how this approach would develop[160]. We confine ourselves to monadic predicates and leave the reader to work out the more general cases. Suitable definitions might be:

$$\text{A. } \psi(\iota x)\phi x =_{\text{Df}} (\exists^1 x)\phi x \,.\, (y)(\phi y \supset \psi y)$$

(i.e. 'The ϕ is ψ' is to mean 'Exactly one thing is ϕ and whatever is ϕ is ψ'); and

$$\text{B. } ((\iota x)\phi x = y) =_{\text{Df}} \phi y \,.\, (x)(\phi x \supset (x = y))$$

(i.e. 'The ϕ is (identical with) y' is to mean 'y is ϕ and nothing else is').

Dealing with terms in this way, however, involves abandoning the project (which underlay the previous semantics) of treating them as straightforward designators of individuals, which could replace individual-variables without destroying validity. For example, with the definitions we have just given, even the simplest instance of the extension to ∀1, viz.

$$(x)\psi x \supset \psi(\iota x)\phi x$$

would become invalid, for it would be an abbreviation of

$$(x)\psi x \supset ((\exists^1 x)\phi x \,.\, (y)(\phi y \supset \psi y))$$

and it is clearly possible for everything to be ψ without there being exactly one thing – or indeed anything at all – that is ϕ. Nor would

$$((\iota x)\phi x = y) \supset L((\iota x)\phi x = y)$$

be valid, even by a semantics which made $(x = y) \supset L(x = y)$ valid. And it seems almost certain that any reasonable alternatives to definitions A and B would have similar results.

One thing this method does do, however, is to enable us to bring out an ambiguity which arises when terms occur within the scope of modal operators, and which is in fact one of the

[160] What follows is essentially an adaptation for the modal case of the method set out in detail in *14 of Whitehead and Russell [1910], which has come to be known as Russell's 'theory of descriptions'. For a fuller account than ours, *vide* Smullyan [1948].

main sources of perplexity in deciding on a rule for value-assignment to terms. What precisely, for example, are we to understand by 'necessarily the ϕ is ψ'?[161] Using the description operator as we did in previous sections, the only way we can express this is as $L\psi(\iota x)\phi x$. If we now apply Definition A to this we obtain:

$$(1)\quad L((\exists^1 x)\phi x \,.\, (y)(\phi y \supset \psi y))$$

which means: it is a necessary truth that exactly one thing is ϕ and that whatever is ϕ (viz. the thing in question) is also ψ. (From this we can easily derive $L(\exists x)\psi x$, but not $(\exists x)L\psi x$.) (1) is one of the things we could take 'necessarily the ϕ is ψ' to mean; another, which our present richer notation enables us to distinguish quite clearly from it is:

$$(2)\quad (\exists^1 x)\phi x \,.\, (y)(\phi y \supset L\psi y)$$ [162]

i.e., exactly one thing is ϕ and it is necessarily ψ. (From this we *can* derive $(\exists x)L\psi x$.)

If now we turn back to the derivation of the troublesome formula (5) – viz. $L(\exists^1 x)\phi x \supset (\exists x)L\phi x$ – on p. 204 we shall see that it proceeds by showing that $L(\exists^1 x)\phi x$ implies $L\phi(\iota x)\phi x$, and that this in turn implies $(\exists x)L\phi x$. But if with our present notation we construe $L\phi(\iota x)\phi x$ in the manner of (1), then we take it as:

$$(1')\quad L((\exists^1 x)\phi x \,.\, (y)(\phi y \supset \phi y))$$

whereas if we construe it in the manner of (2) we take it as:

$$(2')\quad (\exists^1 x)\phi x \,.\, (y)(\phi y \supset L\phi y)$$

Now (1′) is certainly implied by $L(\exists^1 x)\phi x$, but it does not imply $(\exists x)L\phi x$; and (2′) certainly implies $(\exists x)L\phi x$, but it is not implied

[161] The problem is not unlike one which Whitehead and Russell raise in respect of the analysis of $\sim\psi(\iota x)\phi x$ (op. cit. p. 69). In fact the upshot of $*14$ of that work is that provided we have $(\exists^1 x)\phi x$ (in Russell's notation $E!(\iota x)(\phi x)$) and provided all complex predicates of which $(\iota x)(\phi x)$ is an argument are made up using only truth functions, then $(\iota x)(\phi x)$ can be treated as an individual symbol, but that otherwise it cannot be so treated (*ibid.*, p. 83).

[162] (2) in fact is what we should obtain from $L\psi(\iota x)\phi x$ by Definition A if instead of regarding $\psi(\iota x)\phi x$ as the argument of L we were to regard $L\psi$ as a predicate and $(\iota x)\phi x$ as its argument. Thus the ambiguity we are pointing out here reflects the distinction between $L(\phi x)$ and $(L\phi)x$ which we drew in footnote 151.

by $L(\exists^1 x)\phi x$. So the ambiguity we have been considering underlies the derivation of (5).

$\psi(\iota x)\phi x$ can after all be modalized only in one place, the beginning; but $(\exists^1 x)\phi x \,.\, (y)(\phi y \supset \psi y)$ can be modalized in several importantly different ways, of which we have mentioned two (another is $(\exists^1 x)\phi x \,.\, L(y)(\phi y \supset \psi y)$, which may express what we very often mean by 'necessarily the ϕ is ψ').

Sets and attributes

The LPC's we have so far considered have contained predicate variables but only one predicate constant, $=$. The distinction between predicate variables and predicate constants in lower predicate calculi is simply that for the predicate constants there will be special axioms over and above the axioms and rules for truth functions and quantification. For $=$ these axioms are I1 and I2. It is even possible to dispense with the predicate variables and have only a (finite or infinite) number of predicate constants with axioms governing their use. One of the most useful predicate calculi of this kind (especially in mathematics) is set theory[163]. Formally, set theory is a lower predicate calculus with one dyadic predicate symbol, the constant $\in$. We understand $x \in y$ to mean 'x is a member of the set y'. One feature of set theory is that sets having exactly the same members are the same set; i.e.,

$$(x = y) \equiv (z)((z \in x) \equiv (z \in y))$$

is a theorem. This of course means that in some of our modal LPC systems (the ones in which we have $(x = y) \supset L(x = y)$) we should have

$$(z)((z \in x) \equiv (z \in y)) \supset L(z)((z \in x) \equiv (z \in y))$$

which means that if two sets have the same members then they necessarily have the same members. While considerations of the kind mentioned on p. 191 may enable a sense to be given to such a principle which makes it valid[164], it causes difficulties if we

[163] A discussion of set theory lies outside the scope of this book. For references *vide*, e.g., Fraenkel and Bar Hillel [1958], Quine [1963], Stoll [1961].

[164] In Fitch [1966] and [1967] we have a set theory in which $(x \in y) \supset L(x \in y)$ is an axiom (a theory, moreover, which according to Fitch does not involve Russell's contradiction and does not require the elaborate safeguards of non-modal set theories).

wish to speak of the set of all objects which possess such and such a property. For where $\hat{x}\phi x$[165] denotes the set of all x's for which ϕx holds, we define it as $(\imath y)(x)[\phi x \equiv (x \in y)]$ (i.e. the thing which has as members those and only those things which are ϕ). And in fact the description operator will get us into exactly the same trouble here as we have already looked at (p. 203f).

Two properties may *happen* to determine the same set, but we say they are still two distinct properties, or two *attributes*. Corresponding to 'x is a member of set y' one might set up a dyadic relation 'x possesses the attribute y', symbolizing it by $x\eta y$, and attempt to produce an LPC parallel to set theory[166]. The advantage of such a theory over set theory in a modal LPC is this: while two sets are the same if they have the same members, we could say that two attributes are identical only if they *necessarily* hold of the same individuals; we should thus have

$$(x = y) \equiv L(z)((z\eta x) \equiv (z\eta y))$$

as a theorem. We could then define the attribute formed from the predicate ϕ as $(\imath y)L(x)(\phi x \equiv x\eta y)$, and it turns out that we could then use D2 and avoid some (though apparently not all[167]) of the difficulties which arise in a straightforward set theory based on modal logic.

In this chapter we have reached fewer firm conclusions than usual; but the topics of identity and description are among the most difficult in modal logic, and in the present state of the subject are still full of obscurities and unsolved problems.

[165] Most set theories do not have predicate variables and would express this as $\hat{x}\alpha$ (i.e. the set of x's which satisfy α), but $\hat{x}\phi x$ is a little easier to grasp intuitively. (Another common notation in place of $\hat{x}\alpha$ is $\{x : \alpha\}$.)

[166] Lemmon [1963] and Barcan Marcus [1963]. We have used η in place of Lemmon's α, which the notation used in this book might make confusing. For formal details of these systems we refer the reader to the articles cited and to Barcan Marcus [1967] and Parsons [1967].

[167] *Vide* Lemmon, op. cit., p. 96.

PART THREE

A Survey of Modal Logic

CHAPTER TWELVE

The Lewis Systems (I)

Historical preamble

Modal logic was discussed by several ancient authors, notably Aristotle[168], and also by mediaeval logicians; their work, however, lies outside the scope of this book. The subject then appears to have been almost completely neglected until fairly recent times. In fact the first steps towards modern modal logic seem to have been taken by Hugh MacColl towards the end of the 19th century[169]. MacColl introduces the operations of disjunction $(a + b)$, negation (a') and implication $(a : b)$[170]. He then asserts as a valid principle

$$(a : b) : a' + b^{171}$$

but denies the validity of

$$(a : b) = a' + b^{172}$$

on the ground that if a means 'He will persist in his extravagancy' and b means 'He will be ruined', then the negation of $a : b$ is 'He may persist in his extravagancy without necessarily being ruined', while the negation of $a' + b$ is 'He will persist in his extravagancy and he will not be ruined'. MacColl objects to the identification of these precisely because the first asserts only possibility while the second asserts something more. What this

168 Aristotle [350 BC], 29b29–40b16. An attempt to formalize Aristotle's modal logic will be found in McCall [1963]. For a general history of ancient and mediaeval modal logic *vide* Kneale [1962] pp. 81–96, 117–138, 212, 232, 236, 243, or Bocheński [1961] pp. 81–88, 101–103, 114–115, 224–230.

169 MacColl [1880].

170 Op. cit. pp. 50–52.

171 Op. cit. p. 53.

172 Op. cit. p. 55.

amounts to is that he regards $a : b$ as expressing strict implication, and $a' + b$ as expressing material implication. In later papers, and in his book entitled *Symbolic Logic and its Applications*[173], this becomes even clearer: for he explicitly denies that his implicational connective can be given a truth-functional interpretation, and he defines $(A : B)$ as $(A' + B)^{\epsilon}$ (or alternatively as $(AB')^{\eta}$), where ϵ and η represent necessity and impossibility respectively[174]. (Although he speaks of several kinds of 'certainty' and 'possibility', presumably the 'formal certainty' and 'formal possibility' he speaks of in this connection are logical necessity and logical possibility.) MacColl's work also contains what appears to be the first formulation in modern times of the so-called 'paradoxes of implication'[175]; and he gives a concept of the degree of a formula which seems to anticipate that of Parry[176].

But MacColl does not give any axioms[177] and his system can hardly be called a modal logic of the distinctively modern kind with which this book is concerned. For that we have to wait until shortly after the publication in 1910 of *Principia Mathematica*[178], a work which did more than any other to establish the axiomatic method in logic. Beginning in 1912 C. I. Lewis published a series of articles and books[179] in which he expressed dissatisfaction with the notion of material implication found in *Principia*. The grounds of his dissatisfaction were very much the same as those of MacColl, but he had the great advantage of being able to use an axiomatic method based on that of *Principia* itself, and he used it to construct a system (or rather a series of systems) in which strict rather than material implication played the dominant role. It is the work of Lewis which marks the beginning of modern modal logic properly so called, and the rest of this chapter is devoted to it.

173 MacColl [1903], [1906a], [1906b].

174 MacColl [1903], pp. 356–7.

175 MacColl [1906b], p. 513.

176 MacColl [1903], p. 356; Parry [1939], p. 144 (and *supra* p. 50n).

177 He does give ([1906a] p. 8) a list of 'self-evident formulae' and it would be interesting to know which of the more recent modal systems is the weakest in which all these are true.

178 Whitehead and Russell [1910].

179 Lewis [1912], [1913], [1914a], [1914b], [1918]. Lewis and Langford [1932].

The 'paradoxes of implication'

In the system of *Principia Mathematica* – indeed in any standard system of PC – there are found the theses:

$$(1)\quad p \supset (q \supset p)$$

$$(2)\quad {\sim}p \supset (p \supset q)$$

The sense of (1) is often expressed by saying that if a proposition is true, any proposition whatsoever implies it: that of (2) by saying that if a proposition is false, it implies any proposition whatsoever. Together they are often called the 'paradoxes of (material) implication'.

Moreover, since for any proposition p, either the antecedent of (1) or the antecedent of (2) must be true, it is easy to derive from (1) and (2) the further thesis:

$$(3)\quad (p \supset q) \vee (q \supset p)$$

i.e. in any pair of propositions, either the first implies the second or the second implies the first.

Lewis did not wish to reject these theses. On the contrary, he argued[180] (and surely correctly) that (1) and (2), when properly understood, are 'neither mysterious sayings, nor great discoveries, nor gross absurdities', but merely reflect the truth-functional sense in which Whitehead and Russell were using the word 'imply'. But he also maintained that there is another, stronger, sense of 'imply', a sense in which when we say that p implies q we mean that *q follows from p*; and that in this sense of 'imply' it is not the case that every true proposition is implied by any proposition whatsoever, or that every false proposition implies any proposition whatsoever. Moreover in this stronger sense of 'imply' there are pairs of propositions neither of which implies the other. Lewis was thus led to draw the distinction between an implication which holds materially and one which holds necessarily or *strictly*, and to make analogous distinctions for disjunction and equivalence[181]. In his early articles he sometimes took strict disjunction as primitive[182], sometimes

[180] Lewis [1912] p. 522.

[181] Lewis [1912], [1913], [1914a], [1914b]. As far as we have been able to discover, the term 'strict implication' first occurs in Lewis [1912], p. 526 n. 1.

[182] Lewis [1912], [1914a].

strict implication[183], sometimes logical impossibility[184]; and in his book *A Survey of Symbolic Logic*[185], he set out an axiomatic system (the *Survey* system) in which he again took logical impossibility as the primitive modal operator (along with conjunction and negation as primitive truth-functional operators). In 1930[186] Oskar Becker proposed some additional axioms for the *Survey* system and showed that they enable all modalities to be reduced (*vide* p. 47) to a small number of non-equivalent ones. But the first comprehensive treatment of systems of strict implication (or indeed of systems of modal logic at all) appeared in 1932 in Lewis and Langford's book *Symbolic Logic*. Here possibility is taken as the primitive modal operator, and two axiomatic systems of strict implication (called S1 and S2 respectively) are developed in considerable detail. In an Appendix several other systems are outlined as well: one of these is the system of the *Survey* (S3); two others, which contain certain of Becker's reduction postulates, are called S4 and S5[187].

We shall set out these systems in the form in which they occur in *Symbolic Logic*, except that we shall use the notation and terminology employed in Part I of this book. An account of Lewis and Langford's own notation is given in Appendix 4, pp. 347–349.

The system S1

Primitive symbols

$p, q, r, \ldots$	[Propositional variables]
$\sim, M$.	[Monadic operators]
$\cdot$	[Dyadic operator]
(,)	[Brackets]

183 Lewis [1913]. 184 Lewis [1914b].

185 Lewis [1918], Ch. V (amended in Lewis [1920]).

186 Becker [1930].

187 These names ('S1' etc.), by which the systems have since become generally known, are given on pp. 500–501 of Appendix II (written by Lewis) in Lewis and Langford [1932]. They do not occur in Chapter VI, where S1 and S2 are developed (unless we count a brief reference to 'System 1' and 'System 2' on pp. 177–178). Lewis' S4 and S5 are deductively equivalent to the S4 and S5 of Part I of this book, though they have different bases. In the present chapter 'S4' and 'S5' refer to these systems as axiomatized in Lewis and Langford. For a more detailed survey of the axioms, theorems and rules of the various Lewis systems *vide* Feys [1965].

Formation rules[188]

1. A propositional variable standing alone is a wff.
2. If α is a wff, so are $\sim\alpha$ and $M\alpha$.
3. If α and β are wffs, so is $(\alpha \,.\, \beta)$.

Definitions[189]

[Def $\vee$] $(\alpha \vee \beta) =_{\text{Df}} \sim(\sim\alpha \,.\, \sim\beta)$
[Def $-\!\!3$] $(\alpha -\!\!3\, \beta) =_{\text{Df}} \sim M(\alpha \,.\, \sim\beta)$
[Def $=$] $(\alpha = \beta) =_{\text{Df}} ((\alpha -\!\!3\, \beta) \,.\, (\beta -\!\!3\, \alpha))$
[Def L][190] $L\alpha =_{\text{Df}} \sim M\sim\alpha$

Axioms[191]

AS1.1 $(p \,.\, q) -\!\!3\, (q \,.\, p)$
AS1.2 $(p \,.\, q) -\!\!3\, p$
AS1.3 $p -\!\!3\, (p \,.\, p)$
AS1.4 $((p \,.\, q) \,.\, r) -\!\!3\, (p \,.\, (q \,.\, r))$
AS1.5 $((p -\!\!3\, q) \,.\, (q -\!\!3\, r)) -\!\!3\, (p -\!\!3\, r)$
AS1.6 $(p \,.\, (p -\!\!3\, q)) -\!\!3\, q$

Transformation rules

1. *Uniform Substitution*, as in the systems in Part I.
2. *Substitution of strict equivalents*: If $\vdash\alpha$, and β differs from α only in having some wff, δ, at one or more places where α has a wff γ, then if $\vdash(\gamma = \delta)$, $\vdash\beta$.
3. *Adjunction*: $\vdash\alpha$, $\vdash\beta \rightarrow \vdash(\alpha \,.\, \beta)$.
4. *Modus Ponens* (*Detachment*): $\vdash\alpha$, $\vdash(\alpha -\!\!3\, \beta) \rightarrow \vdash\beta$.

There is one striking difference between the above basis for S1 and any of the bases discussed in Part I, and that is that it is not constructed as an extension of PC[192]. In fact none of the

[188] Lewis does not state any formation rules, but these are clearly the ones he uses.

[189] We write these definitions in the style adopted in Part I. Lewis writes them as strict equivalences, using propositional variables.

[190] Lewis does not have a single symbol for necessity, but writes $\sim\Diamond\sim$ throughout. (His $\Diamond$ = our M). But the abbreviation provided by this definition is an obvious convenience.

[191] Our numbering of these axioms is not the same as that of Lewis and Langford. Moreover we omit the axiom $p -\!\!3 \sim\sim p$ since this was shown to be non-independent in McKinsey [1934] and is derived by us as TS1.8 below. Instead of AS1.6 we may have $\sim Mp -\!\!3 \sim p$, or $p -\!\!3\, Mp$.

[192] The first axiomatization of modal logic starting from a PC basis and adding extra axioms and rules to it (as in Part I) appears to be that in Gödel [1933].

axioms of S1 is a wff of PC at all. Moreover, while the rule of Uniform Substitution belongs to PC, the S1 Modus Ponens rule is stated for strict implication, not for material implication as in PC. (We shall often call it the rule of *Strict Detachment*, and the corresponding PC rule, the rule of *Material Detachment*.) As a result proofs of theorems in S1 are apt to have a somewhat different 'style' from those in, say, T, since we are not free to help ourselves to any thesis of PC which seems likely to be useful.

Nevertheless, S1 contains PC; i.e., every thesis of PC is a theorem of S1. It is easy to introduce the operators $\supset$ and $\equiv$ (as Lewis himself does[193]) by the definitions:

$$[\text{Def} \supset] \quad (\alpha \supset \beta) =_{\text{Df}} {\sim}(\alpha \cdot {\sim}\beta)$$

$$[\text{Def} \equiv] \quad (\alpha \equiv \beta) =_{\text{Df}} ((\alpha \supset \beta) \cdot (\beta \supset \alpha))$$

Then clearly every wff of PC is a wff of S1. To show that S1 contains PC we then have to take some complete basis for PC and (a) derive its axioms as theorems of S1; (b) establish those of its transformation rules which are not primitive rules of S1 as derived rules of S1; and (c) derive strict equivalences corresponding to any definitions in the basis which do not occur in S1 (for then transformation rule 2 will license the same replacements as the definitions would). Lewis in fact derives the *Principia Mathematica* basis for PC in this way[194]. It is, however, somewhat more convenient to work with a version of PC in which the primitive operators are $\sim$ and $.$ as in S1 and the definitions are the same as those of S1. Given such primitives and definitions, the following axioms and rules are known to be sufficient for PC[195]:

A1 $p \supset (p \cdot p)$
A2 $(p \cdot q) \supset p$
A3 $(p \supset q) \supset ({\sim}(q \cdot r) \supset {\sim}(r \cdot p))$

Rules: 1. Uniform Substitution; 2. Material Detachment.

Since Uniform Substitution is a primitive rule of S1, we can prove that S1 contains PC by deriving A1–3 and the rule of Material Detachment in S1. We shall now do this.

[193] Lewis and Langford [1932], p. 136.
[194] Lewis and Langford [1932], pp. 136 ff.
[195] *Vide*, e.g., Rosser [1953], pp. 14–15, 55 ff.

We first prove a number of preliminary theorems. In the proofs which follow we use 'Eq' and 'MP' for the S1 transformation rules 2 and 4, not for the analogous rules in PC, and 'Adj' for the rule of Adjunction.

TS1.1 $(p \cdot q) = (q \cdot p)$

PROOF

AS1.1 $[q/p, p/q]$: (1) $(q \cdot p) ⥽ (p \cdot q)$

AS1.1, (1) × Adj, × Def =: (2) $(p \cdot q) = (q \cdot p)$ **Q.E.D.**

An easily derived rule is Syll in the form:

$$\vdash(\alpha ⥽ \beta),\ \vdash(\beta ⥽ \gamma) \rightarrow \vdash(\alpha ⥽ \gamma)$$

Clearly if $\vdash(\alpha ⥽ \beta)$ and $\vdash(\beta ⥽ \gamma)$, Adj gives $\vdash((\alpha ⥽ \beta) \cdot (\beta ⥽ \gamma))$. Hence by substitution in AS1.5 and MP we have $\vdash(\alpha ⥽ \gamma)$.

TS1.2 $p ⥽ p$

PROOF

AS1.3: (1) $p ⥽ (p \cdot p)$

AS1.2 $[p/q]$: (2) $(p \cdot p) ⥽ p$

(1), (2) × Syll: (3) $p ⥽ p$ **Q.E.D.**

By adjoining TS1.2 to itself and applying Def = we obtain

TS1.3 $p = p$

TS1.4 $({\sim}p ⥽ q) = ({\sim}q ⥽ p)$

PROOF

TS1.3 $[{\sim}M({\sim}p \cdot {\sim}q)/p]$: (1) ${\sim}M({\sim}p \cdot {\sim}q) = {\sim}M({\sim}p \cdot {\sim}q)$

(1), TS1.1 $[{\sim}p/p, {\sim}q/q]$ × Eq: (2) ${\sim}M({\sim}p \cdot {\sim}q) = {\sim}M({\sim}q \cdot {\sim}p)$

(2) × Def ⥽: (3) $({\sim}p ⥽ q) = ({\sim}q ⥽ p)$ **Q.E.D.**

Clearly if we have a thesis of the form $(\alpha = \beta)$, then by Def = we have $\vdash((\alpha ⥽ \beta) \cdot (\beta ⥽ \alpha))$; whence, by AS1.2, $\vdash(\alpha ⥽ \beta)$. We shall in future use the implicative forms of any equivalential theses when this is convenient.

TS1.5 $(p \cdot (q \cdot r)) = ((p \cdot q) \cdot r)$

PROOF

AS1.4 $[r/p, p/r]$:	(1) $((r \cdot q) \cdot p) \prec (r \cdot (q \cdot p))$
(1), TS1.1 × Eq:	(2) $(p \cdot (q \cdot r)) \prec ((p \cdot q) \cdot r)$
(2), AS1.4 × Adj, × Def =:	(3) $(p \cdot (q \cdot r)) = ((p \cdot q) \cdot r)$

Q.E.D.

TS1.6 $\sim\sim p \prec p$

PROOF

TS1.4 $[\sim p/q]$:	(1) $(\sim p \prec \sim p) \prec (\sim\sim p \prec p)$
(1), TS1.2 $[\sim p/p]$ × MP:	(2) $\sim\sim p \prec p$

Q.E.D.

We remarked earlier that the converse of TS1.6, viz. $p \prec \sim\sim p$, was one of Lewis' axioms for S1, but that it was later discovered not to be independent. We now show how to prove it as a theorem. As a preliminary we prove:

TS1.7 $\sim p = \sim\sim\sim p$

PROOF

TS1.6 $[\sim\sim p/p]$:	(1) $\sim\sim\sim\sim p \prec \sim\sim p$
(1), TS1.6 × Syll:	(2) $\sim\sim\sim\sim p \prec p$
TS1.4 $[\sim\sim\sim p/p, p/q]$:	(3) $(\sim\sim\sim\sim p \prec p) \prec (\sim p \prec \sim\sim\sim p)$
(2), (3) × MP:	(4) $\sim p \prec \sim\sim\sim p$
TS1.6 $[\sim p/p]$:	(5) $\sim\sim\sim p \prec \sim p$
(4), (5) × Adj, × Def =:	(6) $\sim p = \sim\sim\sim p$

Q.E.D.

TS1.8 $p \prec \sim\sim p$

PROOF

TS1.2 $[\sim M(p \cdot \sim p)/p]$:	(1) $\sim M(p \cdot \sim p) \prec \sim M(p \cdot \sim p)$
(1), TS1.7 × Eq:	(2) $\sim M(p \cdot \sim p) \prec \sim M(p \cdot \sim\sim\sim p)$
(2) × Def $\prec$:	(3) $(p \prec p) \prec (p \prec \sim\sim p)$
TS1.2, (3) × MP:	(4) $p \prec \sim\sim p$

Q.E.D.

TS1.6 and TS1.8 yield, by Adj and Def =:

TS1.9 $p = \sim\sim p$

From TS1.4 and TS1.9 are easily derived:

TS1.4a $(p \prec \sim q) = (q \prec \sim p)$
TS1.4b $(p \prec q) = (\sim q \prec \sim p)$

Furthermore from Def $\supset$ and TS1.9 we can easily derive:

$$\vdash (p \supset \sim q) = \sim(p \,.\, q)$$
$$\vdash \sim(p \supset q) = (p \,.\, \sim q) \quad \text{etc.}$$

We shall regard 'Def $\supset$' as a sufficient indication of the replacements licensed by any of these.

TS1.10 $((p \,.\, q) \prec r) = ((p \,.\, \sim r) \prec \sim q)$

PROOF

TS1.3 $[\sim M((p \,.\, q) \,.\, \sim r)/p]$:
(1) $\sim M((p \,.\, q) \,.\, \sim r) = \sim M((p \,.\, q) \,.\, \sim r)$

(1), TS1.5 $[\sim r/r] \times$ Eq:
(2) $= \sim M(p \,.\, (q \,.\, \sim r))$

(3), TS1.1 $[q/p, \sim r/q] \times$ Eq:
(3) $= \sim M(p \,.\, (\sim r \,.\, q))$

(3), TS1.5 $[\sim r/q, q/r] \times$ Eq:
(4) $= \sim M((p \,.\, \sim r) \,.\, q)$

(4), TS1.9 $\times$ Eq: (5) $= \sim M((p \,.\, \sim r) \,.\, \sim\sim q)$

(5) $\times$ Def $\prec$: (6) $((p \,.\, q) \prec r) = ((p \,.\, \sim r) \prec \sim q$

Q.E.D.

In the above proof the left hand side in lines (2)–(5) is the same as in line (1), and has been omitted for ease of reading.

The next theorem provides an important link between S1 and PC.

TS1.11 $(p \prec q) \prec (p \supset q)$

PROOF

TS1.10 $[p \prec q/q, q/r]$: (1) $((p \,.\, (p \prec q)) \prec q) \prec ((p \,.\, \sim q) \prec \sim(p \prec q))$

(1), AS1.6 $\times$ MP: (2) $(p \,.\, \sim q) \prec \sim(p \prec q)$

(2), TS1.4a $\times$ Eq: (3) $(p \prec q) \prec \sim(p \,.\, \sim q)$

(3) $\times$ Def $\supset$: (4) $(p \prec q) \prec (p \supset q)$ **Q.E.D.**

From TS1.11 we have immediately, by AS1.3 and AS1.2, the first two of the PC axioms, viz.

A1 $p \supset (p \,.\, p)$
A2 $(p \,.\, q) \supset p$

We next derive the PC rule of Modus Ponens for material implication, i.e.

$$\vdash\alpha, \vdash(\alpha \supset \beta) \rightarrow \vdash\beta$$

To do this we first prove:

TS1.12 $(p \,.\, (p \supset q)) ⥽ q$

PROOF

TS1.10 [$\sim q/q$, $p \,.\, \sim q/r$]:	(1) $((p \,.\, \sim q) ⥽ (p \,.\, \sim q)) ⥽$ $((p \,.\, \sim(p \,.\, \sim q)) ⥽ \sim\sim q)$
TS1.2 [$p \,.\, \sim q/p$], (1) × MP:	(2) $(p \,.\, \sim(p \,.\, \sim q)) ⥽ \sim\sim q$
(2), TS1.9 × Eq, × Def $\supset$:	(3) $(p \,.\, (p \supset q)) ⥽ q$ **Q.E.D.**

If now we have $\vdash\alpha$ and $\vdash(\alpha \supset \beta)$, then by Adj we have $\vdash(\alpha \,.\, (\alpha \supset \beta))$; hence by substitution in TS1.12 and MP we obtain $\vdash\beta$. I.e. Modus Ponens for material implication is a derived rule of S1[196].

We have one further PC axiom to prove. It is convenient to begin by deriving:

TS1.13 $(p \,.\, (p \supset q)) ⥽ \sim(r \,.\, \sim(q \,.\, r))$

PROOF

TS1.12 [r/p, $\sim q/q$]:	(1) $(r \,.\, (r \supset \sim q)) ⥽ \sim q$
(1), TS1.4a [$r \,.\, (r \supset \sim q)/p$] × Eq:	(2) $q ⥽ \sim(r \,.\, (r \supset \sim q))$
(2) × Def $\supset$:	(3) $q ⥽ \sim(r \,.\, \sim(r \,.\, q))$
(3), TS1.1 × Eq:	(4) $q ⥽ \sim(r \,.\, \sim(q \,.\, r))$
TS1.12, (4) × Syll:	(5) $(p \,.\, (p \supset q)) ⥽$ $\sim(r \,.\, \sim(q \,.\, r))$ **Q.E.D.**

[196] Note that this rule will be provable in any system formed by adding extra axioms to S1 (we shall consider several such extensions of S1 later).

TS1.14 $(p \supset q) \supset (\sim(q \cdot r) \supset \sim(p \cdot r))$

PROOF

TS1.13, TS1.10 $[p \supset q/q, \sim(r \cdot \sim(q \cdot r))/r] \times$ Eq:	(1) $(p \cdot \sim\sim(r \cdot \sim(q \cdot r))) \prec \sim(p \supset q)$
(1), TS1.4a, TS1.9 × Eq:	(2) $(p \supset q) \prec \sim(p \cdot (r \cdot \sim(q \cdot r)))$
(2), TS1.5 × Eq:	(3) $\prec \sim((p \cdot r) \cdot \sim(q \cdot r))$
(3), TS1.1 × Eq:	(4) $\prec \sim(\sim(q \cdot r) \cdot (p \cdot r))$
(4) × Def $\supset$:	(5) $\prec (\sim(q \cdot r) \supset \sim(p \cdot r))$
(5), TS1.11 × MP:	(6) $(p \supset q) \supset (\sim(q \cdot r) \supset \sim(p \cdot r))$

Q.E.D.

This last theorem is the PC axiom A3. Hence S1 contains the propositional calculus.

S1 and T compared

T was the weakest of the systems discussed in Part I. But S1 is weaker still: it contains nearly all of the basis of T but not quite all.

We have already seen that S1 contains the whole of PC. Moreover, by TS1.9, we have

$$Mp = \sim\sim M \sim\sim p$$

and hence, by the definition of L as $\sim M \sim$,

$$Mp = \sim L \sim p$$

By extension of this, the rule we called LMI in Part I (p. 36) holds in S1; i.e. the relations between possibility and necessity are the same as in T. We also have in S1 an equivalence corresponding to the definition of $\prec$ in T:

TS1.15 $(p \prec q) = L(p \supset q)$

PROOF

TS1.3 $[p \prec q/p]$ × Def $\prec$:	(1) $(p \prec q) = \sim M(p \cdot \sim q)$
(1) × Def $\supset$:	(2) $(p \prec q) = \sim M \sim (p \supset q)$
(2) × Def L:	(3) $(p \prec q) = L(p \supset q)$ **Q.E.D.**

Axioms A5 and A6 of T (viz. $Lp \supset p$ and $L(p \supset q) \supset (Lp \supset Lq)$) are also in S1. They are proved below as TS1.18 and TS1.21.

TS1.16 $p = (\sim p \supset p)$

PROOF

AS1.2 $[\sim p/p, \sim p/q]$: (1) $(\sim p \,.\, \sim p) ⥽ \sim p$
AS1.3 $[\sim p/p]$, (1) × Adj, × Def =: (2) $\sim p = (\sim p \,.\, \sim p)$
TS1.9, (2) × Eq: (3) $p = \sim(\sim p \,.\, \sim p)$
(2) × Def ⊃: (4) $p = (\sim p \supset p)$

Q.E.D.

TS1.17 $Lp = (\sim p ⥽ p)$

PROOF

TS1.3 $[Lp/p]$: (1) $Lp = Lp$
(1), TS1.16 × Eq: (2) $Lp = L(\sim p \supset p)$
(2), TS1.15 × Eq: (3) $Lp = (\sim p ⥽ p)$ **Q.E.D.**

TS1.18 $Lp \supset p$ [A5 of T]

PROOF

TS1.11 $[\sim p/p, p/q]$: (1) $(\sim p ⥽ p) ⥽ (\sim p \supset p)$
(1), TS1.17, TS1.16 × Eq: (2) $Lp ⥽ p$
(2), TS1.11 × MP (3) $Lp \supset p$ **Q.E.D.**

It will sometimes be helpful to use $Lp ⥽ p$, derived at step (2), as a theorem. We shall call it **TS1.18a**

The next two theorems are steps towards proving A6 of T.

TS1.19 $((p \,.\, q) ⥽ r) = (p ⥽ (q \supset r))$

PROOF

TS1.10 $[q/p, p/q]$: (1) $((q \,.\, p) ⥽ r) = ((q \,.\, \sim r) ⥽ \sim p)$
(1), TS1.1 × Eq: (2) $((p \,.\, q) ⥽ r) = ((q \,.\, \sim r) ⥽ \sim p)$
(2), TS1.4a × Eq: (3) $= (p ⥽ \sim(q \,.\, \sim r))$
(3) × Def ⊃: (4) $= (p ⥽ (q \supset r))$ **Q.E.D.**

TS1.20 $((p ⥽ q) \,.\, (q ⥽ r) \,.\, (r ⥽ s)) ⥽ (p ⥽ s)$

PROOF

AS1.5 $[r/q, s/r]$: (1) $((p ⥽ r) \,.\, (r ⥽ s)) ⥽ (p ⥽ s)$
(1), TS1.19 $[p ⥽ r/p, r ⥽ s/q, p ⥽ s/r]$ × Eq:
(2) $(p ⥽ r) ⥽ ((r ⥽ s) \supset (p ⥽ s))$
AS1.5, (2) × Syll: (3) $((p ⥽ q) \,.\, (q ⥽ r)) ⥽ ((r ⥽ s) \supset (p ⥽ s))$
(3), TS1.19 $[(p ⥽ q) \,.\, (q ⥽ r)/p, r ⥽ s/q, p ⥽ s/r]$ × Eq:
(4) $((p ⥽ q) \,.\, (q ⥽ r) \,.\, (r ⥽ s)) ⥽ (p ⥽ s)$

Q.E.D.

TS1.21 $L(p \supset q) \supset (Lp \supset Lq)$ [A6 of T]

PROOF

TS1.20 $[{\sim}q/p, {\sim}p/q, p/r, q/s]$:

(1) $(({\sim}q \prec {\sim}p) \cdot ({\sim}p \prec p) \cdot (p \prec q)) \prec ({\sim}q \prec q)$

(1), TS1.1 × Eq: (2) $(({\sim}q \prec {\sim}p) \cdot (p \prec q) \cdot ({\sim}p \prec p)) \prec ({\sim}q \prec q)$

(2), TS1.4b × Eq, TS1.17 × Eq:

(3) $((p \prec q) \cdot (p \prec q) \cdot Lp) \prec Lq$

(3), TS1.19 × Eq: (4) $((p \prec q) \cdot (p \prec q)) \prec (Lp \supset Lq)$

AS1.3 $[p \prec q/p]$, (4) × Syll:

(5) $(p \prec q) \prec (Lp \supset Lq)$

(5), TS1.15 × Eq: (6) $L(p \supset q) \prec (Lp \supset Lq)$

(6), TS1.11 × MP:

(7) $L(p \supset q) \supset (Lp \supset Lq)$ **Q.E.D.**

We shall refer to the formula derived at (5) above, viz. $(p \prec q) \prec (Lp \supset Lq)$, as **TS1.21a**.

The only remaining item in the basis of T is the Rule of Necessitation (TR3), viz. $\vdash\alpha \rightarrow \vdash L\alpha$ (p. 31). This rule does not hold in S1. Or more exactly, it does not hold unrestrictedly, as it does in T: what we do have in S1 is a weaker rule of the same type, restricted to the cases where α is a PC thesis, i.e. a thesis which contains no modal operators. That is to say, we have in S1 the rule:

$$\vdash_{PC}\alpha \rightarrow \vdash_{S1}L\alpha$$

We shall call this rule, N(S1). We now prove that it is a rule of S1[197].

The PC theses of S1 can be taken to be A1–A3 together with all their transforms under the rules of uniform substitution and material detachment. Now for each of A1–A3, N(S1) holds; for each is of the form $\alpha \supset \beta$, and in proving each of them we first obtained (or had as an axiom) $\alpha \prec \beta$; and, by TS1.15, $\vdash((\alpha \prec \beta) = L(\alpha \supset \beta))$. Moreover, the property expressed in N(S1) is preserved by the transformation rules. For (1) if for a wff, α, both $\vdash\alpha$ and $\vdash L\alpha$, and β is obtained from α by uniform substitution, $L\beta$ will be obtained from $L\alpha$ by the same substitution; hence both $\vdash\beta$ and $\vdash L\beta$. And (2) if we have both $\vdash\alpha$

[197] Cf. Wajsberg [1933] and Parry [1939], p. 143.

and $\vdash(\alpha \supset \beta)$ (so that we have β by detachment), then by hypothesis we have $\vdash L\alpha$ and $\vdash L(\alpha \supset \beta)$; and in that case we also have $\vdash L\beta$, since by TS1.21 we have

$$\vdash L(\alpha \supset \beta) \supset (L\alpha \supset L\beta)$$

whence by two detachments we have $\vdash L\beta$.

Some particular applications of N(S1) are the following: Where α and β are PC formulae,

$$(1) \quad \vdash(\alpha \supset \beta) \rightarrow \vdash(\alpha \prec \beta)$$

$$(2) \quad \vdash(\alpha \equiv \beta) \rightarrow \vdash(\alpha = \beta)$$

(1) follows directly from N(S1) by TS1.15. For (2), whenever we have $\vdash(\alpha \equiv \beta)$ we have, by ordinary PC transformations, both $\vdash(\alpha \supset \beta)$ and $\vdash(\beta \supset \alpha)$. Therefore, by (1), since α and β are PC formulae, $\vdash(\alpha \prec \beta)$ and $\vdash(\beta \prec \alpha)$; whence by Adj and Def =, $\vdash(\alpha = \beta)$.

At this point we can easily derive the so-called *paradoxes of strict implication* (*vide* pp. 39f) in the following forms:

TS1.22 $Lp \supset (q \prec p)$

PROOF

PC:	(1) $p \supset (q \supset p)$
(1) × N(S1):	(2) $L(p \supset (q \supset p))$
(2) × TS1.21:	(3) $Lp \supset L(q \supset p)$
(3), TS1.15 × Eq:	(4) $Lp \supset (q \prec p)$ **Q.E.D.**

TS1.23 $\sim Mp \supset (p \prec q)$

PROOF

PC × N(S1):	(1) $L(\sim p \supset (p \supset q))$
(1) × TS1.21:	(2) $L\sim p \supset L(p \supset q)$
(2) × LMI, TS1.15 × Eq:	(3) $\sim Mp \supset (p \prec q)$ **Q.E.D.**

In S1, however, we cannot prove the 'paradoxes' in their strict forms, viz. $Lp \prec (q \prec p)$ and $\sim Mp \prec (p \prec q)$[197a].

[197a] Halldén [1948a].

As we have said, N(S1) cannot be generalized to apply to all theses of S1. If it could, S1 would of course contain T. But whereas in S1 (as in T) we have, e.g.:

$$\vdash (p \cdot q) ⥽ (q \cdot p)$$

we do not have (as we do in T):

$$\vdash L((p \cdot q) ⥽ (q \cdot p))$$

or, of course, its equivalent by TS1.15:

$$\vdash LL((p \cdot q) \supset (q \cdot p))$$

It is in fact possible to prove – and we shall do so later (p. 236) – that S1 contains no theses of the form $L L\alpha$[198]. Here, however, we shall prove the different, though related, metatheorem[199] that if any thesis of the form $LL\alpha$ were in S1, we should be able to derive the unrestricted rule of Necessitation, and as a result S1, thus extended, would contain T. (In fact, provided α is a thesis of T, the system would be deductively equivalent to T[200], since it is not difficult to show that T contains S1.)

PROOF

Assume that $\vdash LL\alpha$. We first of all show that in that case, for any wff β, $\vdash L\beta \rightarrow \vdash LL\beta$. The proof of *this* is as follows.

Given: (1) $LL\alpha$

(2) $L\beta$

(1), TS1.18 × MP: (3) $L\alpha$

TS1.22 $[\alpha/p, \beta/q]$: (4) $L\alpha \supset (\beta ⥽ \alpha)$

(3), (4) × MP: (5) $\beta ⥽ \alpha$

(2), TS1.22 $[\beta/p, \alpha/q]$ × MP: (6) $\alpha ⥽ \beta$

(6), (5) × Adj, × Def =: (7) $\alpha = \beta$

(1), (7) × Eq: (8) $LL\beta$ **Q.E.D.**

198 We cannot, of course, take a PC thesis, α, and obtain $\vdash LL\alpha$ by applying N(S1) to it twice; for even if α is a wff of PC, $L\alpha$ is not.

199 Cf. Yonemitzu [1955].

200 Of course if the formula α in question were not a thesis of T then the addition to S1 of $LL\alpha$ would give a system stronger than T (and if α is inconsistent with T then the addition of $LL\alpha$ to S1 would give an inconsistent system).

Now all the axioms of S1 are *strict* wffs, i.e. are equivalent to wffs of the form $L\beta$. Hence the derivability of $\vdash LL\alpha$ would ensure that for every axiom, γ, of S1, $L\gamma$ was a thesis; i.e. the rule of Necessitation would hold for the axioms. We have therefore only to show that the property expressed in the rule is preserved by the transformation rules of S1. We have already shown (in the proof of N(S1)) that this property is preserved by the rules of uniform substitution and material detachment. Similar proofs for transformation rules 2, 3 and 4 are straightforward (but are omitted here). Hence the addition of $\vdash LL\alpha$ would yield the unrestricted rule of Necessitation, and thus strengthen S1 into an alternative axiomatization of T.

Note that the proof we have just given depends on the particular axiomatization we are using. For we are assuming that adjunction, strict detachment and Eq still hold when $LL\alpha$ is added. It is possible to have a system deductively equivalent to S1 in which these rules are not taken as primitive and in which it is not always the case that they will still hold when further axioms are added. Our proof, however, shows that in any *strict* extension of *Lewis'* S1 (i.e. in any system formed from it by adding strict wffs as axioms), if $\vdash LL\alpha$ then N holds unrestrictedly.

There are several other ways of turning S1 into T. Either of the following would do:

(a) Add to S1 the rule: $\vdash(\alpha \equiv \beta) \rightarrow \vdash(L\alpha \equiv L\beta)$[201].

(b) Add to S1 a rule allowing the substitution of proved *material* equivalents[202].

‘*T*-principles’ in S1

As we have said, the rule of Necessitation does not hold unrestrictedly in S1. In particular there are some S1 theses in which the main operator is $\supset$ but which cease to be theses if the $\supset$ is replaced by $⥽$. As an example, we have

$$\vdash_{S1}(p \supset p) \supset (p ⥽ p)$$

but the corresponding formula,

$$(1)\quad (p \supset p) ⥽ (p ⥽ p)$$

is not a thesis of S1.

201 Åqvist [1964].

202 Cresswell [1960], pp. 60 *et seq.*

(Note that if (1) were a thesis we should have the following result. By substituting $p \supset p/p$ and $p \prec p/q$ in TS1.21 (in the form $(p \prec q) \supset (Lp \supset Lq)$) we obtain

$$((p \supset p) \prec (p \prec p)) \supset (L(p \supset p) \supset L(p \prec p))$$

If (1) were a thesis then, since $L(p \supset p)$ certainly is one, we could detach twice to obtain $L(p \prec p)$ and hence its equivalent $LL(p \supset p)$. But as we have remarked, there are no S1 theses of the form $LL\alpha$.)

Now the fact that formulae such as (1) are not theses of S1 presents Lewis with a certain difficulty. For he intended the relation expressed by $\prec$ to be such that whenever a material implication is 'asserted' (i.e. whenever a formula $(\alpha \supset \beta)$ is a thesis), then the corresponding strict implication can also be asserted (i.e. $(\alpha \prec \beta)$ is a thesis)[203]; and in that case formulae such as (1) – or formulae which will, so to speak, do the same work – *ought* to be theses. To overcome this difficulty, Lewis introduces what he calls *T-principles*[204].

These can be explained as follows. In every case where $(\alpha \supset \beta)$ is a theorem but $(\alpha \prec \beta)$ is not, there is some theorem, which is referred to as 'T', such that $\vdash((\alpha \, . \, T) \prec \beta)$. The proof that this is so is very simple: for T may be simply $(\alpha \supset \beta)$ itself, which by hypothesis is a theorem; and by TS1.12 we have $\vdash((\alpha \, . \, (\alpha \supset \beta)) \prec \beta)$, i.e. $\vdash((\alpha \, . \, T) \prec \beta)$. T need not always be $(\alpha \supset \beta)$, however, and very often there is a shorter thesis which will do the job equally well. Among the most important of the T-principles are these:

TS1.24 $((p \prec q) \, . \, T) \prec ((p \, . \, r) \prec (q \, . \, r))$
TS1.25 $((p \prec q) \, . \, (p \prec r) \, . \, T) \prec (p \prec (q \, . \, r))$
TS1.26 $((p \prec (q \, . \, r)) \, . \, T) \prec ((p \prec q) \, . \, (p \prec r))$

Proofs are omitted here, but will be found in Lewis and Langford[205].

To see the importance of the T-principles in S1, consider TS1.24. Regarded as a principle of inference (and it is Lewis'

[203] Lewis and Langford [1932], pp. 152, 245.

[204] *Ibid*, pp. 147–153. Note that 'T' as used in this context has no connection with the System T.

[205] *Ibid.*, pp. 147, 148, 150 where the theorems are numbered 16.1, 16.2 and 16.5 respectively.

intention that all theses whose main operator is $\prec$ can be so regarded), this assures us that a proposition of the form $((p \cdot r) \prec (q \cdot r))$ can be validly inferred from two premisses: (a) the corresponding proposition of the form $(p \prec q)$, and (b) some proposition which has the form of a thesis of S1. But (b) is *bound* to be true; so the truth of (a) is sufficient to guarantee the truth of the conclusion. Thus as a principle of inference, TS1.24 'can be used exactly as if the T did not occur in it'[206], even though without the T it is not a thesis of S1. Analogous considerations apply to the other T-principles.

The system S2

The System S2 is obtained by adding to S1 the axiom

AS2.1 $M(p \cdot q) \prec Mp$

Lewis calls this the *Consistency Postulate*[207]. Its sense is that only a possible (or *consistent*) proposition can be a term in a consistent conjunction.

In S2 a number of the T-principles of S1 are provable without the T. This applies to all of TS1.24–TS1.26[208]. However, the rule of Necessitation is not provable unrestrictedly even in S2 (*vide* p. 253), and we cannot dispense with T-principles entirely. E.g., the formula numbered (1) in the previous section, viz.

$$(p \supset p) \prec (p \prec p)$$

is not a theorem of S2, though of course we do have

$$\vdash_{S2}((p \supset p) \cdot T) \prec (p \prec p)$$

S2, though stronger than S1, is not as strong as T[209].

[206] *Ibid.*, p. 147. Note that our earlier proof that MP holds in S1 for material implication in effect draws attention to the same result.

[207] *Ibid.*, p. 166–167.

[208] *Ibid.*, pp. 171–172.

[209] In fact if we could dispense with T-principles in S2 (or S1) altogether – i.e., if wherever we had $\vdash((\alpha \cdot T) \prec \beta)$ we had $\vdash(\alpha \prec \beta)$ – we should have the system T, for the following reason. Suppose that $\vdash\gamma$. Then (by PC) $\vdash((p \supset p) \supset \gamma)$. But from TS1.12 we have $\vdash(((p \supset p) \cdot ((p \supset p) \supset \gamma)) \prec \gamma)$; and since $(p \supset p) \supset \gamma$ is a thesis, we can express this as the T-principle $\vdash(((p \supset p) \cdot T) \prec \gamma)$. So if all T-principles could be dispensed with we should, whenever γ is a thesis, have $\vdash((p \supset p) \prec \gamma)$. From this it follows, by TS1.21a, that $\vdash(L(p \supset p) \supset L\gamma)$; and hence, since by TS1.2 and

We now prove a couple of S2 theorems and then an important rule.

TS2.1 $Mp \prec M(p \vee q)$

PROOF

AS2.1 $[p \vee q/p, p/q]$: (1) $M((p \vee q) . p) \prec M(p \vee q)$

PC $\vdash(((p \vee q) . p) \equiv p) \times$ N(S1): (2) $((p \vee q) . p) = p$

(1), (2) $\times$ **Eq**: (3) $Mp \prec M(p \vee q)$ **Q.E.D.**

TS2.2 $L(p . q) \prec Lp$

PROOF

TS2.1 $[\sim p/p, \sim q/q]$: (1) $M\sim p \prec M(\sim p \vee \sim q)$

PC (De Morgan) $\times$ N(S1): (2) $(\sim p \vee \sim q) = \sim(p . q)$

(1), (2) $\times$ **Eq**: (3) $M\sim p \prec M\sim(p . q)$

(3), TS1.4b $\times$ Eq: (4) $\sim M\sim(p . q) \prec \sim M\sim p$

(4) $\times$ Def L: (5) $L(p . q) \prec Lp$ **Q.E.D.**

A simple corollary of TS2.2, by TS1.1, is:

TS2.2a $L(p . q) \prec Lq$

We are now in a position to derive a rule which is one of the distinguishing features of S2. This rule (or some closely related form of it) is often called *Becker's Rule*[210].

TS1.15 we have $\vdash L(p \supset p)$, we should have $\vdash L\gamma$. I.e., the rule of Necessitation would hold unrestrictedly, and a system at least as strong as T would result. Moreover, in T all T-principles can be eliminated, since by obvious steps we have

$$\vdash_T((\alpha . T) \prec \beta) \rightarrow \vdash_T(T \prec (\alpha \supset \beta)) \rightarrow \vdash_T(\alpha \supset \beta) \rightarrow \vdash_T(\alpha \prec \beta)$$

Hence if we could dispense with all T-principles in S2 (or S1) we should have exactly the system T.

[210] Becker [1930] p. 522 states this rule for any 'affirmative' modality (i.e. any string of monadic operators containing an even number of negation signs). Churchman ([1938] p. 79) who gives it the name 'Becker's Rule' proves it in S2 in the form we have given. It should not be confused with the rule $\vdash\alpha \supset \beta \rightarrow \vdash L\alpha \supset L\beta$ (our DR1 of Part I) which is not a rule of S2 and which, if added, would produce T. (From $\vdash(p \supset p) \supset L(p \supset p)$ we would have $\vdash L(p \supset p) \supset LL(p \supset p)$ and hence $\vdash LL(p \supset p)$, which we have shown to yield T.)

BR $\vdash(\alpha \prec \beta) \rightarrow \vdash(L\alpha \prec L\beta)$

PROOF

PC × N(S1): (1) $(p \,.\, {\sim}q) = (p \,.\, {\sim}(p \,.\, q))$

TS1.2 [$\sim M(p \,.\, {\sim}q)/p$], (1) × Eq:

(2) ${\sim}M(p \,.\, {\sim}q) \prec {\sim}M(p \,.\, {\sim}(p \,.\, q))$

(2) × Def $\prec$: (3) $(p \prec q) \prec (p \prec (p \,.\, q))$

Given: (4) $\alpha \prec \beta$

(4), (3)[$\alpha/p, \beta/q$] × MP: (5) $\alpha \prec (\alpha \,.\, \beta)$

AS1.2 [$\alpha/p, \beta/q$]: (6) $(\alpha \,.\, \beta) \prec \alpha$

(5), (6) × Adj, × Def =: (7) $\alpha = (\alpha \,.\, \beta)$

TS1.2 [$L\alpha/p$], (7) × Eq: (8) $L\alpha \prec L(\alpha \,.\, \beta)$

TS2.2a [$\alpha/p, \beta/q$]: (9) $L(\alpha \,.\, \beta) \prec L\beta$

(8), (9) × Syll: (10) $L\alpha \prec L\beta$ **Q.E.D.**

Note that in proving BR we have (at step (9)) used TS2.2a, which in turn depends on the special S2 axiom AS2.1. BR cannot in fact be proved in S1. The nearest we can get to BR in S1 is

$$\vdash(\alpha \prec \beta) \rightarrow \vdash(L\alpha \supset L\beta)$$

We do however have in S1:

$$\vdash(\alpha = \beta) \rightarrow \vdash(L\alpha = L\beta)$$

and also, interestingly enough, since the above proof makes use only of S1 principles up to step (8):

$$\vdash(\alpha \prec \beta) \rightarrow \vdash(L\alpha \prec L(\alpha \,.\, \beta))$$

At first glance this may look like a stronger rule than BR itself – $L(\alpha \,.\, \beta)$ appears to assert more than $L\beta$ does – and it may seem surprising to find it in the weaker system. But this impression is only a superficial one. The point to remember here is that $L(p \,.\, q) \prec Lq$ is not a thesis of S1, however 'natural' a principle it may sound. In fact its absence from S1 is one way of pinpointing the limitations of that system.

BR enables us to prove in S2 the 'paradoxes of strict implication' in their stronger forms (cf. TS1.22 and TS1.23):

TS2.3 $Lp \prec (q \prec p)$

PROOF

PC × N(S1): (1) $p \prec (q \supset p)$

(1) × BR: (2) $Lp \prec L(q \supset p)$

(2), TS1.15 × Eq: (3) $Lp \prec (q \prec p)$ **Q.E.D.**

TS2.4 $\sim Mp \prec (p \prec q)$

The proof is left to the reader.

The system S3

Although Becker's Rule belongs to S2, the formula,

$$(1)\ (p \prec q) \prec (Lp \prec Lq)$$

which might be confused with it, is not a thesis of S2. (The formula in question should not be confused with the rule, nor should its absence from S2 occasion surprise: it bears the same relation to the rule as the S4 axiom $Lp \prec LLp$, which of course is not in T, does to the rule $\vdash \alpha \rightarrow \vdash L\alpha$, which *is* in T.)

Nevertheless we might well wish to have (1) as a thesis, and if we add it to S2 (or to S1, for that matter) we obtain the System S3. Actually Lewis constructs S3 [211] by adding to the basis of S1 not (1) but

AS3.1 $(p \prec q) \prec (\sim Mq \prec \sim Mp)$

(1), however, is then easily derived as follows:

TS3.1 $(p \prec q) \prec (Lp \prec Lq)$

PROOF

AS3.1 $[\sim q/p, \sim p/q]$: (1) $(\sim q \prec \sim p) \prec (\sim M\sim p \prec \sim M\sim q)$

(1), TS1.4b × Eq, × Def L: (2) $(p \prec q) \prec (Lp \prec Lq)$ **Q.E.D.**

(AS3.1 could be derived from TS3.1 by a similar proof, so it is immaterial which of the two we take as the characteristic S3 axiom.)

That S3 contains S2 may be proved by deriving AS2.1 in S3.

[211] Lewis and Langford [1932], pp. 493, 500. S3 is the system of Lewis [1918] pp. 294, 295 (the *Survey* system) where the axioms are labelled 1.1–1.8; 1.8 is given (in the *Survey* notation with $\sim$ as impossibility and $-$ as negation) as $(p \prec q) = (\sim q \prec \sim p)$. This was shown to yield $\sim p = -p$ and in Lewis [1920] was amended to $(p \prec q) \prec (\sim q \prec \sim p)$. In [1932] $\sim$ is replaced by $\sim\Diamond$ and the amended axioms are listed on p. 493 as A1–A8. A8 is our AS3.1. A1–A7 are not identical with our S1 axioms (which are B1–B7 of [1932]), but they form an alternative axiom set for S1. For a study of S3 *vide* Parry [1939].

TS3.2 [AS2.1] $M(p \cdot q) \prec Mp$

PROOF

AS3.1 $[p \cdot q/p, p/q]$: (1) $((p \cdot q) \prec p) \prec (\sim Mp \prec \sim M(p \cdot q))$

AS1.2, (1) × MP: (2) $\sim Mp \prec \sim M(p \cdot q)$

(2), TS1.4b × Eq: (3) $M(p \cdot q) \prec Mp$ **Q.E.D.**

TS3.3 $Lp \prec (Lq \prec Lp)$

PROOF

TS2.3: (1) $Lp \prec (q \prec p)$

TS3.1 $[q/p, p/q]$: (2) $(q \prec p) \prec (Lq \prec Lp)$

(1), (2) × Syll: (3) $Lp \prec (Lq \prec Lp)$ **Q.E.D.**

The formulae obtained from a number of important PC theses by replacing $\supset$ by $\prec$ throughout are provable in S3 though not in S1 or S2. Among these are the analogues of Comp and the two laws of Syllogism:

TS3.4 $(p \prec q) \prec ((p \prec r) \prec (p \prec (q \cdot r)))$

PROOF

Comp × N(S1): (1) $(p \supset q) \prec ((p \supset r) \supset (p \supset (q \cdot r)))$

(1) × BR, × TS1.15: (2) $(p \prec q) \prec ((p \supset r) \prec (p \supset (q \cdot r)))$

TS3.1 $[p \supset r/p, p \supset (q \cdot r)/q]$ × TS1.15:

(3) $((p \supset r) \prec (p \supset (q \cdot r))) \prec ((p \prec r) \prec (p \prec (q \cdot r)))$

(2), (3) × Syll: (4) $(p \prec q) \prec ((p \prec r) \prec (p \prec (q \cdot r)))$ **Q.E.D.**

TS3.5 $(p \prec q) \prec ((q \prec r) \prec (p \prec r))$

TS3.6 $(q \prec r) \prec ((p \prec q) \prec (p \prec r))$

The proofs have exactly the same structure as that for TS3.4.

It is worth noting, however, that the analogue of Perm, viz. $(p \prec (q \prec r)) \prec (q \prec (p \prec r))$, is not a theorem of S3 (or even of S5).

TS3.7 $LLq \prec (Lp \prec LLp)$[212]

PROOF

TS3.3 $[Lq/p, p/q]$: (1) $LLq \prec (Lp \prec LLq)$

TS3.1 $[Lq/p, Lp/q]$: (2) $(Lq \prec Lp) \prec (LLq \prec LLp)$

TS3.3, (2) × Syll: (3) $Lp \prec (LLq \prec LLp)$

TS3.4 $[Lp/p, LLq \prec LLp/q, LLq/r]$:

(4) $(Lp \prec (LLq \prec LLp)) \prec ((Lp \prec LLq) \prec (Lp \prec ((LLq \prec LLp) . LLq)))$

(3), (4) × MP: (5) $(Lp \prec LLq) \prec (Lp \prec ((LLq \prec LLp) . LLq)))$

AS1.6 $[LLq/p, LLp/q]$, TS1.1 × Eq:

(6) $((LLq \prec LLp) . LLq) \prec LLp$

TS3.6 $[(LLq \prec LLp) . LLq/q, LLp/r, Lp/p]$, (6) × MP:

(7) $(Lp \prec ((LLq \prec LLp) . LLq)) \prec (Lp \prec LLp)$

((1), (5) × Syll), (7) × Syll:

(8) $LLq \prec (Lp \prec LLp)$ **Q.E.D.**

An important feature of S3 is that, unlike S1 and S2 – and also unlike T – but like S4 and S5, it contains theorems which enable modalities to be reduced (p. 47). W. T. Parry has in fact shown [213] that there are only 42 distinct modalities in S3.

S3 is, however, like S2 and S1 (and unlike T, S4 and S5) in that it does not contain the rule of Necessitation (N) in an unrestricted form. TS3.7 is important because if S3 did contain N, we could use this theorem to derive $\vdash(Lp \prec LLp)$, which is the characteristic S4 axiom [214]. The derivation would proceed as follows: Let α be any thesis of S3. Then by N, $\vdash L\alpha$; and by N again, $\vdash LL\alpha$. Now in TS3.7 we put α/q, and then MP immediately gives $\vdash(Lp \prec LLp)$. Indeed, if S3, even in the absence of N, contained any thesis of the form $LL\alpha$, we could derive $\vdash(Lp \prec LLp)$ by putting α/q in TS3.7 and detaching. Hence if either N or any thesis of the form $LL\alpha$ were added to S3, we should obtain a system at least as strong as S4.

212 For this theorem and its proof *vide* Parry [1939]. Our TS3.7 is Parry's 32.1 (p. 141).

213 *Ibid.*, esp. pp. 144–148.

214 I.e. Lewis' S4 axiom. The S4 axiom we used in Chapter 3 was $Lp \supset LLp$, but since our basis for S4 contained N as a primitive rule the two are obviously interchangeable.

We shall prove later on (p. 242) that $Lp \strictif LLp$ is not a thesis of S3. In the light of the foregoing argument, this result will constitute a proof that neither N nor any thesis of the form $LL\alpha$ is in S3 – or of course in S2 or S1 either.

The system S4

In Chapter 3 we constructed the system we there called S4 by adding to T the extra axiom $Lp \supset LLp$. Lewis obtains his system S4 – i.e. S4 in its original form – by adding to *S1* the extra axiom[215]:

AS4.1 $Lp \strictif LLp$

$\vdash(Lp \supset LLp)$, of course, follows immediately from this by TS1.11.

We can show that S4 contains S3 (and therefore S2) by deriving AS3.1 in S4.

We first prove:

TS4.1 $(p \strictif q) \strictif L(p \strictif q)$

PROOF

AS4.1 $[p \supset q/p]$: (1) $L(p \supset q) \strictif LL(p \supset q)$

(1), TS1.15 × Eq: (2) $(p \strictif q) \strictif L(p \strictif q)$ **Q.E.D.**

TS4.2 $(p \strictif q) \strictif (Lp \strictif Lq)$

PROOF

TS1.21a $[p \strictif q/p, Lp \supset Lq/q]$:

(1) $((p \strictif q) \strictif (Lp \supset Lq)) \strictif (L(p \strictif q) \supset L(Lp \supset Lq))$

TS1.21a × TS4.1: (2) $L((p \strictif q) \strictif (Lp \supset Lq))$

TS1.21a $[(p \strictif q) \strictif (Lp \supset Lq)/p, L(p \strictif q) \supset L(Lp \supset Lq/q]$:

(3) $(1) \strictif ((2) \supset L(L(p \strictif q) \supset L(Lp \supset Lq)))$

(1), (2), (3) × MP, × MP (for $\supset$):

(4) $L(L(p \strictif q) \supset L(Lp \supset Lq))$

(4), TS1.15 × Eq: (5) $L(p \strictif q) \strictif (Lp \strictif Lq)$

TS4.1, (5) × Syll: (6) $(p \strictif q) \strictif (Lp \strictif Lq)$ **Q.E.D.**

215 The suggestion of taking $Lp \strictif LLp$ as an extra axiom comes from Becker [1930] p. 514. Becker calls it 1.92 to follow (the amended) 1.8 of the *Survey* (p. 295). In the *Survey* notation 1.92 is $\sim-p \strictif \sim-\sim-p$. Lewis and Langford write this as $\sim\Diamond\sim p \strictif \sim\Diamond\sim\sim\Diamond\sim p$ (C10, p. 497), and define S4 as S1 + C10 ([1932] p. 501).

TS4.3 [AS3.1] $(p \strictif q) \strictif (\sim Mq \strictif \sim Mp)$

PROOF

TS4.2 $[\sim q/p, \sim p/q]$,

Def L: (1) $(\sim q \strictif \sim p) \strictif (\sim M\sim\sim q \strictif \sim M\sim\sim p)$

(1), TS1.4b, TS1.9 × Eq: (2) $(p \strictif q) \strictif (\sim Mq \strictif \sim Mp)$ **Q.E.D.**

We have now shown that S4 contains S3. We shall next prove that the unrestricted rule of Necessitation holds in S4[216].

N: $\vdash \alpha \rightarrow \vdash L\alpha$

PROOF

By TS4.1 $[p/q]$ and TS1.15 we have $\vdash_{S4}((p \strictif p) \strictif LL(p \supset p))$, and hence by TS1.2 and MP, $\vdash_{S4} LL(p \supset p)$. But we know (pp. 227f) that the addition to S1 of any thesis of the form $LL\alpha$ gives a system in which the unrestricted rule of Necessitation holds.

Thus all the axioms and rules of the S4 of Part I are provable in Lewis' S4, and since (as it is easy to show) the converse also holds, the two systems are deductively equivalent. This was, of course, our justification for giving the name 'S4' to the system in Part I, in the first place.

The system S5

Lewis' system S5 is S1 with the addition of [217]

AS5.1 $Mp \strictif LMp$

S5 contains S4. We first prove:

TS5.1 $Mp = LMp$

PROOF

AS5.1 (1) $Mp \strictif LMp$

TS1.18a $[Mp/p]$: (2) $LMp \strictif Mp$

(1), (2) × Adj × Def =: (3) $Mp = LMp$ **Q.E.D.**

216 The first proof that N is a rule of Lewis' S4 appears to be that given in McKinsey and Tarski [1948], p. 5 (theorem 2.1).

217 This comes from Becker's 1.9 ([1930] p. 511): $-\sim p \strictif \sim\sim p$. Lewis and Langford write it as $\Diamond p \strictif \sim\Diamond \sim\Diamond p$ (C11, p. 497).

TS5.2 $Lp = LLp$

PROOF

TS1.3 $[Lp/p]$ × Def L:	(1) $Lp = \sim M\sim p$
(1), TS5.1 $[\sim p/p]$ × Eq:	(2) $Lp = \sim LM\sim p$
(2) × Def L:	(3) $Lp = \sim\sim M\sim M\sim p$
(3), TS1.9 × Eq, × Def L:	(4) $Lp = MLp$
(4), TS5.1 $[Lp/p]$ × Eq:	(5) $Lp = LMLp$
(5), (4) × Eq:	(6) $Lp = LLp$ **Q.E.D.**

The S4 axiom, $Lp \prec LLp$, obviously follows easily from this last theorem. Hence S5 contains S4, and of course S3 and S2 as well[218].

AS5.1 is simply the strict form of the S5 axiom of Chapter 4, $Mp \supset LMp$, and in virtue of N derivable from it. Hence Lewis' S5 is the same system as our earlier S5.

[218] Sobociński [1962a] shows that if we have AS5.1 and AS3.1 we may drop AS1.6 as an axiom. Churchman [1938] (pp. 78–81) obtains S5 from S2 by adding $\vdash(p \prec \sim MM\sim Mp)$.

CHAPTER THIRTEEN

The Lewis Systems (II)

The distinctness of S1–S5 [219]

In our exposition of S1–S5 in the last chapter we proved that each succeeding system contains all the earlier ones. We shall now prove that none of the systems is contained in any of the earlier ones, and hence that none is deductively equivalent to any of the others.

To do this we prove (1) that the characteristic S2 axiom:

AS2.1 $M(p \,.\, q) \prec Mp$

is not a theorem of S1; (2) that similarly

AS3.1 $(p \prec q) \prec (\sim Mq \prec \sim Mp)$

is not a theorem of S2; (3) that

AS4.1 $Lp \prec LLp$

is not a theorem of S3; and finally (4) that

AS5.1 $Mp \prec LMp$

is not a theorem of S4 [220].

Our general method (a common one for propositional logics) will be the same in each case. For each system we give a set, K, of values (represented by numerals) over which the variables are to be taken to range. From this set we select one or more as the *designated* values. We then give a set of *matrices* (value-tables), one for each of the primitive operators, which enable us to calculate the value of any wff, once we have assigned a value to

[219] The results in this section are all found in Appendices II and III of Lewis and Langford [1932], (pp. 492–514), where their various sources are indicated. (Appendix III is contained in the 1959 reprint, but not in the original 1932 edition.)

[220] We have of course already proved by another method (pp. 66–70) that S4 does not contain S5.

each of the variables in it. The set of matrices will have the following characteristics:

(a) Each axiom of the system in question will take one of the designated values for every assignment of values to the variables in it.

(b) If any transformation rule of the system is applied to a formula or formulae which take only the designated value(s), then the resulting formula will take only the designated value(s). (A matrix which fulfils condition (b) is said to be *hereditary* with respect to the transformation rules in question.)

(c) The characteristic axiom of the next system will take a non-designated value for some assignment of values to its variables.

Clearly from (a) and (b) it follows that every theorem of the system takes only one of the designated values; whence from (c) it follows that the characteristic axiom of the next system is not a theorem of the previous one.

(1) We prove that $M(p \,.\, q) \prec Mp$ is not a theorem of S1.

Let $K = \{1, 2, 3, 4\}$. (I.e. every variable can take any of the values 1, 2, 3, 4, but no others.) The designated values are 1 and 2. The matrices are as follows[221]:

p	$\sim p$	Mp
1	4	1
2	3	2
3	2	1
4	1	3

	$p \,.\, q$	q = 1	2	3	4
p	1	1	2	3	4
	2	2	2	4	4
	3	3	4	3	4
	4	4	4	4	4

	$p \prec q$	q = 1	2	3	4
p	1	2	4	3	4
	2	2	2	3	3
	3	2	4	2	4
	4	2	2	2	2

(The matrix for $\prec$ is a theoretical luxury only, though it is convenient to have it: it is simply calculated from the others by Def $\prec$ – i.e. it sets out the values of $\sim M(p \,.\, \sim q)$ for each assignment of values to p and q.)

It is a lengthy and tedious, but simple and mechanical, matter to check that every axiom of S1 takes only one of the designated values 1 and 2 for every assignment of values to its variables.

221 Lewis and Langford [1932], pp. 493–494. These matrices are their Group V.

We have now to show that this characteristic is preserved by the four transformation rules of S1.

Let α be some formula to which *uniform substitution* is applied, and let p be the variable in α for which the wff β is substituted. Then by hypothesis, α takes one of the designated values no matter what value (in K) is assigned to p. But whatever values are assigned to the variables in β, β must take some value in K. Hence the uniform replacement of p by β cannot lead to any non-designated value for the resulting formula as a whole.

Next, consider *substitution of strict equivalents*. If a formula $(\alpha = \beta)$ has the value 1 or 2, then by Def = and the matrix for ., each of $(\alpha \prec \beta)$ and $(\beta \prec \alpha)$ must have the value 1 or 2: and by the matrix for $\prec$ this can be so only when α and β have the same value. Hence the replacement of α by β in any formula cannot alter its value.

For *Adjunction*, it is clear from the table for . that if both α and β take a designated value, so does $(\alpha \, . \, \beta)$.

Finally, for *Modus Ponens*, the table for $\prec$ shows that the only cases in which α and $(\alpha \prec \beta)$ both take either the value 1 or the value 2 are those in which β likewise takes either 1 or 2.

Since every theorem of S1 is a transform of the S1 axioms by one or more of these rules, every theorem takes a designated value for every assignment of values (in K) to its variables.

Now consider AS2.1 $(M(p \, . \, q) \prec Mp)$.

When $p = 2$ and $q = 4$, AS2.1 $= (M(2.4) \prec M2) = (M4 \prec M2) = (3 \prec 2) = 4$. I.e. AS2.1 can take a non-designated value. Hence it is not a theorem of S1.

(2) To prove that AS3.1 is not a theorem of S2 we use an 8-valued matrix discovered by W. T. Parry[222]. For this we have

$$K = \{0, 1, 2, 3, 4, 5, 6, 7\}$$

[222] This matrix is given in Parry [1934], p. 79, and cited in the second edition of Lewis and Langford [1932] on p. 507. Lewis in 1932 had not been able to find a proof of the non-derivability of AS3.1 in S2 (*vide* p. 496).

and the designated values are 6 and 7. The matrices are:

p	$\sim p$	Mp
0	7	1
1	6	5
2	5	7
3	4	7
4	3	7
5	2	7
6	1	7
7	0	7

	$p \,.\, q$	q: 0	1	2	3	4	5	6	7
p	0	0	0	0	0	0	0	0	0
	1	0	1	0	1	0	1	0	1
	2	0	0	2	2	0	0	2	2
	3	0	1	2	3	0	1	2	3
	4	0	0	0	0	4	4	4	4
	5	0	1	0	1	4	5	4	5
	6	0	0	2	2	4	4	6	6
	7	0	1	2	3	4	5	6	7

These matrices are hereditary with respect to the transformation rules, and every axiom (and therefore every theorem) of S2 takes a designated value. But for AS3.1, when $p = 1$ and $q = 0$, $(p \prec q) \prec (\sim Mq \prec \sim Mp) = 0$.

(3) We next prove that AS4.1 is not a theorem of S3. As in the proof that AS2.1 is not in S1 we have

$$\mathrm{K} = \{1, 2, 3, 4\}$$

with 1 and 2 as the designated values. We also have the same (4-valued) matrices for $\sim$ and $.$. But for M we have[223]:

p	Mp
1	1
2	1
3	1
4	3

and hence for $\prec$ and L:

	$p \prec q$	q: 1	2	3	4	Lp
p	1	2	4	4	4	2
	2	2	2	4	4	4
	3	2	4	2	4	4
	4	2	2	2	2	4

These matrices are again hereditary, and all the axioms of S3 always take a designated value. But when $p = 1$, AS4.1 evaluates as follows:

$$(L1 \prec LL1) = (2 \prec L2) = (2 \prec 4) = 4$$ [224]

(4) For the non-derivability of AS5.1 in S4 we again use the same 4-valued matrices for $\sim$ and $.$ and the same designated values. But for M, and consequently for $\prec$, we use[225]:

223 Lewis and Langford [1932], p. 493 (Group I); and *vide* also p. 498.

224 Lewis (*ibid* p. 498) says that AS4.1 fails for these matrices when $p = 3$, but this is presumably a slip.

225 *Ibid*, p. 493 (Group II).

p	Mp
1	1
2	2
3	1
4	4

	$p \prec q$	q: 1	2	3	4
p	1	1	4	3	4
	2	1	1	3	3
	3	1	4	1	4
	4	1	1	1	1

The matrices are again hereditary and all the axioms of S4 always take a designated value. But for AS5.1 we have, when $p = 2$[226]:

$$(M2 \prec {\sim}M{\sim}M2) = (2 \prec {\sim}M{\sim}2) = (2 \prec {\sim}M3)$$
$$= (2 \prec {\sim}1) = (2 \prec 4) = 3$$

S1–S5 are therefore five distinct systems, in ascending order of strength.

There is a further result of some importance which we can establish by our present techniques. For the following matrices[227] ($\sim$ and ., and designated values, as before), every theorem of S5 always takes a designated value:

p	Mp
1	1
2	1
3	1
4	4

	$p \prec q$	q: 1	2	3	4
p	1	1	4	4	4
	2	1	1	4	4
	3	1	4	1	4
	4	1	1	1	1

But for $Mp \prec p$, when $p = 2$ we have:

$$(M2 \prec 2) = (1 \prec 2) = 4$$

(and also, for that matter, if $p = 3$, $(Mp \prec p) = 4$). Hence $Mp \prec p$ is not a theorem even of S5[228].

226 Lewis (*ibid*, p. 498) says that AS5.1 also fails for these matrices when $p = 4$, but this again appears to be a slip.

227 Lewis (*ibid* p. 493), Group III.

228 Lewis (*ibid* p. 498) uses these matrices to show also that MMp is not a thesis of S5.

The importance of this result lies in the fact that if $Mp \prec p$ were a thesis of S5, then, since its converse, $p \prec Mp$ is a thesis even of S1, we should have $\vdash_{S5}(Mp = p)$. In that case S5 would collapse into PC, in the sense explained on p. 59. It is the absence from S5 of theses such as $Mp \prec p$ which prevents this happening and preserves for S5 its genuinely modal status.

Alternative bases for the Lewis systems

In 1953 Leo Simons[229] proved that S3 can be derived from the following axiom set:

H1 $p \prec (p \cdot p)$
H2 $(p \cdot q) \prec p$
H3 $((r \cdot p) \cdot \sim (q \cdot r)) \prec (p \cdot \sim q)$
H4 $\sim Mp \supset \sim p$
H5 $p \prec Mp$
H6 $(p \prec q) \prec (\sim Mq \prec \sim Mp)$

and the rules of uniform substitution and *material* detachment[230].

It is not difficult to show that this system is contained in Lewis' S3: H1, H2 and H6 are among Lewis' axioms; H5 is easily derivable from TS1.18a, H3 from PC by N(S1), and H4 from H5 with the help of TS1.4b and TS1.11; and material detachment holds even in S1. Simons shows that his system contains S3 by deriving in it the S3 axioms and rules[231]:

Simons also shows that each of his axioms is independent, and furthermore that the enlarged set of axioms formed by adding

H7 $MMp \prec Mp$

remains independent. Now H7 is a simple variant of the special S4 axiom $Lp \prec LLp$; hence we have here a set of independent axioms for S4.

[229] Simons [1953].

[230] Simons actually uses axiom-schemata and so does not require a rule of uniform substitution; but we set out his system here in the style we have previously used for propositional logics.

[231] He proves the primitive rules of Lewis' S3 in such a way that they will still hold in any extensions. Thus adding the same extra axioms to his S3 and Lewis' will give the same system.

Another axiom set for S4 consists of H1–H4, H6 and $\vdash \sim M \sim (p \prec Mp)$, with the same two rules[232]. Feys also gives an axiomatization of S4 with only these two rules as primitive, but his axiom set is somewhat more complicated than Simons'[233].

We shall have something to say about S4 based on other primitives than M or L on pp. 296f.

Simons also suggests adding to H1–H6 the axiom

H8 $Mp \prec \sim M \sim Mp$

which would give S5; but he gives no proof that all the axioms would then be independent. Anderson[234] gives an axiomatization of S5 by adding to H1–H6:

$$(\sim M \sim p \prec Mp) \prec (p \prec \sim M \sim Mp)$$

and he shows this to be independent by using Simons' matrices. Sobociński[235] shows that if we have H8 we may omit H5 and still have S5.

The bases we have so far mentioned in this section have been like Lewis' own in that they have not been constructed by the method of subjoining extra axioms and/or rules to a PC basis. It is, however, possible to use this method for each of the Lewis systems, as of course we did for S4 and S5 in Part I. There are several ways of basing a modal system explicitly on PC. One is to include in the basis for the modal system some set of axioms and rules (say those of PM) known to be sufficient for PC. Another, notationally simpler, procedure is to have the rule:

PC1 If α is a valid wff of PC, then $\vdash \alpha$

[232] Simons [1962]. Anderson [1957] gives a basis for T by modifying Simons' axioms for S3 and S4 and changing the rule of detachment to $\vdash \alpha$, $\vdash L\alpha \supset \beta$, $\rightarrow \vdash \beta$. Kripke [1965b] p. 219 shows that T and S2 (unlike S3 and S4) cannot have a basis containing only a finite number of schemata with material detachment as the only primitive transformation rule. (A basis of this kind is sometimes called a 'rule simplified' basis.) Lemmon [1965b] also proves this by algebraic means (*vide* Chapter 17) and states some general conditions under which finite axiomatizability in this sense holds. (It is possible to have finitely axiomatizable modal systems weaker than S3 but they are not any of the standard ones. *Vide* Canty [1965].)

[233] Feys [1965], p. 97.

[234] Anderson [1956].

[235] Sobociński [1962a].

(If we adopt this second procedure the rules of uniform substitution and detachment will not be needed for pure PC derivations. If the *modal* axioms are stated as schemata, the uniform substitution rule will not be needed at all, but if – as in the systems set out below – they are not, it will be needed for derivations involving them. And whether or not the modal axioms are stated as schemata, some form of the detachment rule will be required for modal derivations.)

The systems we shall now develop are due to E. J. Lemmon[236]. As far as the PC basis is concerned, we shall use a modified version of the second procedure mentioned above[237]. PC1 is not used as a primitive rule, but it follows directly from one of the primitive rules and one of the axioms. (Handling things in this way makes the statement of the whole basis somewhat more economical.) Uniform substitution and material detachment (MP) are primitive rules. Of the modal operators, L is taken as primitive, and M and $\prec$ are defined (as in Part I) by:

$$M\alpha =_{\mathrm{Df}} \sim L \sim \alpha$$

$$(\alpha \prec \beta) =_{\mathrm{Df}} L(\alpha \supset \beta)$$

Lemmon then constructs four systems, which he calls P1–P4, each with an explicit PC basis, which are deductively equivalent to S1–S4 respectively; and there is no difficulty in extending the process to S5.

For P1 Lemmon has (in addition to uniform substitution and material detachment, as mentioned above) the following axioms and rules (our numbering):

AP1.1 $((p \prec q) \cdot (q \prec r)) \supset (p \prec r)$
AP1.2 $Lp \supset p$

[236] Lemmon [1957]. Lemmon also considers various subsystems of the Lewis systems. *Vide infra*, pp. 301–303.
[237] Lemmon (op. cit., p. 177) refers by 'PC' to three rules:

PCa. If α is valid (PC), then $\vdash \alpha$
PCb. Uniform substitution
PCc. Modus Ponens

He includes PC in the bases for all his systems. We make use of the fact that (in virtue of our RP1.1 below) PCb and PCc are the only PC rules needed.

RP1.1 If α is a valid wff of PC or an axiom, then $\vdash L\alpha$[238].

RP1.2 (Substitution of Equivalents). If α differs from β only in having a wff γ in some of the places where β has δ, then if $\vdash(\gamma = \delta)$, then $\vdash(\alpha = \beta)$.

Clearly, if α is a valid wff of PC, then by RP1.1, $\vdash L\alpha$, and then by AP1.2 and MP, $\vdash\alpha$. I.e., PC1 follows immediately from RP1.1 and AP1.2.

We now show that Lewis' S1 contains P1.

Uniform substitution and RP1.2 are primitive rules of S1. AP1.1 is derivable from AS1.5 by TS1.11. AP1.2 is TS1.18a. One part of RP1.1 is provided by N(S1); for the other part, if we apply TS1.15 to AS1.5 and TS1.18a we obtain

$$\vdash_{S1} L(((p \prec q) \,.\, (q \prec r)) \supset (p \prec r)) \text{ and } \vdash_{S1} L(Lp \supset p)$$

respectively (i.e. the result of prefixing L to the P1 axioms). Hence RP1.1 holds in S1. Finally we have in S1 strict equivalences corresponding to those definitions in P1 which are not in S1: *vide* p. 223.

The proof that P1 contains S1 is as follows.

AS1.1–AS1.5 follow from standard PC theses and AP1.1 by RP1.1 and Def $\prec$. The rule of Adjunction holds in P1 because by PC1 we have $\vdash(p \supset (q \supset (p \,.\, q)))$; whence if $\vdash\alpha$ and $\vdash\beta$, two detachments yield $\vdash(\alpha \,.\, \beta)$. For strict detachment in P1: if $\vdash(\alpha \prec \beta)$, then by Def $\prec$ and AP1.2 we have $\vdash(\alpha \supset \beta)$; hence if $\vdash\alpha$ and $\vdash(\alpha \prec \beta)$, then $\vdash\alpha$ and $\vdash(\alpha \supset \beta)$, and therefore, by material detachment, $\vdash\beta$.

It remains to prove in P1 strict equivalences corresponding to the S1 definitions of L and $\prec$, and also AS1.6. We first note that the following rule holds in P1:

RP1.3 Where α and β are wffs of PC, $\vdash(\alpha \equiv \beta) \rightarrow \vdash(\alpha = \beta)$.
(The proof is as on p. 226, since RP1.1 includes N(S1).)
We can now prove:

[238] If the $\supset$ in AP1.1 and AP1.2 were replaced by $\prec$, and we had MP for *strict* implication, the phrase 'or an axiom' in RP1.1 could be omitted.

TP1.1 $Lp = \sim M\sim p$ [For Def L in S1]

PROOF

$(p \equiv p) \times$ RP1.3: (1) $p = p$

$(p \equiv \sim\sim p) \times$ RP1.3: (2) $p = \sim\sim p$

(1) $[Lp/p]$, (2) $\times$ Eq: (3) $Lp = L\sim\sim p$

(3), (2) $[L\sim\sim p/p] \times$ Eq: (4) $Lp = \sim\sim L\sim\sim p$

(4) $\times$ Def M: (5) $Lp = \sim M\sim p$ **Q.E.D.**

TP1.2 $(p \prec q) = \sim M(p \,.\, \sim q)$ [For Def $\prec$ in S1]

PROOF

TP1.1 $[p \supset q/p]$: (1) $L(p \supset q) = \sim M\sim(p \supset q)$

$\vdash\sim(p \supset q) \equiv (p \,.\, \sim q) \times$ RP1.3: (2) $\sim(p \supset q) = (p \,.\, \sim q)$

(1), (2) $\times$ Eq: (3) $L(p \supset q) = \sim M(p \,.\, \sim q)$

(3) $\times$ Def $\prec$: (4) $(p \prec q) = \sim M(p \,.\, \sim q)$

Q.E.D.

TP1.3 $(p \,.\, (p \prec q)) \prec q$ [AS1.6]

PROOF

PC $\times$ RP1.3: (1) $(p \supset (q \supset r)) = ((q \,.\, p) \supset r)$

$\vdash(p \supset p)[L(p \supset (q \supset r))/p]$: (2) $L(p \supset (q \supset r)) \supset L(p \supset (q \supset r))$

(2), (1) $\times$ Eq: (3) $L(p \supset (q \supset r)) \supset L((q \,.\, p) \supset r)$

AP1.2, $\times$ RP1.1: (4) $L(Lp \supset p)$

(4) $[p \supset q/p]$: (5) $L(L(p \supset q) \supset (p \supset q))$

(3) $[L(p \supset q)/p, p/q, q/r]$: (6) $L(L(p \supset q) \supset (p \supset q))$
$\supset L((p \,.\, L(p \supset q)) \supset q)$

(5), (6) $\times$ MP, $\times$ Def $\prec$: (7) $(p \,.\, (p \prec q)) \prec q$ **Q.E.D.**

This completes the proof that P1 contains S1. Hence P1 and S1 are deductively equivalent.

Lemmon's P2 is formed from P1 by replacing AP1.1 by

AP2.1 $L(p \supset q) \supset (Lp \supset Lq)$

and RP1.2 (the substitution of equivalents rule) by

RP2.1 $\vdash L(\alpha \supset \beta) \rightarrow \vdash L(L\alpha \supset L\beta)$

The other axioms and rules remain unchanged.

It is easy to show that S2 contains P2: AP2.1 is TS1.21; the result of prefixing L to AP2.1 (which RP1.1 allows us to do) becomes, by Def $\strictif$, TS1.21a; and RP2.1, by Def $\strictif$, is BR (p. 232).

Proving that P2 contains S2 is more complicated. As a first step we prove that P2 contains P1. To do this we have of course to derive in P2 both AP1.1 and the substitution of equivalents rule, RP1.2. But there is an additional point which we must not overlook. RP1.1, though it is still with us, entitles us to prefix an L *only to the axioms of the system in which it occurs*; and since AP1.1 is not an axiom of P2, RP1.1 does not automatically license putting an L before it in P2, as it does in P1. Therefore we also have to derive the formula which consists of AP1.1 preceded by L, i.e. (by Def $\strictif$)

$$((p \strictif q) \cdot (q \strictif r)) \strictif (p \strictif r)$$

Once we have this, of course, AP1.1 will follow by AP1.2.

We now turn to these derivations. In the following proofs we use $\times$ AP2.1 to indicate the use of the rule

$$\vdash L(\alpha \supset \beta) \rightarrow \vdash (L\alpha \supset L\beta)$$

which is immediately derivable from AP2.1. The first two theorems are steps towards the thesis mentioned above.

TP2.1 $(p \strictif q) \supset ((q \strictif r) \supset (p \strictif r))$

PROOF

PC $\times$ RP1.1: (1) $L[(p \supset q) \supset ((q \supset r) \supset (p \supset r))]$

(1) $\times$ AP2.1: (2) $L(p \supset q) \supset L((q \supset r) \supset (p \supset r))$

AP2.1 $[q \supset r/p, p \supset r/q]$: (3) $L((q \supset r) \supset (p \supset r)) \supset (L(q \supset r) \supset L(p \supset r))$

(2), (3) $\times$ Syll, $\times$ Def $\strictif$: (4) $(p \strictif q) \supset ((q \strictif r) \supset (p \strictif r))$

Q.E.D.

TP2.1 gives us the rule Syll in the form:

$$\vdash (\alpha \strictif \beta), \vdash (\beta \strictif \gamma) \rightarrow \vdash (\alpha \strictif \gamma)$$

To distinguish this from Syll for material implication, we refer to it as *Syll'*.

TP2.2 $(p \prec (q \supset r)) \supset ((p \cdot q) \prec r)$

PROOF

PC × RP1.1: (1) $L[(p \supset (q \supset r)) \supset ((p \cdot q) \supset r)]$

(1) × AP2.1: (2) $L(p \supset (q \supset r)) \supset L((p \cdot q) \supset r)$

(2) × Def $\prec$: (3) $(p \prec (q \supset r)) \supset ((p \cdot q) \prec r)$ **Q.E.D.**

TP2.3 $((p \prec q) \cdot (q \prec r)) \prec (p \prec r)$

PROOF

PC × RP1.1: (1) $L[(p \supset q) \supset ((q \supset r) \supset (p \supset r))]$

(1) × RP2.1: (2) $L[L(p \supset q) \supset L((q \supset r) \supset (p \supset r))]$

(2) × Def $\prec$: (3) $(p \prec q) \prec L((q \supset r) \supset (p \supset r))$

AP2.1 × RP1.1: (4) $L[L(p \supset q) \supset (Lp \supset Lq)]$

(4) $[q \supset r/p, p \supset r/q]$, × Def $\prec$:

(5) $L((q \supset r) \supset (p \supset r)) \prec ((q \prec r) \supset (p \prec r))$

(3), (5) × Syll′: (6) $(p \prec q) \prec ((q \prec r) \supset (p \prec r))$

(6), TP2.2 $[p \prec q/p, q \prec r/q, p \prec r/r]$ × MP:

(7) $((p \prec q) \cdot (q \prec r)) \prec (p \prec r)$ **Q.E.D.**

We have now to derive the rule of substitution of equivalents (RP1.2) in P2. Lemmon uses a version of PC in which the primitive operators are $\sim$ and $\supset$; so all formulae in his modal systems may be written with $\sim$, $\supset$ and L as the only operators. Therefore to establish RP1.2 we only need to derive the rules:

(a) $\vdash(\alpha = \beta) \rightarrow \vdash(\sim\alpha = \sim\beta)$

(b) $\vdash(\alpha = \beta) \rightarrow \vdash((\alpha \supset \gamma) = (\beta \supset \gamma))$

(c) $\vdash(\alpha = \beta) \rightarrow \vdash((\gamma \supset \alpha) = (\gamma \supset \beta))$

(d) $\vdash(\alpha = \beta) \rightarrow \vdash(L\alpha = L\beta)$

and RP1.2 will follow by induction on the construction of formulae (and by the easily established rule: $\vdash\alpha$, $\vdash(\alpha = \beta) \rightarrow \vdash\beta$).

TP2.4 $(p = q) \supset (\sim p = \sim q)$

PROOF

PC × RP1.1: (1) $L[(p \supset q) \supset (\sim q \supset \sim p)]$

(1) × AP2.1, × Def $\prec$: (2) $(p \prec q) \supset (\sim q \prec \sim p)$

(2) $[q/p, p/q]$: (3) $(q \prec p) \supset (\sim p \prec \sim q)$

PC: (4) $(p \supset q) \supset ((r \supset s) \supset ((p \,.\, r) \supset (s \,.\, q)))$

(2), (3) × (4): (5) $((p \prec q) \,.\, (q \prec p)) \supset ((\sim p \prec \sim q) \,.\, (\sim q \prec \sim p))$

(5) × Def =: (6) $(p = q) \supset (\sim p = \sim q)$ **Q.E.D.**

((2) above, viz. $(p \prec q) \supset (\sim q \prec \sim p)$, and its associated rule $\vdash(\alpha \prec \beta) \rightarrow \vdash(\sim\beta \prec \sim\alpha)$, we shall call **TP2.4a.**)

TP2.5 $(p = q) \supset ((p \supset r) = (q \supset r))$
TP2.6 $(p = q) \supset ((r \supset p) = (r \supset q))$

These are proved similarly, but starting from

$$\vdash((p \supset q) \supset ((q \supset r) \supset (p \supset r)))$$

and

$$\vdash((p \supset q) \supset ((r \supset p) \supset (r \supset q)))$$

respectively.

Clearly TP2.4–6 immediately yield rules (a)–(c). For (d) we have: $\vdash(\alpha = \beta)$ yields, by Def = and PC, both $\vdash(\alpha \prec \beta)$ and $\vdash(\beta \prec \alpha)$; whence, by RP2.1 and Def $\prec$, both $\vdash(L\alpha \prec L\beta)$ and $\vdash(L\beta \prec L\alpha)$. Adj and Def = then give $\vdash(L\alpha = L\beta)$.

This completes the proof that P2 contains P1. In particular since all the axioms of S1 are in P1, they are in P2 as well. We now prove that AS2.1, which when added to S1 gives S2, is a theorem of P2.

TP2.7 [AS2.1] $M(p \,.\, q) \prec Mp$

PROOF

PC × RP1.1 (1) $L(\sim p \supset \sim(p \,.\, q))$

(1) × RP2.1, × Def $\prec$: (2) $L\sim p \prec L\sim(p \,.\, q)$

(2) × TP2.4a: (3) $\sim L\sim(p \,.\, q) \prec \sim L\sim p$

(3) × Def M: (4) $M(p \,.\, q) \prec Mp$ **Q.E.D.**

We have now proved that P2 contains all the axioms of S2. The primitive transformation rules of S2 are uniform substitution, adjunction, strict detachment and substitution of strict equivalents. Of these, uniform substitution is a primitive rule of P2; we have proved that substitution of strict equivalents holds in P2; and adjunction and strict detachment hold in P2 for the same reasons as we gave in the case of P1 on p. 247. Moreover, since the equivalences corresponding to the S1 and

S2 definitions of L and ⥽ (viz. TP1.1 and TP1.2) are theorems of P1, they are also theorems of P2. Hence P2 contains S2.

To obtain P3 we replace both AP2.1 and RP2.1 by:

AP3.1 $L(p \supset q) \supset L(Lp \supset Lq)$

(I.e. P3 has only AP1.2 and RP1.1 – and of course substitution and detachment – in addition to AP3.1.)

We show first that P3 contains P2. To do this we have to derive RP2.1 and $\vdash(L(p \supset q)$ ⥽ $(Lp \supset Lq))$[239]. Clearly RP2.1 follows directly from AP3.1. For the other we begin by proving:

TP3.1 $(p$ ⥽ $q) \supset ((q$ ⥽ $r) \supset (p$ ⥽ $r))$

PROOF

AP1.2 $[Lp \supset Lq/p]$: (1) $L(Lp \supset Lq) \supset (Lp \supset Lq)$

AP3.1, (1) × Syll: (2) $L(p \supset q) \supset (Lp \supset Lq)$

Since this last formula, (2), is AP2.1, the rest of the proof is as for TP2.1[240], which is the same formula as TP3.1. Thus we have the rule Syll′ in P3 as well as in P2.

TP3.2 $L(p \supset q)$ ⥽ $(Lp \supset Lq)$

PROOF

AP3.1 × RP1.1, × Def ⥽: (1) $L(p \supset q)$ ⥽ $L(Lp \supset Lq)$

AP1.2 × RP1.1, × Def ⥽: (2) Lp ⥽ p

(2) $[Lp \supset Lq/p]$: (3) $L(Lp \supset Lq)$ ⥽ $(Lp \supset Lq)$

(1), (3) × Syll′: (4) $L(p \supset q)$ ⥽ $(Lp \supset Lq)$

Q.E.D.

So P3 contains P2, and therefore S1 as well. Now S3 is S1 plus AS3.1, or alternatively (as we showed on p. 233) S1 plus TS3.1 – i.e. $(p$ ⥽ $q)$ ⥽ $(Lp$ ⥽ $Lq)$; and this is easily derivable in P3 by applying RP1.1 and Def ⥽ to AP3.1. Hence P3 contains S3.

The proof that S3 contains P3 is straightforward. AP1.2 is TS1.18. RP1.1, so far as it applies to PC formulae, is N(S1). So far as it applies to axioms, it will hold in S3 if we can prove the

[239] I.e. the strict form of AP2.1 – for the reason given on p. 249.

[240] P. 249. The proof given there does not involve applying RP1.1 to AP2.1, so it can be followed through in P3.

strict forms of AP1.2 and AP3.1. The former is TS1.18a; the latter (by Def $\prec$) is TS3.1.

P4 has the same axioms as P3, but in place of RP1.1 (i.e. as the sole rule apart from substitution and detachment) we have:

RP4.1 If α is a valid wff of PC or a thesis of the system, then $\vdash L\alpha$[241].

Clearly P4 is P3 with the addition of the rule: $\vdash\alpha \rightarrow \vdash L\alpha$ – i.e. the unrestricted rule of necessitation (N). But P3 is deductively equivalent to S3, and we proved on p. 235 that the addition of this rule to S3 yields a system which contains S4. Hence P4 contains S4.

Moreover S4 contains P4, for we proved on p. 236f both that S4 contains S3 (and therefore P3) and that N holds in S4.

The two systems are therefore deductively equivalent[242].

Lemmon does not continue his series of systems beyond P4. But clearly we could form a system P5, deductively equivalent to S5, by adding to P4:

AP5.1 $Mp \supset LMp$

If we did this we could then weaken AP3.1 to AP2.1, and then we should have the S5 axiomatization of Chapter 3 (apart from the different style of setting out the PC part of the basis).

The Lewis systems and T

None of the Lewis systems S1–S5 is deductively equivalent to T.

We proved in Part I that T is contained in, but does not contain, S4 (p. 70). Moreover, it is not difficult to see that T contains P2, and therefore S2; for the basis of P2 (apart from RP2.1, which is an obvious rule of T in any case) is simply that of T with certain restrictions on the rule of necessitation (N). S2, however, does not contain T; for by N we have $\vdash_T LL(p \supset p)$, but we have shown that S2 has no theses of the form $LL\alpha$ (p. 236). Since S3 too has no theses of this form, it cannot contain

[241] Lemmon actually gives PC1 and the rule $\vdash\alpha \rightarrow \vdash L\alpha$; (op. cit. p. 178) but our RP4.1, which has the same effect as these combined, is more in line with the rule RP1.1 which we had in P1–P3.

[242] For alternative ways of obtaining S4 (and also T) *vide* Åqvist [1964], pp. 79–81. Our S4 basis of p.125 is in fact the same as that of P4.

T either. But although T contains S2, it does not contain S3; for T has the unrestricted rule of necessitation; and as we proved on p. 235, if this rule were added to S3 we should obtain S4, which as we know is *not* contained in T. T and S3 are therefore independent systems; i.e., neither contains the other.

The relations holding among S1–S5 and T can be represented in the following diagram[243]:

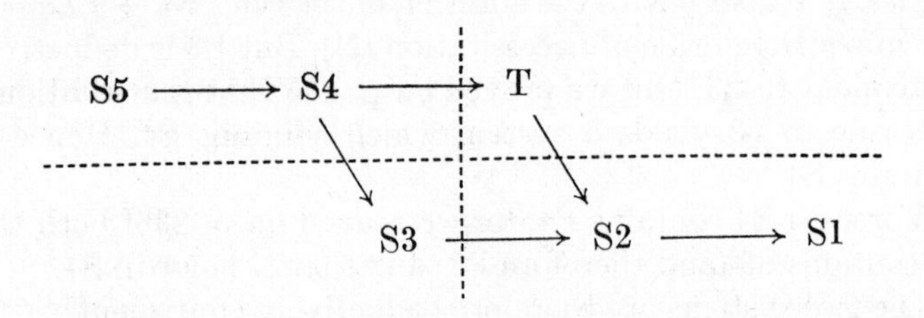

A → B means that every theorem of B is also a theorem of A (but not *vice versa*). Systems above the horizontal line have the unrestricted rule of Necessitation. Those to the left of the vertical line have only a finite number of distinct modalities.

[243] Cf. Prior [1957], pp. 123–124.

CHAPTER FOURTEEN

Other Modal Propositional Systems

So far we have confined our attention to the six systems, S1–S5 and T. There are, however, many other modal systems than these – so many, in fact, that all we can attempt here is to give a brief summary of the main ones and to indicate their relations to each other and to the systems already discussed.

To make our topic more manageable, we shall restrict ourselves for the present to L- or M-based systems which have all of the following features[244].

1. The relations between the modal operators L, M, $\prec$ and $=$ are such that the following are theorems:

$$\text{(a)}\ \sim L\sim p \equiv Mp$$

$$\text{(b)}\ \sim M\sim p \equiv Lp$$

$$\text{(c)}\ (p \prec q) \equiv L(p \supset q)$$

$$\text{(d)}\ (p = q) \equiv ((p \prec q) \cdot (q \prec p))$$

2. The rule of uniform substitution holds (either as primitive or, if axiom-schemata are employed, as derivable).
3. The rule of material detachment holds.
4. If α is a valid wff of PC then $\vdash L\alpha$.
5. The following are theorems:

$$\text{(a)}\ Lp \supset p$$

$$\text{(b)}\ L(p \supset q) \supset (Lp \supset Lq)$$

6. $p \supset Lp$ is not a theorem.

This list of conditions is neither arbitrary nor unduly restrictive. From the intuitive point of view, it is based on two principles. One is that the systems to be considered should be capable of being interpreted as logics of necessity and possibility in some fairly natural sense; the other is that they should include

[244] Some of these conditions are given in Łukasiewicz [1953]. Examples of systems not satisfying all of them will be found in Chapter 16.

orthodox two-valued propositional calculus, both in the sense that every valid PC formula will be a theorem and also in the sense that every proposition with the form of a valid PC formula shall count as a necessary truth.

As far as restrictiveness is concerned, we drew up a somewhat analogous list of conditions in Chapter 2, and the weakest system which satisfied *them* was T. T also satisfies our present set of conditions, but so do the weaker systems S1 and S2, in which it certainly seems possible to think of L and M as necessity and possibility operators. In fact the present set can be satisfied by systems which are even weaker than S1, as we shall see in a moment. So perhaps the list in Chapter 2 was unduly restrictive; but it is at least arguable that a system which did not comply with all of our present set of conditions would involve a strained sense of 'necessity' or a departure from standard PC, or both. At any rate, their common possession of the features we have listed gives a certain coherence to the group of systems we are about to study.

The system S0.5

The weakest system satisfying all of conditions 1–6 would be one equivalent to the following:

L is taken as primitive, and the other modal operators are defined so as to satisfy condition 1.

The axioms are:

LA1 $Lp \supset p$
LA2 $L(p \supset q) \supset (Lp \supset Lq)$

The transformation rules are: uniform substitution; material detachment; and

PCL If α is a valid wff of PC, then $\vdash L\alpha$.

This system is called S0.5 by E. J. Lemmon[245].

LA1 and LA2 are the axioms A5 and A6 of the System T (p. 31). Since by PCL and LA1 we can derive all PC theses in

[245] Lemmon [1957], p. 181. S0.5 should not be confused with systems such as the S.1 of Parry [1953] (*vide* also Hamblin [1959]) which only consider first-degree formulae. Since such systems (as noted by Lemmon, *loc. cit.*) have to restrict either the class of formulae or the rule of uniform substitution they will not be discussed in this book. Cf Pollock [1967b].

S0.5, we can think of S0.5 as T but with the rule of Necessitation weakened to PCL. In fact, not merely is S0.5 weaker than T, it is even weaker than S1. Lemmon proves this[246] by showing that it is weaker than a system which he calls S0.9 and which he shows to be contained in S1 (and possibly weaker than S1). S0.9 is Lemmon's P2, but with RP2.1 replaced by the weaker rule: $\vdash(\alpha = \beta) \rightarrow \vdash(L\alpha = L\beta)$.

As far as interpretation is concerned, Lemmon suggests that the operator L in S0.5 can be thought of as meaning 'it is tautologous by truth tables that . . .'[247].

Systems with the Brouwerian axiom

In Chapter 3 we referred briefly to the *Brouwerian System*[248], obtaining by adding to T the extra axiom

B $p \supset LMp$

and we saw that this system is intermediate between T and S5 but independent of S4.

Lewis, following Becker, cites the strict form of the Brouwerian axiom[249], viz.

B′ $p \prec LMp$

(When we add the Brouwerian axiom to T, the presence in T of the rule of Necessitation makes it immaterial whether we use B′ or B; but when we add it to some other systems it may well make a difference which form we choose.)

We show that the addition of B′ to S1 also yields the Brouwerian system. The proof is as follows:

TS1.21a	(1) $(p \prec q) \prec (Lp \supset Lq)$
B′, (1) $[LMp/q] \times$ MP:	(2) $Lp \supset LLMp$
(2) $[p \supset p/p]$:	(3) $L(p \supset p) \supset LLM(p \supset p)$
PC $\times$ N(S1):	(4) $L(p \supset p)$
(4), (3) $\times$ MP:	(5) $LLM(p \supset p)$

[246] Op. cit., p. 181.

[247] Lemmon [1959], p. 31.

[248] *Supra* pp. 57f. *Vide* also Kripke [1963a], and the system T⁺ of Thomas [1964b].

[249] Becker [1930] (p. 509) using $\sim$ for impossibility has simply $p \prec \sim\sim p$. (Becker of course was thinking of it as a possible addition to S3.) Lewis and Langford [1932] write it as $p \prec \sim\Diamond \sim\Diamond p$ (C12, p. 497).

Now (5) is of the form $LL\alpha$, and so any strict extension of S1 in which it is derivable contains T (*vide* pp. 227f)[250]. Hence $S1 + B' = T + B$[251]. Moreover, the proof also shows that $S3 + B' = S5$[252]. For from B′ we have derived (using S1 theses) a formula of the form $LL\alpha$ and we know[253] that the addition of such a formula to S3 gives S4 and that the addition of B′ to S4 gives S5; so $S3 + B'$ contains S5. And the converse holds, since we have already shown that S5 contains S3 and that B′ is a theorem of S5.

Although B′ does not strengthen S1 to S5, there is a closely related formula which does, viz.

B″ $p \prec LLMp$

To prove this, we first note that B′ is a theorem of $S1 + B''$. For by substitution in $\vdash(Lp \prec p)$ we have $\vdash(LLMp \prec LMp)$; whence by B″ and Syll we obtain $\vdash(p \prec LMp)$. Now we have already shown that $S1 + B'$ contains $LLM(p \supset p)$ and hence that it and any strict extensions (in particular $S1 + B''$) contain the unrestricted rule of Necessitation and so the whole axiomatic basis of T. So we need only show that T with the addition of B″ yields S5, and this will give us our desired result.

Working with the material forms of theses (which of course are sufficient in T and stronger systems since we have the rule of Necessitation), we have the following proof of the characteristic S5 axiom:

B $[ML{\sim}p/p]$:	(1) $ML{\sim}p \supset LMML{\sim}p$
(1) × Transp, × LMI	(2) $MLLMp \supset LMp$
B″ (material form) × DR3 (p. 37):	(3) $Mp \supset MLLMp$
(3), (2) × Syll:	(4) $Mp \supset LMp$ **Q.E.D.**

So the addition of B″ to S1 gives us at least S5. That it gives us no more follows from the fact that B″ is a thesis of S5. This is

[250] In this chapter, unless otherwise stated, S1–S3 are understood as axiomatized by Lewis, though the results also hold for Lemmon's axiomatizations.

[251] *Vide supra*, p. 58. The result also follows from an observation in Sobociński [1962a] (p. 59).

[252] Parry [1939], pp. 150–152.

[253] *Vide supra*, p. 235 and Lewis and Langford [1932], p. 498.

easy to prove, since we already know that B′ is in S5; whence by $\vdash(Lp \prec LLp)$ and Syll we obtain B″.

B″ was obtained from B′ by replacing the L by two L's. If we decide to have as a thesis a formula obtainable by replacing the L in B′ by a number of L's, we have what Becker[254] calls a 'generalized' Brouwerian axiom, which can be written as:

$$p \prec L_n Mp$$

where L_n represents a sequence of n consecutive L's and n is a natural number greater than 1.

Now clearly, for $n > 1$, $\vdash_{S1} L_n Mp \prec LLMp$; and hence the addition of any generalized Brouwerian axiom to S1 will, by Syll, give $\vdash p \prec LLMp$ (B″) and thus yield S5. Moreover every generalized Brouwerian axiom is itself in S5, since the above argument to show that B″ is in S5 can be extended to cover any number of L's. Hence the addition of any generalized Brouwerian axiom to S1 gives a system deductively equivalent to S5[255].

Systems with generalized S4 axioms

We turn now to a series of systems stronger than T but weaker than S4. In this section we shall be thinking of S4 as axiomatized by adding $\vdash(Lp \supset LLp)$ to T.

Clearly if we have, as in S4, $\vdash(Lp \supset LLp)$, we also have, by substitution, $\vdash(LLp \supset LLLp)$, $\vdash(LLLp \supset LLLLp)$, and in general, where L_n is a sequence of n consecutive L's, $\vdash(L_n p \supset L_{n+1} p)$. But we cannot work in the reverse direction: if, for example, our only addition to T is $\vdash(LLp \supset LLLp)$, we cannot derive $\vdash(Lp \supset LLp)$.

We can state this result more generally. Let S be a system formed by adding $\vdash(L_n p \supset L_{n+1} p)$ to T. Then if $m \geqslant n$, $\vdash_S(L_m p \supset L_{m+1} p)$. But if $m < n$, $(L_m p \supset L_{m+1} p)$ is not a thesis of S.

[254] Becker [1930], p. 524.

[255] Churchman [1938] has proved that adding generalized Brouwerian axioms to S2 yields S5. Sobociński [1962c] has proved the same result for a certain proper sub-system of S1. For systems without the theorem $Lp \prec p$, however, which generalized Brouwerian axioms we add can make a difference. *Vide* Sobociński [1962c]; Thomas [1963a], [1963c] and [1964b].

The first half of this result follows immediately from the rule of uniform substitution, as we have already remarked in the case of S4. The second half can be proved from a consideration of models of the kind explained in Chapter 4[256].

The upshot of this is that we have an infinite series of non-equivalent systems between T and S4, each obtained by adding to T an axiom of the form $L_n p \supset L_{n+1} p$. We can call these[257], $S4_n$ systems.

Any particular $S4_n$ system will have exactly n L's in the antecedent of its characteristic axiom. (Thus, $S4_1$ is simply S4.) Each $S4_n$ system ($n > 1$) is stronger than T but weaker than S4; and the larger n is, the weaker the $S4_n$ system.

One feature of $S4_n$ systems is that while, as we saw in Chapter 3 (p. 48), when $n = 1$ (i.e. in S4) the number of non-equivalent modalities is finite, when $n > 1$ there are an infinite number of non-equivalent modalities in $S4_n$. For example it has been proved[258] that in $S4_2$ there are no equivalent modalities in the sequence:

$$Lp;\ MLp;\ LMLp;\ MLMLp;\ \text{etc.}$$

Each $S4_n$ system can be combined with the Brouwerian axiom to yield a series of systems $S4_n + B$. Thomas[259] gives the name T_n^+ to these systems and proves that each T_n^+ is independent of T_{n+1}^+.

Systems between S4 and S5

A considerable number of systems have been proposed which contain S4. Of these some are contained in S5, but others are not. In the present Section we shall be concerned only with the former kind; we shall consider the others later on.

Each of these systems is formed by adding an extra axiom or axioms to S4. For convenience we shall assume that S4 has been axiomatized (as in Chapter 3) with $\vdash \alpha \rightarrow \vdash L\alpha$ as a primitive rule. It will then be sufficient to add the new axioms in a form which

256 This result is also cited in Thomas [1964a] as an unpublished result of Sobociński.

257 Following Feys [1965], p. 127.

258 Sugihara [1964].

259 Thomas [1964a].

has a material implication (or disjunction) sign as the main operator[260].

Dummett and Lemmon[261] credit P. T. Geach with investigating a system which they call S4.2, whose proper axiom is:

G1[262] $MLp \supset LMp$

They prove that S4.2 has exactly ten non-equivalent modalities, the ones indicated in the following diagram and their negations. (The arrows represent implications.)

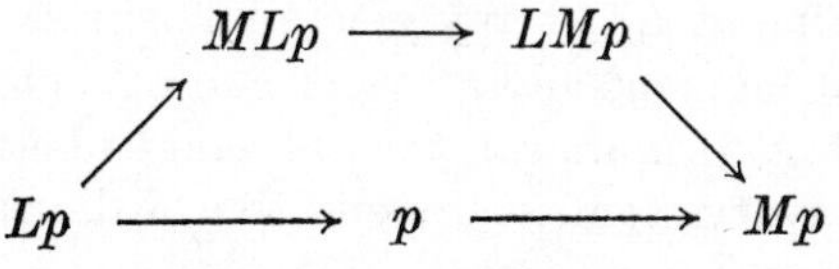

$$\begin{array}{ccccc} & MLp & \longrightarrow & LMp & \\ \nearrow & & & & \searrow \\ Lp & \longrightarrow & p & \longrightarrow & Mp \end{array}$$

S4.2 is contained in S5, since in S5 we have $\vdash(MLp \supset Lp)$, $\vdash(Lp \supset Mp)$ and $\vdash(Mp \supset LMp)$, whence by Syll we obtain G1.

The same authors also consider a system S4.3, with the proper axiom:

D2 $L(Lp \supset Lq) \vee L(Lq \supset Lp)$

They show[263] that S4.3 contains S4.2.

[260] This might not be sufficient if S4 were axiomatized differently. A version of S4 in which the rule of Necessitation is not primitive would admittedly be bound to contain the derived rule that whenever $\vdash_{S4}\alpha$, $\vdash_{S4}L\alpha$; but this would give us no guarantee that an unrestricted rule of Necessitation would still hold if we were to add further axioms. It is for this reason that some authors, working with other axiomatizations of S4, have added their new axioms in strict form (i.e. preceded by L, or with ⥽ instead of $\supset$ as the main operator). This applies to many of the axioms listed below.

[261] Dummett and Lemmon [1959], p. 252.

[262] The numberings in this Section are those given in Sobociński [1964b], where a series of systems related to S4.2 and S4.3 are considered. In Parry [1939], p. 152, we find $MLp = LMp$ (his C16), which is G1 strengthened to a strict equivalence. But since this yields $LMp \supset MLp$, which is our Kb, the system of which it is the proper axiom is not contained in S5 (*vide infra*, p. 264n.)

[263] Op. cit., p. 252. Further results concerning S4.2, S4.3 and their extensions can be found in Bull [1964a], [1965c] and [1966a]. Like many of the results of this section they are obtained using Boolean Algebra. (For some discussion of the algebraic study of modal systems *vide* Chapter 17.)

S4.3 and another system, D, which we shall mention in a moment, have a special interest in connection with an issue raised by A. N. Prior in *Time and Modality*[264]. Prior was thinking of propositions as things which could change their truth-values (could *become* true or *become* false) with the passage of time, and he wanted to be able to interpret Lp to mean 'It is and always will be the case that p'. He therefore suggested that we think of time as a series of moments, at each of which a given proposition could have the value 1 (true) or the value 0 (false), without prejudice to its value at any other moment; and he defined the value of Lp at any given moment as 1 if p has the value 1 at that moment and at every subsequent moment, and otherwise as 0. A formula can then be said to be valid iff it has the value 1 at every moment, irrespective of the values assigned to its variables at any moment.

The problem is to find an axiomatic system whose theorems shall coincide with the formulae which are valid by this criterion. In *Time and Modality* Prior made the conjecture[265] that S4 was the required system, but this was discovered to be incorrect and he abandoned it shortly afterwards[266]. The problem is further complicated by the fact that the criterion of validity can be taken in either of two ways, according as time is regarded as discrete or continuous. To regard time as discrete is to think of it as consisting of a series of distinct moments, so that given one moment we can speak of the next moment, the next again, and so forth. To regard time as continuous is to suppose that between any two moments there is a third, and then it will make no sense to speak of the *next* moment after a given one. This distinction is important here, since it turns out that there are formulae which are not valid when time is taken to be continuous but which are valid when time is taken to be discrete. Thus two systems are required, a weaker one for the interpretation with time as continuous, and a stronger one (D) for the interpretation with time as discrete.

[264] Prior [1957], Chapter II. For a later and fuller introduction to the whole topic of the temporal interpretation of modal logics *vide* Prior [1967].

[265] Prior [1957], p. 23. *Vide* also Prior [1955b].

[266] Prior [1958].

Dummett and Lemmon show[267] that the stronger system (D) contains S4.3 and also the following formula (which is not in S4.3)

M1 $((p \strictif Lp) \strictif Lp) \supset (MLp \supset Lp)$

It has since been proved that S4.3 is precisely the weaker system required (when time is taken as continuous) and that the stronger system, D (for time taken as discrete) is precisely S4.3 + M1[268].

M1 can in fact be replaced in the basis of D by the shorter[269] formula:

N1 $((p \strictif Lp) \strictif p) \supset (MLp \supset p)$

Moreover, Sobociński has shown that the addition to S4 of

R1 $p \supset (MLp \supset Lp)$

produces a system which contains, but is not contained in, D, and which he calls S4.4[270].

On a temporal interpretation R1 would mean that if any proposition which is now true is ever going to be always true from that moment on, then that proposition will always be true from the present moment on. It can hardly be claimed that this has much intuitive plausibility (except perhaps when restricted to certain special kinds of propositions), but of course S4.4 is stronger than D and R1 is not a theorem of D.

We have been considering M1 (or N1) as added to S4.3 (to produce D), but we could of course add it instead to the weaker systems S4 or S4.2. We may thus set out the following series of systems:

S4.4:	S4 + R1
S4.3.1 (D):	S4.3 + N1
S4.3:	S4 + D2
S4.2.1:	S4.2 + N1

[267] Dummett and Lemmon [1959], p. 264.

[268] Bull [1965a]. The result for D is there credited to Kripke. For a historical account of these developments *vide* Prior [1967], pp. 20–31. An alternative basis for D is S4.3 + $MLp \supset (((p \strictif Lp) \strictif Lp) \strictif Lp)$. (This last formula is an obvious variant of one given by Bull (*op. cit.*, p. 58).)

[269] Prior [1967] (p. 29) credits this shortening, and also (p. 27) an analogous shortening of D2 to $(Lp \strictif q) \vee (Lq \strictif p)$, to P. T. Geach.

[270] Sobociński [1964b], where D (axiomatized with N1) is called S4.3.1.

S4.2:	S4 + G1
S4.1.1:	S4 + M1
S4.1[271]:	S4 + N1

Sobociński shows that these systems are related as indicated in the following diagram[272]. The arrows indicate that the system on the left contains the system on the right.

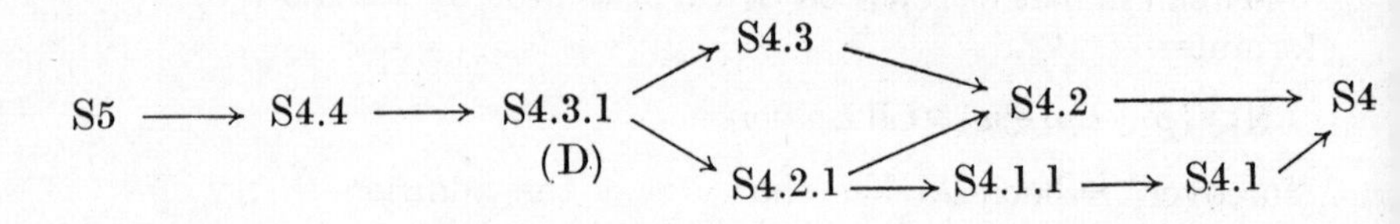

Another extension of S4 is the system S4.5 of Parry[273], obtained by adding the axiom

$$LMLp \supset Lp$$

This system, however, is equivalent to S5[274].

Systems between S3 and S5

We can also extend S3 to obtain systems which are contained in S5. Of these some are also contained in S4, but others are not.

Two axioms which could be added to S3 are:

$$M(Lp ⥽ LLp)$$ [275]

and

$$LMMp ⥽ LMp$$ [276]

[271] This system is not the same as the one called S4.1 in McKinsey [1945], p. 92. We follow Sobociński in defying historical usage here, on the ground that McKinsey's S4.1 (to which we shall give Sobociński's name, K1 – *vide* p. 265) is not contained in S5 and hence cannot be, as its name would suggest, a proper sub-system of S4.2.

[272] Sobociński [1964b], p. 307. He also considers a system V1 which is not contained in S5.

[273] Parry [1939], p. 150.

[274] For a proof *vide* Feys [1965], pp. 130–131. This result was also obtained by Dummett and Lemmon [1959], p. 251, and by Sobociński [1964a], p. 74.

[275] Halldén [1949a], p. 233.

[276] Feys [1965], p. 128 and Parry [1939], p. 149.

Each of the systems thus obtained is contained in S4. They may be called S3.1 and S3.2 respectively.

A system contained in S5 but not in S4, which we may call S3.3, results from adding to S3:

$$LMLp \prec Lp$$ [277]

Another extension of S3 which is not contained in S4 has as its proper axiom:

$$\sim Lp \supset L\sim Lp$$

(or the obvious alternative, $Mp \supset LMp$). Åqvist calls this system, S3.5, and we shall have more to say about it later on[278]. Here we merely call attention to the fact that the strict form of this axiom is the proper axiom of S5, and its addition to S3 in that form would of course produce S5. In its material form, however, it results, when added to S3, in a system weaker than S5.

Extensions of S4 not contained in S5

All the systems we have so far dealt with have been contained in S5, but some additions to S4 result in systems which are not, and we now turn to these. They are not, however, *stronger* than S5; rather they are independent of it, in the sense that they have theorems which are not in S5 but lack some which are[279].

Sobociński shows[280] that the addition to S4 of any of the following axioms will produce the same system, which he calls K1:

Ka $(LMp \,.\, LMq) \supset M(p \,.\, q)$[281]
Kb $LMp \supset MLp$
Kc $LM(Mp \supset Lp)$
Kd $LML(p \supset Lp)$
Ke $LML(Mp \supset p)$

Other axioms he considers are two we have mentioned previously (p. 261) viz.

[277] Feys [1965], pp. 128–129.

[278] Åqvist [1964], p. 79. Also *vide infra*, p. 272.

[279] Sobociński [1964c] calls these systems, 'non-Lewis systems'.

[280] Sobociński [1964a], p. 77. Our Ka–Ke are his K1–K5. We have made this change to avoid any confusion with the names of the *systems*.

[281] This is the axiom added by McKinsey [1945], p. 92, to S4 to obtain the system he calls S4.1, which is thus Sobociński's K1.

G1 $MLp \supset LMp$
D2 $L(Lp \supset Lq) \vee L(Lq \supset Lp)$

as well as two further ones:

H1 $p \supset L(Mp \supset p)$
J1 $((p \prec Lp) \prec p) \supset p$

By adding these, or various combinations of them, either to S4 or to one of the extensions of S4 already mentioned, he produces the following systems:

K2: K1 + G1 [282]
K3: K1 + D2
K1.1: S4 + J1
K1.2: S4 + H1
K2.1: S4.2 + J1
K3.1: S4.3 + J1
K4: S4.4 + K2

He shows [283] that the K systems are related to each other and to some of the systems previously mentioned in the ways illustrated in the following diagram:

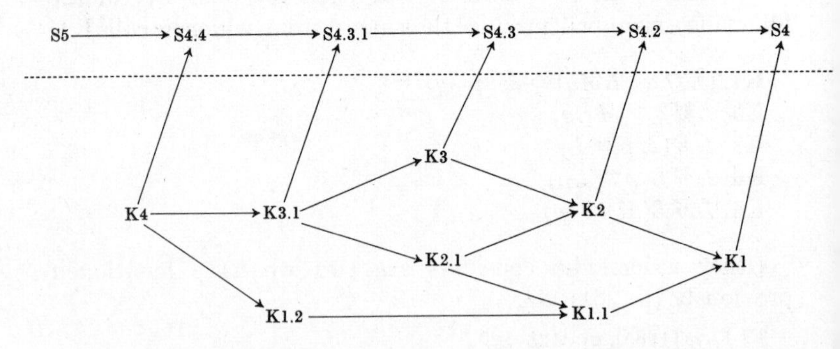

[282] Sobociński [1964a], p. 78. If we take the proper axiom for K1 in the form Kb ($LMp \supset MLp$), then K2 is S4 + ($LMp \equiv MLp$). K2 is the system mentioned in Parry [1939], p. 152 (*vide supra* p. 261 n). Sobociński shows that it is equivalent to the system S3 + ($LMp = MLp$). Prior [1964b] shows that K1 is contained in, but weaker than, K2.

[283] Sobociński [1964c], p. 317 and [1964b], p. 307.

The K systems neither contain nor are contained in S5, and in fact the addition of any of them to S5 would cause it to collapse into PC in the sense explained on p. 59.

We note in passing that there are formulae which if added to various other modal calculi would similarly collapse them into PC. We mention the following[284]:

$p \prec Lp$, $Mp \prec p$, and $LMp = Lp$ would each collapse S1 into PC.

$LMp \prec p$ (and hence $LMp \prec Lp$) would collapse S2 into PC.

Non-regular systems

All the systems we have so far considered, without exception, have had this feature, that the PC-transforms of all their theses are valid PC formulae. By 'the PC-transform of a formula' we mean that formula which results from (a) eliminating all occurrences of $\prec$ and $=$ by definitions or equivalences, and then (b) deleting all occurrences of L and M[285]. Systems possessing this feature have been called *regular*[286]. Establishing that a system is regular, of course, constitutes a consistency proof for that system (*vide* pp. 41f).

In this Section we consider a number of systems which are not regular in this sense.

We noticed in the previous chapter (p. 253) that none of the Lewis systems S1–S3 has any theses of the form $LL\alpha$, and that the addition to them of a thesis of that form leads to T or S4. But although no formula of the form $LL\alpha$ is a thesis of these systems, it is not the case that the negation of every such formula *is* a thesis. It has, however, been suggested that we might strengthen the systems so that this would be the case. The simplest way of doing so would be to add an axiom which Lewis[287] numbers C13:

C13 MMp

– for then if we want to derive any formula of the form $\sim LL\alpha$, we have only to substitute $\sim\alpha$ for p in C13 and apply LMI.

284 Cf. Feys [1965], pp. 131–132.

285 Or, equivalently, by replacing $\prec$ by $\supset$ and $=$ by $\equiv$ everywhere and deleting all occurrences of L and M.

286 Sobociński [1962b].

287 Lewis and Langford [1932], p. 497.

Lewis points out[288] that C13 is independent of each of the systems S1–S5, and incompatible with S4 and S5, but he does not develop any systems containing it. Still, as we have remarked, and as Lewis implicitly acknowledges, it could be consistently added to S1, S2 or S3; and it would certainly yield a non-regular system, for the deletion of its modal operators leaves us simply with p.

One thing which might lead us to add C13 is a wish to remove a disquieting feature of S1–S3 as they stand. Halldén has pointed out[289] that each of these systems contains theses of the form $(\alpha \vee \beta)$ which are such that (i) neither α nor β is itself a thesis, and (ii) α and β have no common variable. Now it would certainly seem strange to have $(\alpha \vee \beta)$ *valid* when neither α nor β is valid and they have no common variable; for by normal criteria of validity this would mean that (a) we could make a value-assignment to the variables in α which would falsify it, (b) this would not commit us to any particular assignment of values to the variables in β, (c) we could make an assignment which would falsify β, and yet (d) we could not falsify both α and β at the same time (which is what we should have to do to falsify $(\alpha \vee \beta)$). This seems paradoxical at the very least. In fact we could, following Halldén[290], define a *normal* interpretation of a system as one in which if any formula $(\alpha \vee \beta)$ is valid, and α and β have no common variable, then either α is valid or β is valid. It follows from all this that for any normal interpretation, if every theorem of S1–S3 is valid then these systems are incomplete. The incompleteness arises in this way. In the case of formulae of the type we have mentioned, we have $\vdash(\alpha \vee \beta)$; hence $(\alpha \vee \beta)$ is valid; hence (by normal interpretation) either α is valid or β is valid; but neither $\vdash\alpha$ nor $\vdash\beta$; therefore some valid formula is not a thesis.

An example[291] is:

$$(1)\quad MM(p \,.\, {\sim}p) \vee LL(q \supset q)$$

[288] *Ibid*, pp. 498–501. C13 is also incompatible with T.

[289] Halldén [1951].

[290] *Ibid.*, p. 127, though our account differs slightly from his. Halldén claims that Lewis' own interpretations of his calculi are normal.

[291] Halldén's own example is ${\sim}(M(p \,.\, {\sim}p) ⥽ (p \,.\, {\sim}p)) \vee (M(q \,.\, {\sim}q) ⥽ (q \,.\, {\sim}q))$.

We can prove this even in S1 as follows:

$\sim p \vee p\,[LL(p \supset p)/p]$:	(a) $\sim LL(p \supset p) \vee LL(p \supset p)$
PC × N(S1):	(b) $(p \supset p) = (q \supset q)$
(a), (b) × Eq:	(c) $\sim LL(p \supset p) \vee LL(q \supset q)$
(c) × Def ⊃, × LMI:	(d) $MM(p\,.\sim p) \vee LL(q \supset q)$

Q.E.D.

Neither disjunct of (1), however, is a thesis even of S3: the first because S3 is regular and hence cannot have any thesis whose PC-transform is $p\,.\sim p$; the second because S3 has no theses of the form $LL\alpha$.

If we are dissatisfied with this situation, there are two obvious ways out. One is to abandon S1–S3 in favour of T or some stronger system; for $LL(q \supset q)$ – the second disjunct in (1) – is an easily derived thesis of T, and hence the paradox will not arise in T[292]. The other is to extend S1, S2 or S3 by adding MMp (C13) as an axiom, for then $MM(p\,.\sim p)$ – the first disjunct in (1) – will be a thesis, and once more the paradox will not arise.

At any rate, we now have a plausible motive for adding C13 to the Lewis systems with which it is compatible. The system S2 + C13 has been called S6[293], and S3 + C13, S7[294]. S3 + $LMMp$ is known as S8.

Special interest attaches to the relations between S3, S7 and S4. We form S7 by adding C13 (which it is convenient here to take as MMq) to S3; and one way at least of obtaining S4 is by adding AS4.1 ($Lp \prec LLp$) to S3. Now C13 and AS4.1 represent what Lewis calls 'contrary assumptions'[295]. C13, as we have already observed, amounts to the assertion that no propositions are necessarily necessary. AS4.1, however, commits us to saying that some propositions *are* necessarily necessary; for it asserts that all necessary propositions are necessarily necessary, and in any system containing AS4.1 there must be some necessary propositions at least. So clearly we cannot add both C13 and AS4.1 to S3, though we could add either of them.

292 Cf. McKinsey [1953].

293 Alban [1943] p. 25. In S2 we have $MM(p\,.\sim p) \prec MMp$, and so $MM(p\,.\sim p)$ would do equally well as the special axiom of S6.

294 Halldén [1949a] p. 230. Lewis' ([1932] p. 498) Group I matrices (*vide* p. 242 *supra*) can be used to show the consistency of these systems.

295 Lewis and Langford [1932], p. 499.

S3, then, is neutral between two conflicting views: (a) the one reflected in S7, that no propositions are necessarily necessary (or that every proposition is possibly possible), and (b) the one reflected in S4, that if a proposition is necessary it is necessarily necessary. But although S3 does not commit us to holding either of these views it does commit us to holding that one of them is correct; for in S3 (though not in weaker systems) we have the following theorem, which is a simple consequence of TS3.7 (p. 235):

TS3.8 $MMq \vee (Lp \prec LLp)$

What this amounts to is that S3 requires that either view (a) or view (b) is correct, but does not tell us which is correct. (And there need be no harm in this omission, for we may well be hesitating between (a) and (b) and want a logic which says: 'The following are truths about logical necessity *whichever* of these two views is the correct one'.)[296]

All this suggests that the theorems of S3 are the formulae which hold independently of which of these two views we take. And in fact we can prove that the theorems of S3 are precisely those formulae which are theorems both of S7 and of S4; i.e. that $\vdash_{S3}\alpha$ iff both $\vdash_{S7}\alpha$ and $\vdash_{S4}\alpha$[297]. It is of course obvious that since S7 and S4 are both stronger than S3, if $\vdash_{S3}\alpha$ then $\vdash_{S7}\alpha$ and $\vdash_{S4}\alpha$; what needs to be proved is the converse, and we do so as follows.

We shall work with an axiomatization of S3 (such as that of Simons[298]) in which uniform substitution and material detachment are the only primitive transformation rules. We saw on p. 235 that the addition to S3 of any axiom of the form $LL\alpha$ produces a system at least as strong as S4; and if we choose as the extra axiom a formula which is in S4 anyway (such as

[296] The fact that TS3.8 is a thesis of S3 does something to explain Halldén's paradox mentioned above, at least as far as that system is concerned. We illustrated the difficulty by pointing out that $MM(p \,.\, {\sim}p) \vee LL(q \supset q)$ is a thesis, yet it consists of two non-theses with no variable in common. We can now see that if view (a) is correct, $MM(p \,.\, {\sim}p)$ holds, while if view (b) is correct, $LL(q \supset q)$ holds; hence it is not surprising that in S3, where we are committed to saying that one or other of these views is correct, the disjunction is a thesis.

[297] This result is due to Halldén [1949a], who however derives it in a slightly more general form.

[298] Simons [1953] and *vide supra* p. 244.

$LL(p \supset p)$), we obtain exactly S4. We can therefore take S4 as S3 + $LL(p \supset p)$, and this will prove more convenient here than our earlier axiomatizations. S7 is of course S3 + MMp.

We first prove two lemmas.

LEMMA I

If $\vdash_{S7}\alpha$ then $\vdash_{S3}{\sim}LL(p \supset p) \supset \alpha$

PROOF

The following are easily proved theses of S3 [299];

A. ${\sim}LL(p \supset p) \supset MMp$

B. ${\sim}LL(p \supset p) \supset {\sim}LL(q \supset q)$

The proof now proceeds by induction on the method of proving theorems in S7.

1. If α is an S3 axiom then by $p \supset (q \supset p)$ we have $\vdash_{S3}{\sim}LL(p \supset p) \supset \alpha$.

2. If α is the proper S7 axiom MMp then ${\sim}LL(p \supset p) \supset \alpha$ is simply A and therefore a theorem of S3.

3. For Modus Ponens: If α is obtained from two already proved wffs β and $\beta \supset \alpha$ then by the induction hypothesis $\vdash_{S3}{\sim}LL(p \supset p) \supset \beta$ and $\vdash_{S3}{\sim}LL(p \supset p) \supset (\beta \supset \alpha)$; whence by standard PC theorems $\vdash_{S3}{\sim}LL(p \supset p) \supset \alpha$.

4. For uniform substitution: Suppose that α is obtained by substitution of a wff, β, for some propositional variable in an S7 theorem, γ. By the induction hypothesis, $\vdash_{S3}{\sim}LL(p \supset p) \supset \gamma$. Now (i), if the variable for which the substitution is made is any other variable than p, the same substitution in ${\sim}LL(p \supset p) \supset \gamma$ will give $\vdash_{S3}{\sim}LL(p \supset p) \supset \alpha$. And (ii) if the substitution is made for p, the same substitution in ${\sim}LL(p \supset p) \supset \gamma$ will give $\vdash_{S3}{\sim}LL(\beta \supset \beta) \supset \alpha$; but by B we have $\vdash_{S3}{\sim}LL(p \supset p) \supset {\sim}LL(\beta \supset \beta)$; hence by Syll we have $\vdash_{S3}{\sim}LL(p \supset p) \supset \alpha$ in this case also.

Therefore the lemma holds for all theses of S7.

LEMMA II

If $\vdash_{S4}\alpha$, then $\vdash_{S3}LL(p \supset p) \supset \alpha$.

The proof is exactly parallel except that in place of A and B we use respectively:

C. $LL(p \supset p) \supset LL(p \supset p)$

D. $LL(p \supset p) \supset LL(q \supset q)$

[299] Actually A is a thesis of S2 and B of S1.

From these two lemmas we have the result that if $\vdash_{S7}\alpha$ and $\vdash_{S4}\alpha$ then $\vdash_{S3}\sim LL(p \supset p) \supset \alpha$ and $\vdash_{S3}LL(p \supset p) \supset \alpha$. Hence by $(\sim p \supset q) \supset ((p \supset q) \supset q)$, we have $\vdash_{S3}\alpha$. **Q.E.D.**

Two other related results are worth mentioning here though we shall not prove either of them.

1. We mentioned on p. 264 a system S3.1 obtained by adding to S3 the axiom $M(Lp \prec LLp)$. Halldén shows[300] that α is a thesis of S3 iff it is a thesis both of S8 and of S3.1.

2. In the same section (p. 265) we also referred to Åqvist's system S3.5, which is S3 + $Mp \supset LMp$; and we remarked that it is weaker than S5 but that the addition of the rule of necessitation would strengthen it to S5. Åqvist also forms a system which we shall call S9[301] by adding MMp to S3.5, and he proves that this system is consistent. S9 and S5 are related to S3.5 in the way S7 and S4 are related to S3; i.e., α is a thesis of S3.5 iff it is a thesis both of S5 and of S9.

At this point we might be inclined to make a similar conjecture about S6, T and S2; i.e., we might conjecture that α is a thesis of S2 iff it is a thesis both of S6 and of T. This is not so, however, for Åqvist shows[302] that the formula;

$$(1)\quad \sim LL(p \supset p) \vee LLL(q \supset q)$$

is not a thesis of S2, yet since $\vdash_{S6}\sim LL(p \supset p)$ and $\vdash_{T}LLL(q \supset q)$, (1) belongs both to S6 and to T.

Other modal predicate calculi

In Part II we considered predicate calculi based in various ways on T, S4 and S5 (and the Brouwerian system). It is of course possible to have predicate calculi based on the systems we have considered in the present chapter. In fact the first developments in modal predicate logic[303] were based on S2 and S4.

Such LPC's will of course differ from one another in the

[300] Halldén [1949a].

[301] Åqvist ([1964], p. 79) calls this system S7.5, but S9 seems a better name for it since it contains S8. This can be easily proved, since $[Mp/p]$ in $Mp \supset LMp$ gives $MMp \supset LMMp$ and this with MMp gives the S8 axiom $LMMp$.

[302] Op. cit., p. 82.

[303] Barcan [1946].

obvious respect that LPC substitution-instances of a propositional formula will sometimes be theorems and sometimes not, depending on the underlying propositional logic. But also, as we saw in the case of T, S4 and S5, the differences may sometimes amount to more than this.

For example, Ruth Barcan added the formula which bears her name to S2 in its strict form (viz., $M(\exists x)\alpha \prec (\exists x)M\alpha$), whereas it was sufficient to add its material form to T and stronger systems (pp. 142f) since the strict form could then be easily derived. Moreover, as we have seen (p. 145), this formula will be a theorem of an LPC based on any modal system which contains the Brouwerian system.

Again, when axioms for identity are added to LPC + S2[304], we can derive LI (p. 190) in its material form (viz. $(x = y) \supset L(x = y)$), though not in its strict form ($(x = y) \prec L(x = y)$).

Little work has, however, been done in this area and few results can be cited.

[304] Barcan [1947], [1950].

CHAPTER FIFTEEN

Validity and Decision Procedures for Various Systems

In Part I we gave definitions of validity for T, S4 and S5 in terms of semantic models, and on the basis of these models we were able to give decision procedures for these systems, using the method of semantic diagrams. In this we were for the most part either following or adapting the work of Kripke. Kripke has also shown how to apply essentially the same kind of analysis to various other modal systems, notably S2 and S3. In the present chapter we shall see how this is done. We shall set out the methods in fair detail for S2 and S3, and then indicate briefly how they can be modified to deal with S6–S8, S3.5, S9, S0.5, S4.3 and D. For many of the other systems we have mentioned, however – including even S1 – no satisfactory semantics seems yet to have been devised.

Validity in S2 and S3

One reason why we shall have to make a substantial change in our earlier definitions of validity if we are to deal with S2 and S3 is that these definitions – i.e. those for T, S4 and S5 – all satisfied the rule of Necessitation. That is to say, if any wff, α, is valid in terms of any of these definitions, $L\alpha$ is also valid. But as we have observed, the rule of Necessitation does not hold, at least unrestrictedly, in S2 and S3 (and in a number of other systems).

Another, related, feature of S2 and S3 which we have also noted is that they are compatible with (though they do not contain) the axiom MMp. This means that they are compatible with (though they do not commit us to) the view that every proposition is 'possibly possible'. And this suggests an idea which is in fact the key to Kripke's S2- and S3-models, that there might be some 'worlds' in which every proposition without exception – even one of the form $p \,.\, {\sim}p$ – is possible. Kripke[305] calls such worlds, *non-normal* worlds, and we shall follow him in this terminology. Worlds of the kind that occur in T-models are by contrast called *normal* worlds. The rules for evaluating non-modal formulae in non-normal world are the same as in normal worlds – thus even in a non-normal world we never have $V(p \,.\, {\sim}p) = 1$, for example – but for modal formulae in non-normal worlds $V(M\alpha)$ always has the value 1, and by consequence $V(L\alpha)$ always has the value 0 (i.e. in a non-normal world no propositions are necessary).

In an S2-model there must be at least one normal world, and there may (but need not) be one or more non-normal worlds. Every normal world is accessible to itself. Every non-normal world is accessible to at least one normal world. But no world (not even itself) is accessible to any non-normal world. Otherwise the accessibility relations can be as we please. In an S3-model there is the additional requirement that the accessibility relation be transitive. Thus S3-models stand to S2-models as S4-models do to T-models. A formula will be said to be S2-(S3-) valid iff it has the value 1 in every *normal* world in every S2-(S3-) model[306].

The point of the stipulation that a non-normal world should not be accessible to itself is this: we want to be able to say that in any world the value of $L\alpha$ or $M\alpha$ is determined by the value of α in the worlds accessible to it; but in a non-normal world we always have $V(L\alpha) = 0$ and $V(M\alpha) = 1$, irrespective of the value of α even in that world itself. In an S2- or S3-model, therefore, the accessibility relation is not unrestrictedly reflexive. It is, however, reflexive in the weaker sense that if any world whatever is accessible to a given world, w_i, then w_i is accessible to itself. Such a relation is often called a *quasi-reflexive* relation[307]. More

[305] Kripke [1965b], pp. 210–211. Some suggestions for interpreting the semantics which follow are given in Cresswell [1967b].

[306] If we define validity as truth in *every* world (whether normal or not) in every model, then we have a semantics for systems which contain no theses of the form $L\alpha$ (*vide infra*, pp. 302f).

[307] An everyday example of a quasi-reflexive relation is 'of the same length as'. Everything has the same length as itself – provided it has the same length as something (i.e. provided it has a length at all). Some things do not have a length at all, and *they* are not of the same length as themselves. So 'of the same length as' is not strictly reflexive; but it is quasi-reflexive.

exactly expressed, if W is a set of objects, then R is a quasi-reflexive relation over W iff for every $w_i \in W$, if there is some $w_j \in W$ such that $w_i R w_j$, then $w_i R w_i$. We can then define a normal world in W as a world, w_i, such that $w_i R w_i$.

We now set out the definition of S2-validity formally. $\langle W, R, V \rangle$ is an S2-model iff (a) W is a set of objects (worlds); (b) R is a quasi-reflexive relation over W; (c) if there is any $w_j \in W$ for which it is not the case that $w_j R w_j$, then there is some $w_i \in W$ such that $w_i R w_j$; (d) there is at least one $w_i \in W$ such that $w_i R w_i$; and (e) V is a value-assignment satisfying the following conditions:

1. For any propositional variable, p_j, and any $w_i \in W$, either $V(p_j, w_i) = 1$ or $V(p_j, w_i) = 0$.

2. For any wff, α, and any $w_i \in W$, $V(\sim\alpha, w_i) = 1$ if $V(\alpha, w_i) = 0$. Otherwise $V(\sim\alpha, w_i) = 0$.

3. For any wffs, α and β, and any $w_i \in W$, $V((\alpha \vee \beta), w_i) = 1$ if either $V(\alpha, w_i) = 1$ or $V(\beta, w_i) = 1$. Otherwise $V((\alpha \vee \beta), w_i) = 0$.

4. For any wff, α, and for any $w_i \in W$, $V(L\alpha, w_i) = 1$ if $w_i R w_i$ and for every w_j such that $w_i R w_j$, $V(\alpha, w_j) = 1$. Otherwise $V(L\alpha, w_i) = 0$. (The effect of this is that if w_i is normal, $V(L\alpha, w_i)$ is computed as in a T-model; but if w_i is non-normal, $V(L\alpha, w_i) = 0$ in every case – and hence, incidentally, $V(M\alpha, w_i) = 1$ in every case).

A wff, α, is S2-valid iff in every S2-model $\langle W, R, V \rangle$ $V(\alpha, w_i) = 1$ for every $w_i \in W$ such that $w_i R w_i$.

An S3-model is defined in the same way as an S2-model, except that we add the condition that R is transitive. A wff α is S3-valid iff in every S3-model $\langle W, R, V \rangle$ $V(\alpha, w_i) = 1$ for every normal $w_i \in W$.

For those who prefer the approach via the parlour games of Chapter 4, the S2-game is like the T-game but with the following modifications. Some of the sheets of paper are, say, white, others pink. In every S2-setting at least one player must have a white sheet; but some of the players may have pink sheets instead. No player with a pink sheet may see any other player, but every such player must be seen by at least one player with a white sheet. The rules for responding to calls are, for players with white sheets, exactly as in the T-game. For players with pink sheets, rules 1–3 are as in the T-game, but rules 4 and 5 (those covering calls with L and M) are replaced by the following:

4′. If a call is of the form $L\alpha$, do not raise your hand.

5′. If a call is of the form $M\alpha$, raise your hand.

A call is an S2-successful call iff in every S2-setting it would lead every player with a white sheet to raise his hand. A formula is S2-valid iff it would form an S2-successful call.

For the S3-game these same modifications are made to the S4-game; it follows that the S3-game will be the S2-game with the added rule that in every setting the seeing-relation must be transitive. S3-successful calls and S3-validity are then defined as above, with 'S3' replacing 'S2' throughout.

We now show that every thesis of the axiomatic system S2 is S2-valid by the above definition. It is convenient to use Lemmon's axiomatization of S2 (his system P2[308]).

We first prove that if α is a valid PC formula then $L\alpha$ is S2-valid. By conditions 1–3 for V, in every S2-model $V(\alpha, w_i) = 1$ for every $w_i \in W$. Now the only situation which would give $V(L\alpha) = 0$ in any normal world w_i is that there should be some world w_j such that $w_i R w_j$ and $V(\alpha, w_j) = 0$. But there is no such world. Hence for every normal world, w_i, $V(L\alpha, w_i) = 1$.

We have next to show that if α is one of the other axioms of P2, $L\alpha$ is S2-valid. Clearly it will be sufficient to show that in every model, $V(\alpha) = 1$ in every world, whether normal or not, for this will give $V(L\alpha) = 1$ in every normal world. These other axioms are $Lp \supset p$ and $L(p \supset q) \supset (Lp \supset Lq)$. For non-normal worlds we may simply note that in each case the antecedent is false (since it begins with L) and that therefore the whole implication is true. For normal worlds the proof then proceeds as for T (p. 68), with simple and obvious modifications.

We next prove that Becker's Rule, i.e.

$$\vdash L(\alpha \supset \beta) \rightarrow \vdash L(L\alpha \supset L\beta)$$

is S2-validity-preserving. We do this by showing that if $L(L\alpha \supset L\beta)$ is not S2-valid, neither is $L(\alpha \supset \beta)$. Suppose then

[308] *Vide supra*, p. 248. Kripke (op. cit., p. 208) uses a slightly different axiomatization based on an infinite (though effectively specifiable) set of axioms with material detachment as the only primitive rule of inference. His axiomatic basis may be easily shown equivalent to Lemmon's and for our purposes there is nothing to choose between them though, unlike the bases we are using, Kripke's basis allows the addition of $LL(p \supset p)$ to S2 without obtaining the unrestricted rule of Necessitation and permits an infinity of systems to be generated by the axioms $L_n(p \supset p)$ (for each n).

that $L(L\alpha \supset L\beta)$ is not S2-valid. This means that in some S2-model there is some normal w_i such that $V(L(L\alpha \supset L\beta), w_i) = 0$. Hence there is some w_j such that $w_i R w_j$ and $V(L\alpha \supset L\beta), w_j) = 0$. It follows that $V(L\alpha, w_j) = 1$ and $V(L\beta, w_j) = 0$. And it follows from this in turn, (a) that w_j is normal, and (b) that there is some world, w_k, such that $w_j R w_k$ and $V(\alpha, w_k) = 1$ and $V(\beta, w_k) = 0$. Hence $V((\alpha \supset \beta), w_k) = 0$, and so, since $w_j R w_k$, $V(L(\alpha \supset \beta), w_j) = 0$. But w_j is normal. Hence in some normal world $V(L(\alpha \supset \beta)) = 0$; i.e. $L(\alpha \supset \beta)$ is not S2-valid.

It is easy to see that the remaining primitive rules of P2, viz. uniform substitution and Modus Ponens, also preserve S2-validity. Hence every thesis of P2 (or S2) is S2-valid.

We may note in passing that provided there is at least one non-normal world (say w_j) in an S2-model then $LL(p \supset p)$ will be false in some normal world (say w_i) in that model. For since w_j is not normal, $V(L(p \supset p), w_j) = 0$; but there must be some normal w_i such that $w_i R w_j$, and hence $V(LL(p \supset p), w_i) = 0$. This shows that $LL(p \supset p)$ is not S2-valid and hence not an S2 theorem.

We now show that every thesis of S3 is S3-valid. We shall again use Lemmon's axiomatization (P3, p. 252). Clearly the additional requirement that R be transitive would not affect any of the above proofs for S2, so every S2 thesis is S3-valid. Therefore we have only to prove that $L(p \supset q) \supset L(Lp \supset Lq)$ is S3-valid and remains S3-valid when prefixed by L. As in the case of the S2 axioms, it will be sufficient if we prove that in every S3-model $V((L(p \supset q) \supset L(Lp \supset Lq)), w_i) = 1$ for every w_i, whether normal or non-normal. If w_i is non-normal, the result follows immediately from the fact that $V(L(p \supset q), w_i) = 0$. If w_i is normal we argue as follows. Suppose $V((L(p \supset q) \supset L(Lp \supset Lq)), w_i) \neq 1$. Then (a) $V(L(p \supset q), w_i) = 1$ and (b) $V(L(Lp \supset Lq), w_i) = 0$. From (b), since w_i is normal, there is some w_j ($w_i R w_j$) for which $V((Lp \supset Lq), w_j) = 0$, and therefore $V(Lp, w_j) = 1$ and $V(Lq, w_j) = 0$. Hence w_j is normal, and there is some w_k ($w_j R w_k$) for which $V(p, w_k) = 1$ and $V(q, w_k) = 0$, and hence (c) $V((p \supset q), w_k) = 0$. But R is transitive, so we have $w_i R w_k$. Hence from (c), $V(L(p \supset q), w_i) = 0$, and this contradicts (a). So the hypothesis that $V((L(p \supset q) \supset L(Lp \supset Lq)), w_i) \neq 1$ has been falsified.

We have therefore shown that every thesis of P3 (or S3) is S3-valid.

Decision procedures for S2 and S3

We now show how to modify the diagrams described in Chapters 5 and 6 in order to give decision procedures for S2 and S3[309].

In these diagrams we represented worlds by rectangles. In S2- and S3-diagrams we may have two types of worlds to represent, normal ones and non-normal ones. We shall use ordinary rectangles as before for normal worlds, and double-lined rectangles for non-normal ones.

In an S2-diagram for a formula α we are trying to construct a model in which α is false in some *normal* world. We therefore begin by placing α in an ordinary rectangle with 0 under its main operator, and evaluating and constructing subsidiary rectangles as in a T-diagram. The difference between an S2-diagram and a T-diagram is this: whenever we reach a rectangle (other than the first one) in which every formula beginning with L has the value 0 and every formula beginning with M has the value 1, we count the world in question as non-normal. Since in a non-normal world $L\alpha = 0$ and $M\alpha = 1$, irrespective of the value of α in that world or any other, there need be no asterisks in any double-lined rectangle, and no arrows go from any such rectangle to any other. As a result we are able to find falsifying S2-models for many formulae for which there are no falsifying T-models; such formulae are T-valid but not S2-valid.

A simple example will illustrate the method. Take the formula $LL(p \supset p)$. Its T-diagram is:

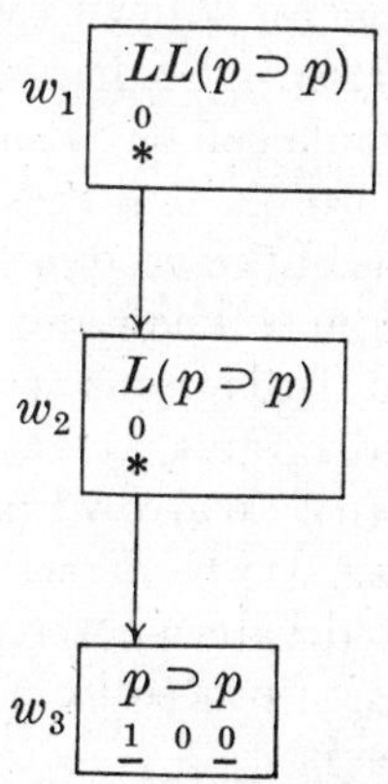

309 Material about the related semantic tableaux for S2 and S3 will be found in Kripke [1965b], p. 212 ff. Decision procedures of quite a different kind occur in Matzumoto [1960], Ohnishi [1961a], [1961b].

The inconsistency in w_3 shows $LL(p \supset p)$ to be T-valid. But its S2-diagram is:

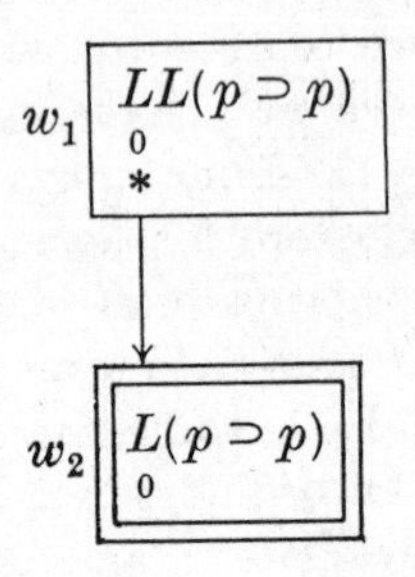

Here w_2 is non-normal, so no third rectangle is required and no inconsistency is reached. Hence $LL(p \supset p)$ is not S2-valid.

For the same kind of reason as in the case of T-diagrams, every S2-diagram will come to an end in a finite number of steps; and either an inconsistency will develop (in which case the original formula is valid), or else it will not (in which case the formula is not valid). Of course some formulae will require a number of alternative diagrams, but these are dealt with as explained for T on pp. 91 ff. Therefore the method can be applied to every wff.

Moreover this decision procedure can be used as the basis of a completeness proof for (axiomatic) S2. For by an argument of the same type as we gave for T in Chapter 5 we can show that whenever a complete system of S2-diagrams for a given formula is inconsistent, that formula can be obtained as a theorem of S2. Hence every S2-valid formula is a thesis of S2. We shall not, however, give the details of the proof here[310].

S3-diagrams differ from S2-diagrams in the same way that S4-diagrams differ from T-diagrams (*vide* Chapter 6). That is, S3-diagrams are like S2-diagrams, except that for any rectangles, w_i, w_j, w_k, whenever there is an arrow from w_i to w_j and one from w_j to w_k, an arrow must also be drawn from w_i to w_k and the appropriate formulae and values entered in w_k. Once more the method gives us a decision procedure, and we can base a completeness proof for S3 on it.

310 Completeness proofs for S2 and the immediately following systems, based on semantic tableaux, will be found in Kripke [1965b], pp. 214–217.

Validity in S6–S8

A number of other modal systems can be dealt with along similar lines. Of these we begin with S6(S2 + MMp), S7(S3 + MMp) and S8(S3 + $LMMp$)[311].

In an S2-model there may be non-normal worlds but there need not be any; in an S6-model there must be at least one. Every such world will of course be accessible to some normal world. If we define a non-normal world as before, we can then say that ⟨W, R, V⟩ is an S6-model iff (a) it is an S2-model and (b) there is some $w_j \in$ W such that it is not the case that w_jRw_j. It follows that in every S6-model ⟨W, R, V⟩ there is some normal $w_i \in$ W and some non-normal $w_j \in$ W, such that w_iRw_j. A wff, α, is S6-valid iff, in every S6-model, V(α, w_i) = 1 for every such w_i – i.e. for every (normal) world to which any non-normal world is accessible.

By the definitions of S2- and S6-models, S6-models are a subset of S2-models. Moreover, the w_i's in which it is required for S6-validity that V(α, w_i) = 1, being all normal, are a subset of those in which this is required for S2-validity. Hence if a formula is S2-valid it is also S6-valid, and so every thesis of S2 is S6-valid. We now show that the special S6 axiom MMp is S6-valid. Let ⟨W, R, V⟩ be any S6-model and w_j any non-normal world in W. Then since w_j is non-normal, V(Mp, w_j) = 1; hence for any w_i such that w_iRw_j, V(MMp, w_i) = 1. It is not difficult to show that the transformation rules preserve S6-validity, and this gives the result that every S6 thesis is S6-valid.

Models and validity for S7 are defined as for S6, but reading 'S3' for 'S2'. The proof that every S7 thesis is S7-valid proceeds analogously.

For S8 we have the following. ⟨W, R, V⟩ is an S8-model iff (a) it is an S3-model and (b) for *every* normal $w_i \in$ W there is a non-normal w_j such that w_iRw_j. Once more we say that a formula, α, is S8-valid iff, in every S8-model, V(α, w_i) = 1 for every normal w_i to which any non-normal w_j is accessible. Since, however, in an S8-model *every* normal world has a non-normal world accessible to it, we can simply revert to the formulation we used for S2 and S3, and say that a formula is S8-valid iff, in every S8-model, it has the value 1 in every normal world.

For the same reason as before, every S3-valid formula (and

[311] *Vide* Kripke, op. cit., p. 220.

hence every thesis of S3) is S8-valid. We now show that the special S8 axiom $LMMp$ is S8-valid. Clearly it will be sufficient if we prove that $V(MMp, w_i) = 1$ for every $w_i \in W$, whether normal or non-normal. If w_i is non-normal, it follows immediately that $V(MMp, w_i) = 1$. If w_i is normal, then there is some non-normal world w_j such that $w_i R w_j$. Since w_j is non-normal, $V(Mp, w_j) = 1$; hence $V(MMp, w_i) = 1$. Therefore $LMMp$ is S8-valid. Once more the transformation rules preserve validity, so every thesis of S8 is S8-valid.

Decision procedures for S6–S8

An S6-diagram is an S2-diagram with the proviso that an arrow must go from the first rectangle to a double-lined rectangle (representing a non-normal world). An S6-diagram need contain no more than one double-lined rectangle, but one must be connected to the *first* rectangle, since otherwise the value of the formula in the first rectangle will not count as far as S6-validity is concerned. A double-lined rectangle can be thought of as implicitly containing all those well-formed parts of the formulae in it which begin with L or M, evaluated as 0 or 1 respectively; but these values need not be written in if we remember that if the other rules force us to place a 1 under an L or a 0 under an M in a double-lined rectangle this counts as an inconsistency and shows the original formula to be valid.

An S7-diagram is an S3-diagram with a similar proviso. An S8-diagram is an S3-diagram but with the stronger proviso that an arrow must go from *every* single-lined rectangle to a double-lined one.

As simple illustrations we show by diagrams that MMp is not S2-valid but is S6-valid, and that $LMMp$ is not S6-valid but is S8-valid.

MMp is immediately falsified by the one-rectangle S2-diagram:

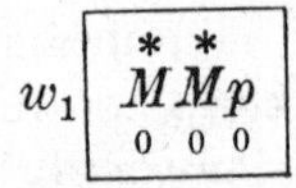

	* *
w_1	MMp
	0 0 0

This diagram is complete for S2, since no asterisks occur below any of the values and double-lined rectangles are not *required* in S2-diagrams. In an S6-diagram, however, we must have a double-lined rectangle connected to w_1. Hence the S6-diagram is:

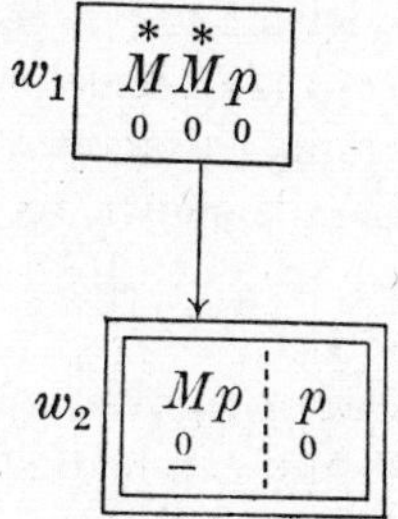

Here the asterisk above the first M in w_1 requires us to put $Mp = 0$ in w_2, but in a double-lined rectangle this counts as an inconsistency. Hence MMp is S6-valid.

Turning now to $LMMp$, we have the S6- and S7-diagram:

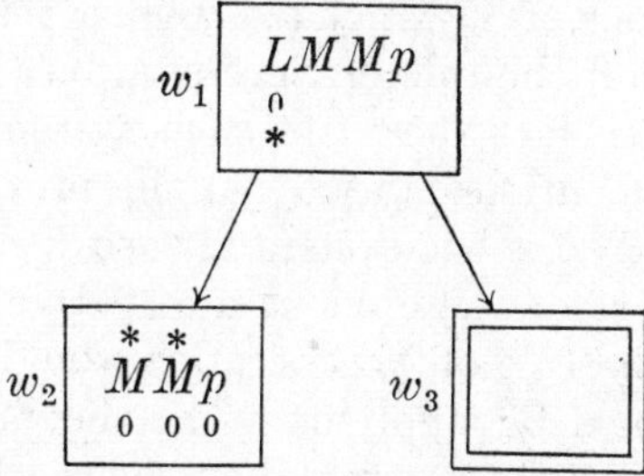

Here no inconsistency occurs. Having $MMp = 0$ in w_2 is sufficient to give $LMMp = 0$ in w_1; we do not have to have it in w_3 as well. Hence $LMMp$ is not S6-valid. In an S8-diagram, however, a double-lined rectangle must be connected not merely to w_1 but to w_2 as well. Hence we have for S8:

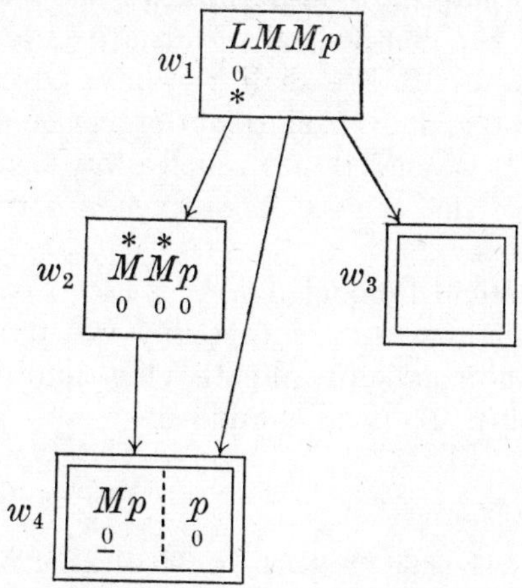

(Note that the transitivity of the accessibility relation in S8 requires an arrow from w_1 to w_4; this in fact makes w_3 unnecessary, but we have put it in to facilitate comparison with the previous diagram.) $LMMp$ is shown to be S8-valid because an asterisk in w_2 forces us to put $Mp = 0$ in the double-lined rectangle w_4.

The method of semantic diagrams provides a decision procedure for S6, S7 and S8, and can be developed into a completeness proof for these systems.

The systems S3.5 and S9

Modal logicians have from time to time speculated about whether there are extensions of S3 which are weaker than S5 but which would become S5 if the rule of Necessitation ($\vdash\alpha \rightarrow \vdash L\alpha$) were added to them. In fact there are such systems, and Åqvist's S3.5, which we mentioned briefly in the previous chapter, is one of them[312]. S3.5 is the system formed by adding to S3 the axiom $Mp \supset LMp$ (or the obvious alternative, $\sim Lp \supset L\sim Lp$). It is clear that the rule of Necessitation would strengthen this system to S5, since by applying it to the proper axiom we obtain $Mp \prec LMp$, which strengthens even S1 to S5. We are now in a position to investigate S3.5 from a semantical point of view.

The simplest way of defining S3.5-models is to dispense with the relation R altogether, and consequently with our earlier definition of normal and non-normal worlds. Instead we shall simply say that the non-normal worlds are an arbitrary subset (possibly empty) of W. We shall also have to give a different value-assignment rule for L, since our earlier one was expressed in terms of R, but we shall do so in such a way that once more $L\alpha$ will always have the value 0 in every non-normal world. We proceed as follows.

$\langle$W, N, V$\rangle$ is an S3.5-model iff (a) W is a set of objects (worlds); (b) N is a non-empty subset (possibly not proper[313]) of W; and (c) V is a value-assignment satisfying conditions 1, 2 and 3 for an S2-model (p. 276) and in addition:

[312] Åqvist [1964], pp. 82–84. *Vide* also *supra*, p. 265, and Cresswell [1967a].

[313] I.e. N may (though it need not) be the whole of W.

4′. For any wff α and any $w_i \in W$, $V(L\alpha, w_i) = 1$ if $w_i \in N$ and, for every $w_j \in W$, $V(\alpha, w_j) = 1$. Otherwise $V(L\alpha, w_i) = 0$.

(N is of course the set of normal worlds on our new account of them, and this last condition ensures that if w_i is non-normal, $V(L\alpha, w_i) = 0$, and consequently $V(M\alpha, w_i) = 1$.)

A formula, α, is S3.5-valid iff for every S3.5-model $\langle W, N, V\rangle$, $V(\alpha, w_i) = 1$ for every $w_i \in N$.

It is not difficult to show (though we omit the details here) that if it is impossible to falsify a formula, α, in any S3-model, it is also impossible to falsify α in any S3.5-model; i.e. that all S3-valid formulae (and hence all theses of S3) are S3.5-valid. We now show that the proper axiom of S3.5 – viz. $Mp \supset LMp$ – is S3.5-valid.

If $Mp \supset LMp$ is not S3.5-valid, then in some S3.5-model $\langle W, N, V\rangle$, there is some $w_i \in N$ for which (a) $V(Mp, w_i) = 1$ and (b) $V(LMp, w_i) = 0$. From (b), since $w_i \in N$, we have $V(Mp, w_j) = 0$ for some $w_j \in W$, hence $V(p, w_k) = 0$ for every $w_k \in W$. It follows that $V(Mp, w_i) = 0$, contrary to (a), and this falsifies the supposition that $Mp \supset LMp$ is not S3.5-valid.

Since the rules again preserve validity, we have the result that every thesis of S3.5 is S3.5-valid. A decision procedure and a completeness proof for S3.5 can be constructed along the same lines as before.

The proper S5 axiom $L(Mp \supset LMp)$, however, is not S3.5-valid, and hence S3.5 is weaker than S5. Let $\langle W, N, V\rangle$ be an S3.5-model in which $w_j \notin N$. Then $V(Mp, w_j) = 1$ and $V(LMp, w_j) = 0$, and hence $V((Mp \supset LMp), w_j) = 0$. Hence for any $w_i \in N$, $V(L(Mp \supset LMp), w_i) = 0$. To express this less formally: although in such a model $Mp \supset LMp$ is true in every normal world, it is false in every non-normal world, and this falsifies $L(Mp \supset LMp)$ in normal worlds.

S9 is S3.5 + MMp. As with the other systems containing MMp, we require that in every model there shall be some non-normal world. Thus we define an S9-model in the same way as an S3.5-model except that we replace (b) by (b′): N is a non-empty *proper* subset of W. Validity is then defined as for S3.5.

We show that MMp is S9-valid. In every S9-model there must be at least one $w_i \in N$ and at least one $w_j \notin N$. If $w_j \notin N$, then $V(Mp, w_j) = 1$; hence for every $w_i \in N$, $V(MMp, w_i) = 1$; i.e.,

MMp is S9-valid. Completeness can be proved as for the preceding systems.

The system S0.5

We can give an account of validity for S0.5 (p. 256) by modifying the account for S2. A non-normal world is again a world to which no world is accessible; in an S0.5-model, however, the value of $L\alpha$ in a non-normal world need not be 0, but can be either 1 or 0 (and hence so can the value of $M\alpha$). Otherwise everything is as for S2. We therefore define an S0.5-model $\langle W, R, V \rangle$ in the same way as an S2-model except that the value-assignment rule for L is replaced by:

For any wff, α, and any $w_i \in W$: (1) if w_i is normal, $V(L\alpha, w_i) = 1$ if for every $w_j \in W$ such that $w_i R w_j$, $V(\alpha, w_j) = 1$; otherwise $V(L\alpha, w_i) = 0$; (ii) if w_i is non-normal, either $V(L\alpha, w_i) = 1$ or $V(L\alpha, w_i) = 0$.

S0.5-validity is defined in the same way as S2-validity; i.e., α is S0.5-valid iff in every S0.5-model $\langle W, R, V \rangle$, $V(\alpha, w_i) = 1$ for every normal $w_i \in W$.

We show briefly that every thesis of S0.5 is S0.5-valid. If α is a valid wff of PC, clearly $V(\alpha, w_j) = 1$ for every $w_j \in W$; hence $V(L\alpha, w_i) = 1$ for every normal $w_i \in W$. For $Lp \supset p$: if w_i is normal and $V(Lp, w_i) = 1$, then since $w_i R w_i$, $V(p, w_i) = 1$; hence $V((Lp \supset p), w_i) = 1$. For $L(p \supset q) \supset (Lp \supset Lq)$: if for any normal $w_i \in W$, $V((L(p \supset q) \supset (Lp \supset Lq)), w_i) = 0$, then $V(L(p \supset q), w_i) = 1$, $V(Lp, w_i) = 1$ and $V(Lq, w_i) = 0$; hence there is some $w_j \in W$ for which $V((p \supset q), w_j) = 1$, $V(p, w_j) = 1$ and $V(q, w_j) = 0$; but this is impossible, even in a non-normal world. Since substitution and Modus Ponens clearly preserve validity, every thesis of S0.5 is S0.5-valid.

For S0.5-diagrams the simplest method is to write the formula in a single-lined rectangle, evaluate it in the usual way, and then make every other rectangle double-lined. In this way we never need to go farther than the second row of rectangles. In a double-lined rectangle (a non-normal world) we can have $L\alpha = 1$ or $L\alpha = 0$ irrespective of the value of α (and similarly with $M\alpha$). We cannot, however, have *both* $L\alpha = 1$ *and* $L\alpha = 0$ in the same rectangle (where α is exactly the same expression in the two cases); thus we cannot have in the one rectangle both $Lp = 1$ and $Lp = 0$, though we could have

$Lp = 1$ and $L{\sim}{\sim}p = 0$. A completeness proof can be developed without difficulty [314].

Some of the peculiarities of the system S0.5 can be brought out by comparing the three formulae:

(1) $L(p \supset {\sim}{\sim}p)$
(2) $LL(p \supset {\sim}{\sim}p)$
(3) $L(Lp \supset L{\sim}{\sim}p)$

Of these, (1) is valid in S0.5 but (2) and (3) are not. Using diagrams, we have, for (1):

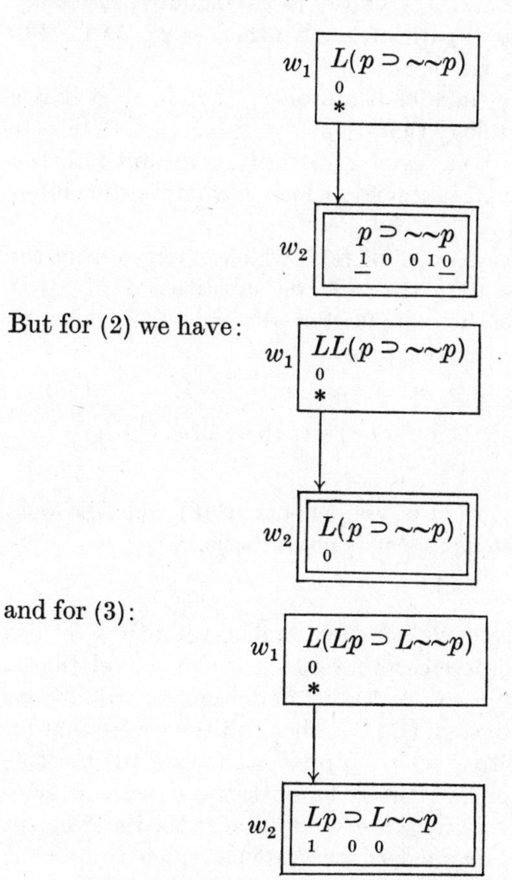

But for (2) we have:

and for (3):

[314] Cresswell [1966a]. We could in fact so define S0.5-models that each S0.5-model would contain only one normal world. If we did so the relation R would become redundant.

We should no doubt expect this result for (2), since S0.5 is weaker than S1 and S1 has no theses of the form $LL\alpha$; but (3) is a formula which *shows* S0.5 to be weaker than S1, since (3) is a thesis of S1. All these results, however, accord well with the interpretation which Lemmon suggests for the operator L in S0.5, namely as 'it is a truth-functional tautology that . . .'. On this interpretation, what (1) means is:

(4) It is a truth-functional tautology that $p \supset \sim\sim p$

—and this is true. But what (2) means is that (4) is itself a truth-functional tautology, and this is false; (4) is no doubt a necessary truth, but a necessary truth of some other kind. As for (3), the part in brackets will mean:

(5) If it is a truth-functional tautology that p, then it is a truth-functional tautology that $\sim\sim p$.

Once again, (5) is true, even necessarily true, but it is not itself a truth-functional tautology, which is what, on this interpretation, (3) says it is.

Another important feature of S0.5 which emerges from the invalidity of (3) is that the rule of substitution of strict equivalents does not hold in it. For the rule PCL gives us $\vdash L(p \supset p)$ and $[Lp/p]$ therein gives

(6) $L(Lp \supset Lp)$

while PCL also gives $\vdash L(p \equiv \sim\sim p)$ and therefore

(7) $(p = \sim\sim p)$

But applying substitution of equivalents to (6) and (7) would yield (3), which as we have seen is not a thesis.

Systems containing T

In Chapter 4 we defined T-models and T-validity. We also showed that if we impose certain restrictions on the relation R in T-models we obtain corresponding definitions of validity for S4, the Brouwerian system and S5. The conditions were that for S4 R must be transitive, for the Brouwerian system it must be symmetrical, and for S5 it must be both transitive and symmetrical. All these conditions are in addition to the requirement of reflexivity for R in models for systems which contain T. Non-normal worlds of course play no part in such models.

At least two of the other systems which contain T can also be dealt with by adding restrictions to R in T-models. We say that a relation R is *connected* (over a set, W) iff it holds, in one direction or the other, between every pair in W: i.e., iff for every $w_i, w_j \in W$, either $w_i R w_j$ or $w_j R w_i$[315].

One kind of model in which it would be natural to think of R as connected over W would be one in which the members of W were moments of time and R was the relation 'either contemporaneous with or earlier than'; for we normally think of the moments of time as all lying, so to speak, on a single straight line. And in fact if we require that R be reflexive, transitive and connected over W we obtain an account of validity for the 'temporal' system S4.3[316].

Having laid it down that R is transitive and connected over W, we may go on to add the stronger requirement that R be *discrete*. The intuitive idea here is that among the things that are related to an object, x, there is one, y, which is the 'next' or 'nearest' to x – i.e. between x and y there is no third thing related to x. A discrete relation we might think of as a step-by-step relation. For the semantic models we can express the additional requirement here as follows: for every $w_i \in W$ there is some $w_j \in W$ such that (a) $w_i R w_j$ and (b) for any $w_k \in W$, if $w_i R w_k$ and $w_i \neq w_k$, then $w_j R w_k$ and $w_j \neq w_i$. (Prior's illustration of time conceived as a series of moments may again help to make the sense of this definition clearer. We can take $w_i R w_j$ to mean that w_j is the same as or later than w_i. Here R is clearly transitive, and connected over the set of moments. To say that R is also discrete then amounts to this: that for any moment there is a second moment which is later than it, and any moment later than the first either is later than the second too, or else *is* the second. This makes precise the idea that after each moment there is a next one.) If we now make this additional requirement that R be discrete, we have a definition of validity for the system D (*vide* pp. 262f).

The method of semantic diagrams can easily be adapted to the system S4.3. As an illustration, we show how to prove the validity

315 A relation of this kind is sometimes called *well-connected* or *strongly-connected*, to distinguish it from the case where we merely make the weaker requirement that the relation holds between any two *different* elements of W – i.e. if we use the definition that for every $w_i, w_j \in W$ if $w_i \neq w_j$ then either $w_i R w_j$ or $w_j R w_i$.

316 *Vide supra* p. 261f and also Kripke [1963a], p. 95. The completeness of S4.3 is proved (algebraically) in Bull [1965a]. *Vide* also Prior [1962].

of $L(Lp \supset Lq) \vee L(Lq \supset Lp)$ when R is connected (as well as reflexive and transitive).

The T-diagram for this formula (which shows it to be invalid in T) is:

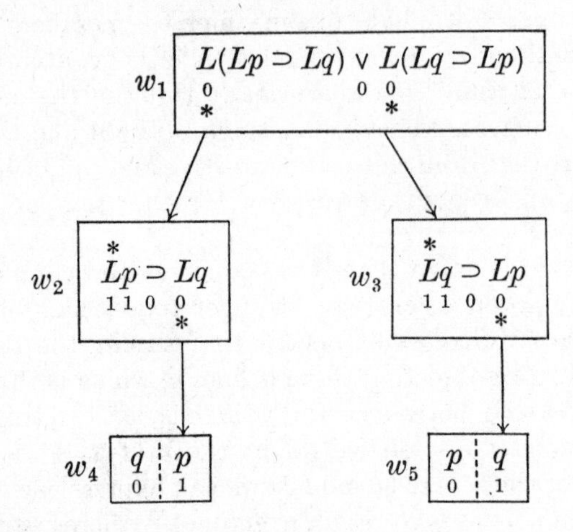

(and clearly this is the only kind of T-diagram which could falsify the formula).

If, however, we require that R be connected, then one thing that will follow is that we must have either w_4Rw_5 or w_5Rw_4. If we have the former then (since R must also be transitive) we have w_2Rw_5, and hence p must be assigned 1 in w_5, thus making it inconsistent. If we have the latter, then for a similar reason q must be assigned 1 in w_4, making *it* inconsistent. So in either case w_1 will count as inconsistent and we therefore have the S4.3-validity of $L(Lp \supset Lq) \vee L(Lq \supset Lp)$.

Semantical study of systems containing T typically proceeds by imposing conditions on the relation R. It should be clear that these conditions on R can frequently be reflected by rules for the construction of the appropriate diagrams (or semantic tableaux).

The existence postulate

For all the systems we have discussed semantically, except for S6–S9, we can have models in which W has only one member. It is this feature which makes them 'regular', in the sense of p. 267, for a one-world model is a model in which Lp and p have the same truth-value and in which $p \supset Lp$ is true. We have generally considered it sufficient for the preservation of the modal status of a system that it should not collapse into PC (p. 59), but we could have insisted on a stronger requirement, viz. that the system should contain a formula which is incompatible with collapsing into PC. The reason for doing so would be the conviction that there are propositions which are true but not necessarily true, and that this is a fact of logic and should find expression in modal systems.

In Lewis and Langford[317] use is made of quantifiers over propositional variables. Thus we could have

$$(1)\ \sim(p)(p \supset Lp)$$

to express the fact that not every true proposition is necessary. In Lewis and Langford the formula

$$(2)\ (\exists p)(\exists q)(\sim(p \prec q) \,.\, \sim(p \prec \sim q))$$

is used and this can easily be shown to be equivalent to (1) (when suitable quantificational axioms are added). Lewis calls (2) the 'existence postulate' and regards it as a possible addition to S1 or S2, though we can also regard it as an addition to any of the other systems. Semantically the rule for the universal quantifier over propositions is:

[V$\forall p$]. For any wff, α, any propositional variable, p, and any $w_i \in W$, $V((p)\alpha, w_i) = 1$ if for every V' differing from V only in assignment (in any $w_i \in W$) to p, $V'(\alpha, w_i) = 1$; otherwise $V((p)\alpha, w_i) = 0$[318].

[317] Lewis and Langford [1932], pp. 178–198.

[318] Quantifiers over propositional variables can be introduced into the ordinary propositional calculus (*vide* Church [1956], pp. 151–154) and into the non-modal LPC (*ibid.*, pp. 296–300). In these systems we can prove $(p)\alpha \equiv (\beta \,.\, \gamma)$, where β is α with some standard true formula, say $(p \supset p)$, everywhere in place of free p in α and γ is α with some standard false formula (say $(p \,.\, \sim p)$) everywhere in place of free p in α. Thus $(p)\alpha$ here may be regarded simply as an abbreviation for $(\beta \,.\, \gamma)$. Systems in which this can be done are often called *extensional* systems and systems in which it cannot, *intensional* systems. Modal systems are intensional since we can easily show that $(p)(p \supset Lp) \equiv (((p \supset p) \supset L(p \supset p)) . ((p \,.\, \sim p) \supset L(p \,.\, \sim p)))$ is not a theorem of any of the modal systems we have discussed. (The right-hand side is a theorem of all of them but if the left-hand side were a theorem we should have, by $\forall 1$, $\vdash p \supset Lp$, and the system would collapse into PC.) For a semantical study of an intensional logic with quantification over propositional variables

It can be seen that [V∀p] reflects the meaning of the universal quantifier when applied to propositions. For it says that $(p)\alpha$ will be true in w_i iff, whatever the truth-value of p in w_i or in other worlds in the model (the value of p in other worlds than w_i can of course affect the value of α in w_i), α is true in w_i.

With this rule $\sim(p)(p \supset Lp)$ will be true in any world w_i in a model provided there is at least one other world, w_j, in that model (such that $w_i R w_j$). For if there is then, given the assignment V in the model we can define an assignment V′ which is like V except that $V'(p, w_i) = 1$ and $V'(p, w_j) = 0$, from which we have $V'((p \supset Lp), w_i) = 0$ and hence by [V∀p], $V((p)(p \supset Lp), w_i) = 0$, i.e. $V(\sim(p)(p \supset Lp), w_i) = 1$. And conversely if there is in the model a world w_i with no other w_j accessible to it the formula will be falsified. Thus the condition (that there be a distinct world accessible to any (normal) world in the model) is equivalent to the validity of the existence postulate.

One could have analogous, more complicated formulae whose validity would be equivalent to insisting on at least 3, 4, . . . or an infinite number of worlds; but an investigation of this would lead us far beyond the present chapter, and indeed far beyond the present stage of development of the subject.

vide Cresswell [1966b], [1967c]. Modality can be treated extensionally if modal operators are regarded as predicates designating properties of sentences or formulae; *vide* Montague [1963], Löb [1966]. cf. also Fitch [1937], [1939], [1948] and Lewis [1968].

CHAPTER SIXTEEN

Non-standard Systems

This chapter contains a hotch-potch of systems, united – if that is the word – only by the fact that in one way or another they fail to satisfy the conditions listed at the beginning of Chapter 14. For the most part we shall content ourselves with outlining the systems and shall not attempt any proofs.

Systems with primitives other than *L* or *M*

As is well known, in ordinary non-modal, two-valued PC we can within certain limits choose which operators we shall take as primitive. Thus $\sim$ and $\vee$, $\sim$ and $.$, and $\sim$ and $\supset$, form equally satisfactory pairs of primitives. We can even take a single operator as primitive and define all the rest in terms of it. The first such single primitive to be discovered was the stroke of 'alternative denial', often called the *Sheffer stroke*[319]: p/q is to be understood as having the value 1 except where p and q both have the value 1; it is thus to be understood in the same way as $\sim p \vee \sim q$ in the standard interpretation.

All the modal systems we have so far examined have had either *L* or *M* as primitive and the other modal operators defined in terms of this one (together with non-modal operators). We can, however, have other modal primitives than these, at least in many systems. It is even possible to have only a single (triadic) operator in terms of which all the other operators, *whether modal or truth-functional*, can be defined. Halldén[320] gives as primitive an operator $[p, q, r]$ which is to be understood as meaning: either p is false, or q is false, or r is impossible – i.e. as what we should ordinarily mean by $(\sim p \vee \sim q \vee \sim Mr)$. He shows that in

[319] Sheffer [1913]. The only other dyadic operator in terms of which *all* other truth-functions can be defined is 'joint denial', which may be written as $\downarrow$. $p \downarrow q = 0$ except where both $p = 0$ and $q = 0$.

[320] Halldén [1949b]. Cf. Alban [1943], Massey [1966], pp. 599–602, [1967].

all systems stronger than S1 the following definitions are then satisfactory:

$$\sim\alpha =_{\mathrm{Df}} [\alpha, \alpha, \alpha]$$

$$(\alpha \,.\, \beta) =_{\mathrm{Df}} \sim[\alpha, \beta, [\alpha, \sim\alpha, \alpha]]$$

$$M\alpha =_{\mathrm{Df}} \sim[[\alpha, \sim\alpha, \alpha], [\alpha, \sim\alpha, \alpha], \alpha]$$

and of course we already know that all other operators in the systems in question can be expressed in terms of the three just defined.

An operator such as Halldén's may, however, be felt to be somewhat artificial; at any rate we shall not pursue this theme any further.

We noted previously that one of Lewis' original proposals was to use strict disjunction as a primitive [321]. A more popular dyadic primitive, however, has been strict implication. If one has a primitive operator of this kind one thing one can do is of course to try to define others in terms of it, but another is to see how much one can do with it on its own. The fragment of (non-modal) PC in which material implication is the only operator is often called the calculus of *pure (material) implication* [322]. Analogously, if we start from a modal system, S, in which $\prec$ occurs, we can form its *pure strict implication* fragment, which will contain all and only those theses of S in which $\prec$ is the sole operator. Of course, if we start from S5 we shall obtain a different calculus of pure strict implication from the one we would obtain from say T or S4.

In 1956 a group of logicians carried out an investigation of pure strict implicational calculi [323]. They gave the names C1–C5 to the implicational fragments of S1–S5 respectively, and CT to the analogous fragment of T, and we shall follow their terminology here. They concentrated mainly on C5, for which one of the group, C. A. Meredith, gave a number of alternative axiomatizations [324]. One of these is the following:

[321] Lewis [1912].

[322] Or sometimes, after the material implication symbol (C) in the Polish notation, *C-pure*.

[323] Lemmon, Meredith, Prior and Thomas [1957].

[324] *Ibid.*, p. 8; also Meredith [1956a], p. 9.4; Meredith and Prior [1964].

Axioms

M1[325] $q \prec (p \prec p)$

M2 $(((p \prec q) \prec r) \prec q) \prec ((q \prec s) \prec (p \prec s))$

Rules: Substitution and (strict) detachment.

It is of some interest that in S5 we can define both L and M, as follows[326]:

$$L\alpha =_{\text{Df}} ((\alpha \prec \alpha) \prec \alpha)$$

$$M\alpha =_{\text{Df}} ((\alpha \prec L\alpha) \prec L\alpha)$$

What we mean by saying that these definitions can be given in S5 is that *in S5* the thesishood or otherwise of a formula remains unaffected if we replace $L\alpha$ anywhere by $(\alpha \prec \alpha) \prec \alpha$, or vice versa; and analogously for M. In S4 the same definition will work for L, but not the one for M. In S3 and weaker systems neither definition will do.

Hacking has shown[327] that the following sets of axioms (again with substitution and detachment as the only rules) suffice for C3–C5.

Axioms common to the three systems:

1. $p \prec p$
2. $(p \prec (q \prec r)) \prec ((p \prec q) \prec (p \prec r))$

For C3, add to 1 and 2:

C3.1 $(q \prec r) \prec ((p \prec q) \prec (p \prec r))$

C3.2 $\alpha \prec (\beta \prec \alpha)$, if α and β are both *strict* formulae, i.e. have the form $(\gamma \prec \delta)$ where γ and δ are wffs.

For C4, add to 1 and 2:

C4.1 $\alpha \prec (p \prec \alpha)$, if α is strict.

For C5, add to 1, 2 and C4.1:

C5.1 $((\alpha \prec p) \prec \alpha) \prec \alpha$, if α is strict.

Having established these C-systems, as we may call them, we can if we wish extend them by adding further operators, together with appropriate axioms. Prior gives the following basis[328] for

[325] The numbering is that of Prior [1961], p. 61.

[326] Meredith [1956a], p. 1.1; Meredith and Prior [1964], p. 204.

[327] Hacking [1963]. *Vide* also Beth and Nieland [1965], pp. 22 and 24.

[328] Prior [1961], [1963a].

C5 + C-pure (i.e. the fragment of S5 in which the only operators are $\prec$ and $\supset$, both being primitive):

Axioms

C′5.1 $(((p \prec q) \prec r) \prec s) \prec ((q \prec s) \prec (p \prec s))$
C′5.2 $p \prec (q \supset p)$
C′5.3 $(p \prec (p \supset q)) \prec (p \prec q)$
C′5.4 $(p \prec q) \prec (p \supset q)$

Rules

R1. Strict detachment.
R2. $\vdash(\alpha \supset \beta) \rightarrow \vdash(\alpha \prec \beta)$.
R3. Uniform substitution for variables.
R4. Uniform substitution of $\supset$ for $\prec$ in any thesis.

We can obtain the full S5 by adding to this basis a constant false (impossible) proposition, 0, and the axiom:

C′5.5 $0 \prec p$

(All the other operators of S5 in its usual form can then be introduced by definitions.)

A similarly constructed basis for S4 [329] is obtained by replacing C′5.1 and C′5.2 by

C′4.1 $(p \prec q) \prec (s \prec ((q \prec r) \prec (p \prec r)))$
C′4.2 $((p \supset q) \supset p) \prec p$

Hacking also axiomatizes C2 and CT [330]. And for each of the full systems S2–S5 and T he gives bases which start with axioms containing only $\prec$ and then add others for $\sim$, . and $\vee$ [331]. (In the case of S3–S5, but not S2 or T, these axioms for $\prec$ yield precisely the corresponding C-systems.)

Another modal operator which can be taken as primitive is strict equivalence. No work appears to have been done on calculi of pure strict equivalence, but as early as 1937 Huntington gave a set of axioms (in algebraic form) for S2 [332], taking strict equivalence as primitive and using the definition:

$$L\alpha =_{\text{Df}} (\alpha = (\alpha \vee \sim\alpha))$$

329 Prior [1963a], p. 3.
330 Hacking [1963], pp. 67–8.
331 Op. cit., pp. 51, 66–67.
332 Huntington [1937].

A system equivalent to S4, with $=$ as the only primitive modal operator and L defined in a way very similar to Huntington's, is the following [333]. We assume an axiomatization of PC containing the constant false proposition 0, and the definition

$$1 =_{\text{Df}} (0 \supset 0)$$

To this we subjoin: $=$ as primitive; the definition:

$$L\alpha =_{\text{Df}} (\alpha = 1)$$

the axiom schema:

I. $(p = q) \supset (\alpha \supset \beta)$, where β differs from α only in having q in some of the places where α has p; and the rule:

R. $\vdash(\alpha \equiv \beta) \rightarrow \vdash(\alpha = \beta)$ [334]

We remarked earlier that the Sheffer stroke ($/$) of alternative denial can be used as the sole primitive operator for PC. Its modal analogue, strict alternative denial, can be symbolized by a double stroke ($//$): $p//q$ is to be understood as we should normally understand $\sim M(p \,.\, q)$. Systems have been put forward in which this has been taken as primitive and the other operators defined as follows:

$$(\alpha \prec \beta) =_{\text{Df}} \alpha//(\beta/\beta)$$
$$(\alpha \supset \beta) =_{\text{Df}} \alpha/(\beta/\beta)$$
$$\sim\alpha =_{\text{Df}} \alpha/\alpha$$
$$(\alpha \vee \beta) =_{\text{Df}} (\alpha/\alpha)/(\beta/\beta)$$
$$(\alpha \,.\, \beta) =_{\text{Df}} (\alpha/\beta)/(\alpha/\beta)$$
$$L\alpha =_{\text{Df}} (\alpha/\alpha)//(\alpha/\alpha)$$
$$M\alpha =_{\text{Df}} (\alpha//\alpha)/(\alpha//\alpha)$$

Using these definitions, Ishimoto gives axioms and rules for S3 and S4 [335].

333 Cresswell [1965], where it is suggested that in this system $=$ could be interpreted as a sign of propositional identity, and Lp taken to mean 'p is identical with some provable truth'.

334 Without the rule R various theorems of S4 – including even $L(p \supset q) \supset (Lp \supset Lq)$ – cannot be derived. An early attempt to define necessity by the method used in this system, that of Bronstein and Tartar [1934] is (as was remarked by McKinsey [1934a]) unsatisfactory for this reason.

335 Ishimoto [1954], [1956a], [1956b]. For a summary of the axioms and rules *vide* Feys [1965], pp. 90–91.

Another possible primitive is *contingency*. Montgomery and Routley[335a] write ∇p for 'it is contingent that p' (i.e. for what would usually be understood by $\sim Lp \,.\, Mp$), and investigate some bases in which this is primitive, as well as some in which non-contingency is primitive.

The system E

The systems so far discussed in this chapter have been attempts to give bases, using different primitives from the usual ones, for already existing systems of modal logic or for fragments of these. They have thus not involved any radical departure from the 'standard' systems we discussed earlier. Some logicians, however, have felt that even the very weakest of these standard systems contain theses which are in some way objectionable, and have set out to construct systems without them. One source of dissatisfaction has been the so-called 'paradoxes of strict implication'. Even in S0.5 we have the following theses:

$$(1)\ q \prec (p \vee \sim p)$$
$$(2)\ (p \,.\, \sim p) \prec q$$
$$(3)\ Lp \supset (q \prec p)$$
$$(4)\ \sim Mp \supset (p \prec q)$$

and in S2 we can also prove

$$(5)\ Lp \prec (q \prec p)$$
$$(6)\ \sim Mp \prec (p \prec q)$$

In the eyes of some logicians and philosophers, theses such as these make it impossible to interpret $\prec$ as 'entails' – i.e. to regard $p \prec q$ as meaning 'q logically follows from p' – or at least prevent the systems in which they occur from being adequate calculi of entailment[336].

Various attempts have been made to construct systems in which (1)–(6) and related formulae are not theses but in which there is a dyadic operator which behaves in other respects as $\prec$ does in one or other of the standard modal calculi. Of these the one with greatest formal interest is one based on a relation which Ackermann called *strenge Implikation*[337], and Anderson and

[335a] Montgomery and Routley [1966].

[336] For a brief account of the philosophical issues raised here, *vide* Appendix 2, pp. 335–339.

[337] Ackermann [1956].

Belnap, who developed the system set out below, simply call *entailment*[338]. Anderson and Belnap use the primitive symbol $\rightarrow$ to represent this relation, and give the following basis for the *pure calculus of entailment*[339]:

Axioms

- **E₁1** $p \rightarrow p$
- **E₁2** $(p \rightarrow q) \rightarrow ((q \rightarrow r) \rightarrow (p \rightarrow r))$
- **E₁3** $(p \rightarrow q) \rightarrow ((r \rightarrow p) \rightarrow (r \rightarrow q))$
- **E₁4** $(p \rightarrow (p \rightarrow q)) \rightarrow (p \rightarrow q)$
- **E₁5** $((p \rightarrow p) \rightarrow q) \rightarrow q$.

Rules: Uniform substitution; Modus Ponens (for $\rightarrow$). These axioms have been shown to be mutually independent in the pure calculus[340].

L can be introduced into this system by a definition analogous to the one in C4 or C5:

$$L\alpha =_{\text{Df}} ((\alpha \rightarrow \alpha) \rightarrow \alpha)$$ [341]

Anderson and Belnap's full *System E* contains truth-functional operators as well as $\rightarrow$ and *L* (the latter defined as above). It is derived from the following fourteen axioms, with substitution, Modus Ponens (for $\rightarrow$) and adjunction as the transformation rules[342].

For entailment

- **E1** $((p \rightarrow p) \rightarrow q) \rightarrow q$
- **E2** $((p \rightarrow q) \rightarrow ((q \rightarrow r) \rightarrow (p \rightarrow r))$
- **E3** $(p \rightarrow (p \rightarrow q)) \rightarrow (p \rightarrow q)$

For conjunction

- **E4** $(p \,.\, q) \rightarrow p$
- **E5** $(p \,.\, q) \rightarrow q$
- **E6** $((p \rightarrow q) \,.\, (p \rightarrow r)) \rightarrow (p \rightarrow (q \,.\, r))$

338 For a bibliographical survey *vide* Grover [1966].

339 Anderson and Belnap [1962b], pp. 39–40, where however the authors use axiom-schemata.

340 Anderson, Belnap, Wallace [1960].

341 Anderson and Belnap [1959a], p. 109. Ackermann [1956] had introduced a constant impossible proposition ('das Absurde'), but Anderson and Belnap show this to be redundant.

342 Anderson and Belnap [1962a], p. 14; Anderson [1963], p. 9.

For necessity and conjunction

E7 $(Lp \,.\, Lq) \rightarrow L(p \,.\, q)$

For disjunction

E8 $p \rightarrow (p \vee q)$
E9 $q \rightarrow (p \vee q)$
E10 $((p \rightarrow r) \,.\, (q \rightarrow r)) \rightarrow ((p \vee q) \rightarrow r)$

For conjunction and disjunction

E11 $(p \,.\, (q \vee r)) \rightarrow ((p \,.\, q) \vee r)$

For negation

E12 $(p \rightarrow \sim p) \rightarrow \sim p$
E13 $(p \rightarrow \sim q) \rightarrow (q \rightarrow \sim p)$
E14 $\sim\sim p \rightarrow p$

In this system the 'paradoxical' formulae (1)–(6) – with $\rightarrow$ replacing ⥽ throughout – are not derivable. Another formula which is not provable is one which expresses the 'disjunctive syllogism', viz.

$$((p \vee q) \,.\, \sim p) \rightarrow q$$

In fact, granted the rest of the system, it could be said that it is the absence of the disjunctive syllogism which prevents the 'paradoxes' from being theorems[343]. E does, however, contain the full propositional calculus, where $\supset$ is defined in the usual manner in terms of other truth-functional operators[344].

Among the important theses of E are[345]:

$$(p \rightarrow q) \rightarrow (Lp \rightarrow Lq)$$
$$(p \rightarrow q) \rightarrow L(p \rightarrow q)$$
$$Lp \rightarrow p$$
$$Lp \rightarrow LLp$$

[343] Those who regard the disjunctive syllogism as a sound principle of entailment will of course regard its absence as telling against the claim of E to be an adequate calculus of entailment: of these some would be prepared to have the 'paradoxes' if that is the price to be paid for the disjunctive syllogism; and some would regard the 'paradoxes' themselves as sound principles of entailment, and so take even *their* absence as a ground of complaint against E. But again, *vide* pp. 335–339.

[344] Anderson and Belnap [1959b], p. 302.

[345] Anderson and Belnap [1962b], p. 44.

It is also possible to derive the rule $\vdash \alpha \rightarrow \vdash L\alpha$[346]. Taken together these are strongly reminiscent of S4. In fact E can be shown to be a proper sub-system of S4, and to have the same irreducible modalities as S4 has[347].

Syntactical definitions of completeness can be given for E (in terms of natural deduction[348]), and its completeness relative to these definitions can be proved.

Systems without the axiom of necessity

One way in which a system could deviate from the conditions at the beginning of Chapter 14 is by not having $Lp \supset p$ as a thesis. Of course, so long as we wish to interpret L as a necessity operator we are not likely to want to dispense with this thesis, since one of the basic things we want to say about necessity is that whatever is necessarily true *is* true. There are, however, other possible interpretations for L, some of which would make it reasonable to abandon $Lp \supset p$. For example, it has been suggested that we might take Lp to mean 'it ought to be the case that p' or 'it is obligatory that p'. The companion interpretation for Mp would be 'it is permissible that p'. The standard relations between L and M would then be preserved; for to say that something is obligatory is to say that it is not permissible that it be omitted, and hence we have $Lp \equiv {\sim}M{\sim}p$. But on this interpretation we should not want to have either $Lp \supset p$, which would mean that whatever is obligatory is the case, or $p \supset Mp$, which would mean that whatever is the case is permissible. On the other hand, $Lp \supset Mp$ – 'whatever is obligatory is permissible' – remains a sound principle; and in fact many systems of modal logic which omit $Lp \supset p$ replace it by the weaker thesis $Lp \supset Mp$[349]. A logic in which the modal operators (or some of them) are given a

[346] Grover [1966], p. 30.

[347] Anderson and Belnap [1959a], p. 108. For a version of E which corresponds to T, *vide* Goble [1966], pp. 205–207.

[348] Anderson [1960]. For a brief account of natural deduction and its application to modal logic *vide* Appendix 1 *infra* pp. 331–334.

[349] Cf., e.g., Lemmon [1957]. Systems which omit $Lp \supset p$ can be studied semantically by following a suggestion of Kripke [1963a] (p. 95) and dropping reflexiveness from the conditions for R. Provided that we then require that for every world there shall always be one which is accessible to it (though this need not be itself), we shall obtain semantics for systems in which $Lp \supset p$ is not simply omitted but replaced by $Lp \supset Mp$.

moral interpretation of this kind is often called a *deontic* logic[350]. Another meaning we might give to Lp which would for obvious reasons also lead us to reject $Lp \supset p$ is 'x believes that p'.

Systems obtained from S1–S5 by simply dropping the axiom of necessity (and not replacing it by any weaker thesis) have been studied by Feys[351] and called by him S1⁰–S5⁰. Their formal properties are very similar to those of the full S1–S5, except as regards the relations between modalized and unmodalized propositions.

Systems without the axiom of necessity have also been studied by Sobociński[352], Thomas[353] and others.

Systems without any theses of the form $L\alpha$

In 1957 E. J. Lemmon[354] proposed a series of systems in which he intended Lp to be interpreted as 'it is scientifically but not logically certain that p'. These systems differ from the ones we have so far considered in that $L(p \supset p)$, for example, is not a thesis of any of them. This, of course, is what the intended interpretation would lead one to expect; for, since $p \supset p$ *is* logically certain, it must be false to say that it is 'scientifically but not logically certain'[355]. In fact these systems of Lemmon's have no theses of the form $L\alpha$ at all[356].

[350] More elaborate deontic logics often contain two distinct pairs of modal operators, one (say L and M) interpreted in the usual way in terms of necessity and possibility, the other (say O and P) in terms of obligatoriness and permissibility. We can then have $Lp \supset p$ and $p \supset Mp$ as theses, but not their deontic counterparts $Op \supset p$ and $p \supset Pp$ – though we shall have $Op \supset Pp$. We can also formulate theses connecting the two types of operators, such as $Op \supset Mp$ ('whatever is obligatory is possible'). Deontic logic is, however, a topic in itself, which we regard as outside the scope of this book. Some bibliographical references will be found in Lemmon [1965a].

[351] Feys [1950], [1965].

[352] Sobociński [1962a], [1962b], [1962c], [1963].

[353] Thomas [1962], [1963a], [1963b], [1963c], [1964b].

[354] Lemmon [1957].

[355] This interpretation does not, however, appear to fit Lemmon's systems adequately. For in all of them $Lq \supset L(p \supset p)$ is a theorem; and on the interpretation in question this would mean, 'If anything is scientifically but not logically certain then $p \supset p$ is scientifically but not logically certain' – which is surely false.

[356] A feature they share with Q and the Ł-modal system, which are discussed below.

One of Lemmon's systems, which he calls E2, is formed by adding to PC the axioms:

A1 $Lp \supset p$
A2 $L(p \supset q) \supset (Lp \supset Lq)$

and the rule:

R1 $\vdash(\alpha \supset \beta) \rightarrow \vdash(L\alpha \supset L\beta)$

(in addition to substitution and detachment).

E2 is contained in, but does not contain, S2. It is, however, worth noting that the rule R1 of E2 is not a rule of the stronger S2.

Another of Lemmon's systems (E3) is the same as E2 but with

A3 $L(p \supset q) \supset L(Lp \supset Lq)$
replacing both A2 and R1.

From a semantical point of view we can characterize E2 and E3 as follows. If we define models as for S2 (S3) but alter the definition of validity so that α is valid iff in every S2-(S3-) model $\langle W, R, V\rangle$, $V(\alpha, w_i) = 1$ for *every* $w_i \in W$ whether w_i is normal or not, then we have a definition of validity for E2 (E3)[357].

A system without the interdefinability of *L* and *M*

All the systems we have so far discussed (provided they have contained L, M and $\sim$) have had the theses $Lp = \sim M \sim p$ and $Mp = \sim L \sim p$. We shall say that in such systems L and M are *interdefinable*, whether or not one of them is explicitly defined in terms of the other. We now turn to a system which does not have this feature. This is Prior's system Q, which he defined in terms of a matrix in 1957[358] but which was not axiomatized till later.

It will be easier to understand Prior's matrix if we have in mind his own interpretation of it. He thinks of a proposition as something which at a given time may be either true, or false, or not statable at all; and a given proposition may change from one status to another with the passage of time. A proposition might be true at all times; or it might have the weaker property of being true *whenever it is statable* – i.e., of being at every moment

[357] *Vide supra* p. 276. *Vide* also Kripke [1965b] for the completeness of these systems.

[358] Prior [1957], pp. 41–54.

either unstatable or true – and such a proposition, though not always true, is at least never false. In this logic a formula, α, will be said to be valid iff every concrete replacement for its variables results in a proposition which is never false, i.e. which is always either unstatable or true.

In the matrices which follow we can think of 1 as representing truth, 3 as representing falsity and 2 as representing unstatability. Each propositional variable is associated with an infinite sequence of numbers, each of which is 1, 2 or 3; these can be thought of as representing the status of the proposition at successive points of time. Given the sequences for a set of propositions, we calculate the sequences for the various functions of them as follows.

For truth-functions we follow the ordinary rules as far as 1 and 3 are concerned, and in the case of 2 we have the rules that the negation of 2 is 2 and that the conjunction of 2 and anything is 2. The other truth-functions can be defined in terms of negation and conjunction in the usual way.

For L the rules are[359]:

(a) If the sequence for α consists entirely of 1's so does that for $L\alpha$.

(b) If the sequence for α contains any 2's, then $L\alpha$ has 2 wherever α has 2, but 3 in every other place.

(c) If the sequence for α contains no 2's but does contain some 3's, then the sequence for $L\alpha$ consists entirely of 3's.

For M the rules are:

(a) If the sequence for α contains no 1's at all, then the sequence for $M\alpha$ is the same as that for α.

(b) If the sequence for α contains some 1's *and* some 2's (whether or not it contains 3's as well), then the sequence for $M\alpha$ has 2 wherever α has 2, but 1 everywhere else.

(c) If the sequence for α contains some 1's but no 2's, then the sequence for $M\alpha$ consists entirely of 1's.

A wff, α, is said to be valid in Q iff, no matter what sequences are associated with the variables in α, the sequence for α itself contains no 3's.

[359] Op. cit. p. 44.

In Q, $Lp \supset {\sim}M{\sim}p$ and $Mp \supset {\sim}L{\sim}p$ are valid but their converses are not, so we do not have the equivalence (either material or strict) of Lp and ${\sim}M{\sim}p$, or of Mp and ${\sim}L{\sim}p$. To see why $Lp \equiv {\sim}M{\sim}p$ is not valid, consider the case in which

	p =	1, 2, and only 2 thereafter
then	${\sim}p$ =	3, 2 ,, ,, ,, ,,
hence	$M{\sim}p$ =	3, 2 ,, ,, ,, ,,
hence	${\sim}M{\sim}p$ =	1, 2 ,, ,, ,, ,,
but	Lp =	3, 2 ,, ,, ,, ,,
hence	$Lp \equiv {\sim}M{\sim}p$ =	3, 2 ,, ,, ,, ,,

Q has been axiomatized by R. A. Bull as follows[360]:
Some complete basis for PC, plus:

Q1 ${\sim}M{\sim}p \supset p$
Q2 $Lp \supset p$
Q3 $(Lp \, . \, Lq) \supset L(p \, . \, q)$

RQLa. $\vdash(\beta \supset \gamma) \rightarrow \vdash(\beta \supset {\sim}M{\sim}\gamma)$, provided β is fully modalized and each variable in β occurs in γ.

RQLb. $\vdash(L\alpha \supset (\beta \supset \gamma)) \rightarrow \vdash(L\alpha \supset (\beta \supset {\sim}M{\sim}\gamma))$, provided β is fully modalized and each variable in β occurs in either α or γ.

RQLc. $\vdash(L\alpha \supset (\beta \supset \gamma)) \rightarrow \vdash(L\alpha \supset (\beta \supset L\gamma))$, provided β is fully modalized and each variable in β occurs in either α or γ.

Prior has shown[361] that this axiomatization is equivalent to one which he had himself previously suggested as sufficient for Q.

Q is of course not equivalent to any of the Lewis systems; it is however contained in S5, though not in any of S1–S4.

Systems based on non-classical propositional calculi

The systems so far considered have contained, as their non-modal part, the classical two-valued propositional calculus – i.e. PC as described in Chapter 1. There are, however, other propositional calculi than this, and modal systems have been based on at least one of them, the *intuitionist calculus* (IC).

In IC, $\sim$, $\vee$, $.$ and $\supset$ are all primitive, and the theses are the formulae derivable from the following axioms by substitution and detachment[362]:

[360] Bull [1964b].
[361] Prior [1964a]. The earlier axiomatization is in Prior [1959].
[362] Heyting [1930].

H1 $p \supset (p \,.\, p)$
H2 $(p \,.\, q) \supset (q \,.\, p)$
H3 $(p \supset q) \supset ((p \,.\, r) \supset (q \,.\, r))$
H4 $((p \supset q) \,.\, (q \supset r)) \supset (p \supset r)$
H5 $p \supset (q \supset p)$
H6 $(p \,.\, (p \supset q)) \supset q$
H7 $p \supset (p \vee q)$
H8 $(p \vee q) \supset (q \vee p)$
H9 $((p \supset r) \,.\, (q \supset r)) \supset ((p \vee q) \supset r)$
H10 $\sim p \supset (p \supset q)$
H11 $((p \supset q) \,.\, (p \supset \sim q)) \supset \sim p$

IC is weaker than the classical two-valued PC. Its most striking feature is that neither $p \vee \sim p$ nor $\sim\sim p \supset p$ is a thesis. We do however have $\vdash_{IC}(p \supset \sim\sim p)$.

Intuitionist S5 – i.e. a system analogous to S5 but with PC replaced by IC – can be axiomatized[363] by adding to IC the operators L and M and the following axioms and rules:

LH1 $Lp \supset p$
MH1 $p \supset Mp$
LH2 $\vdash(\alpha \supset \beta) \rightarrow \vdash(\alpha \supset L\beta)$, provided α is fully modalized.
MH2 $\vdash(\alpha \supset \beta) \rightarrow \vdash(M\alpha \supset \beta)$, provided β is fully modalized.

There are also connections of a different kind between non-classical propositional calculi and modal systems. In particular it has been shown that IC bears a close structural similarity to S4. In 1948 McKinsey and Tarski showed how to translate any formula, α, of IC into a modal formula, $T(\alpha)$, which is a thesis of S4 iff α is a thesis of IC[364]. The rules for this translation are as follows:

T1 If α is a propositional variable, then $T(\alpha)$ is $L\alpha$.
T2 $T(\alpha \vee \beta) = T(\alpha) \vee T(\beta)$
T3 $T(\alpha \,.\, \beta) = T(\alpha) \,.\, T(\beta)$
T4 $T(\alpha \supset \beta) = T(\alpha) \prec T(\beta)$
T5 $T(\sim\alpha) = \sim M(T(\alpha))$

[363] Prior [1957], p. 38; Bull [1965b], p. 142 [1966b]. For a modal system based on a three-valued logic *vide* Segerberg [1967].

[364] McKinsey and Tarski [1948], pp. 13–14. This sort of connection was also explored in Becker [1930] and Gödel [1933]. (*Vide* Rescher [1966].) For the intuitionist predicate calculus *vide* Kripke [1965a].

The correspondence thus established between S4 and IC suggests the project of starting from other modal systems and trying to correlate propositional calculi with them by the same rules T1–T5. Such propositional calculi have been constructed and investigated for S4.2 and S4.3 [365]. It has also been shown that the propositional calculus which corresponds in this way to S5 is the ordinary two-valued PC [366].

The Ł-modal system

One way of axiomatizing PC is to use the single axiom [367]:

$$(1)\ \delta p \supset (\delta{\sim}p \supset \delta q)$$

Here δ is a 'functorial variable': the values of δp are all those well-formed expressions which have p as a constituent – e.g. p itself, $p \supset q$, $(p \supset p) \supset p$, and so forth. By using a rule of uniform substitution for functorial as well as for propositional variables, and the ordinary rule of detachment, the whole of the ordinary two-valued PC can be derived from (1) [368].

In developing PC in this way, of course, we take the range of values of δ to be truth-functions only. But in 1953 Łukasiewicz proposed a system, with (1) as an axiom, in which δ could take modal as well as truth-functional values [369]. This he called the Ł-modal system. Its axioms are:

Ł1 $\delta p \supset (\delta{\sim}p \supset \delta q)$
Ł2 $p \supset Mp$

and its rules are uniform substitution for propositional and functorial variables, and material detachment. The operator L is introduced by the familiar definition:

$$L\alpha =_{\mathrm{Df}} {\sim}M{\sim}\alpha$$

so that L and M have the same relation to each other as in the other systems we have studied (apart from Q).

365 Dummett and Lemmon [1959].
366 *Ibid.*, pp. 263–264.
367 Łukasiewicz [1951].
368 As an alternative to (1) we could use the axiom-schema:

$$\alpha \supset (\beta \supset \gamma)$$

where α, β and γ differ only in that in some places where α has p, β has ${\sim}p$ and γ has q. In this way we could dispense with functorial variables. Cf. Smiley [1963], p. 128.
369 Łukasiewicz [1953].

In virtue of Ł1, the Ł-modal system contains the full propositional calculus, but it has some remarkable properties which make it contrast sharply with other modal systems. We mention a few of these.

1. Unlike any of the Lewis systems (and also unlike Q), it has a finite characteristic matrix. In fact a wff is a theorem of the Ł-modal system iff it satisfies the 4-valued matrix given below[370]. The designated value is 1, and we have for convenience calculated the table for L from those for $\sim$ and M.

	$\supset$	β: 1	2	3	4	$\sim$	M	L
α	1	1	2	3	4	4	1	2
	2	1	1	3	3	3	1	2
	3	1	2	1	2	2	3	4
	4	1	1	1	1	1	3	4

2. Like Q and Lemmon's E systems, but unlike the other systems, the Ł-modal system has no theses of the form $L\alpha$. In fact it goes even further than Q in this matter, in that no proposition of the form Lp ever takes the designated value at all, as the matrix given above makes clear. Prior put this by saying that in the Ł-modal system 'there are no individual true propositions of the form Lp'[371].

3. The Ł-modal system has precisely six irreducible modalities. These are the same six that S5 has, but the reduction laws are different: whereas in S5 all the operators in an unbroken sequence of L's and/or M's can be deleted except the last, in the Ł-modal system all can be deleted except the *first*[372]. Thus $LM\alpha$ is equivalent not to $M\alpha$, but to $L\alpha$, and so forth.

4. As we have shown in earlier chapters, T, S4 and S5 contain the rule N($\vdash\alpha \rightarrow \vdash L\alpha$); S6–S8 contain the thesis MMp, which is incompatible with N; and S1–S3 contain neither N nor

370 This matrix is given in Łukasiewicz [1953], and proved to be characteristic for the Ł-modal system in Smiley [1961]. For a proof that none of the Lewis systems has a finite characteristic matrix, *vide* Dugundji [1940].

371 Prior [1957], p. 42.

372 Prior [1957], p. 126.

$\vdash MMp$, but are compatible with either of them. The Ł-modal system is like S1–S3 in containing neither N nor $\vdash MMp$; but it is unlike all these systems in being positively inconsistent with each of them[373].

5. The full 'unorthodoxy' of the Ł-modal system does not, however, emerge until we consider some of its theses. One of these is:

$$(2)\ (Mp \,.\, Mq) \supset M(p \,.\, q)$$

which means that any two possible propositions are consistent with each other. Another, obtained straightforwardly by replacing δ in Ł1 by M, is:

$$(3)\ Mp \supset (M{\sim}p \supset Mq)$$

which means that if a proposition and its negation are both possible, then any proposition whatever is possible. A third is:

$$(4)\ (p \equiv q) \supset (Mp \supset Mq)$$

which means that if two propositions have the same truth-value, then if one is possible so is the other. Theses such as these may well seem counter-intuitive[374], – though perhaps the more correct conclusion to draw from their presence is that in this system M represents a different sense of 'possible' from any we have encountered in interpreting other systems.

What might this sense of 'possible' be? Łukasiewicz[375] says that he regards Mp and Lp as *truth-functions* of p, and this certainly seems in harmony with the use to which he puts Ł1 in his system. But they are not, as he himself admits, *two-valued* truth-functions: for there are only four monadic two-valued truth-functional operators, and neither M nor L can be identified with any of them. In the Ł-modal system Mp is not p, because $Mp \supset p$ is not a theorem; Mp is not ${\sim}p$, because $Mp \supset {\sim}p$ is not a theorem; Mp is not $(p \supset p)$, since $(p \supset p) \supset Mp$ is not a theorem; and Mp is not $(p \,.\, {\sim}p)$, since $Mp \supset (p \,.\, {\sim}p)$ is not a theorem. (These results can easily be checked by the matrix.)

[373] Prior [1957], p. 126.

[374] Though not, apparently, to Łukasiewicz. For in [1953], p. 122, he claims that (4) is 'intuitively evident'. And in [1951] – when not working within the Ł-modal system at all – he claims that (3) is a sound principle because it is derivable from (1).

[375] Łukasiewicz [1953], p. 113.

For similar reasons Lp is not any of these truth-functions either. Łukasiewicz' own view is that M and L are truth-functional operators in a four-valued logic. Prior, however, has pointed out[376] that we can regard them as operator-*variables*, each ranging over two of the monadic operators in ordinary two-valued PC, Mp ranging over p and $(p \supset p)$ and Lp over p and $(p \,.\, {\sim}p)$. What this amounts to is this: Let α be any wff containing one or more occurrences of M, (any occurrences of L having been eliminated by Def L); let α' be α with M deleted everywhere; and let α'' be α with every sub-formula of the form $M\beta$ replaced by $(\beta \supset \beta)$; then α is a thesis of the Ł-modal system iff both $\vdash_{PC}\alpha'$ and $\vdash_{PC}\alpha''$.

So M and L in the Ł-modal system do seem to express senses of 'possibility' and 'necessity' considerably different from any of the usual ones; and perhaps, if by a 'modal logic' we mean a logic of possibility and necessity, this system takes us to the limit of what we should regard as a modal logic at all. Nevertheless it shares with 'ordinary' modal systems, such as those of Lewis, the following basic formal features: (a) it contains a standard propositional logic (PC, in fact); and (b) it has two monadic operators, M and L, which are such that (i) they cannot be identified with any operators of PC, (ii) Mp is weaker than p, in the sense that $p \supset Mp$ is a thesis but $Mp \supset p$ is not, and Lp is stronger than p in the sense that $Lp \supset p$ is a thesis but $p \supset Lp$ is not, and (iii) $Lp \equiv {\sim}M{\sim}p$ and $Mp \equiv {\sim}L{\sim}p$ are theses.

[376] Prior [1954], p. 204; [1957], p. 4. (*Vide* also Kripke [1965b] p. 210.)

CHAPTER SEVENTEEN

Boolean Algebra and Modal Logic

We remarked earlier that modal systems have been studied in three main ways: axiomatically, semantically and algebraically. This book has concentrated on the first two of these methods, but we now give a brief account of the third [377]. We shall confine our attention almost wholly to a single system, T; several other systems have, however, been investigated in this way, and in some cases the modifications required in order to deal with them will be quite simple and straightforward.

Boolean algebras

The algebraic approach to modal logic is based on a correspondence between various modal systems and certain mathematical structures known as *Boolean algebras* [378]. A Boolean algebra consists of a set (K) of objects, $a, b, c, \ldots$, called *elements*, together with certain operations (or functions) whose arguments are members of K and whose values are likewise members of K. The members of K may be anything we choose, but it is common to think of them as themselves sets of objects, and of the operations as ways of constructing new sets out of old ones. A Boolean algebra must contain a monadic operator, which we shall write as $-$, and a dyadic one, which we shall write as $\times$, and must satisfy the following conditions (often called *postulates*) [379]:

1.1 K contains at least two elements.

1.2 If $a, b \in \mathrm{K}$ [380], then $-a \in \mathrm{K}$ and $(a \times b) \in \mathrm{K}$.

[377] A fuller survey is found in Lemmon [1966a] and [1966b].

[378] Boolean algebras are named after the 19th-century logician and mathematician, George Boole, who first systematically investigated them. The publication in 1847 of his *Mathematical Analysis of Logic* is often regarded as marking the beginning of modern formal logic.

[379] These conditions are given in Stoll [1961], pp. 176–7. Stoll derives them from the postulates in Huntington [1904], given the definitions of + and 0. Several other, equivalent, sets of postulates are known.

[380] '$a, b \in \mathrm{K}$' means that a and b are both members of K.

1.3 If $a, b \in \mathrm{K}$, then $(a \times b) = (b \times a)$[381].

1.4 If $a, b, c \in \mathrm{K}$, then $a \times (b \times c) = (a \times b) \times c$.

1.5 For all $a, b \in \mathrm{K}$, if there is some $c \in \mathrm{K}$ such that $(a \times -b) = (c \times -c)$, then $(a \times b) = a$.

1.6 For all $a, b, c \in \mathrm{K}$, if $(a \times b) = a$, then $(a \times -b) = (c \times -c)$.

It is convenient to add the following definitions:

[0] $\quad 0 =_{\mathrm{Df}} (a \times -a)$

[1] $\quad 1 =_{\mathrm{Df}} -0$[382]

[+] $\quad (a + b) =_{\mathrm{Df}} -(-a \times -b)$

[$\subset$] $\quad (a \subset b) =_{\mathrm{Df}} (a \times b) = a$

We can therefore define a Boolean algebra as an ordered three-membered set $\langle \mathrm{K}, -, \times \rangle$ where K is a set of elements, $-$ and $\times$ represent respectively monadic and dyadic operations on those elements, and the postulates stated above are satisfied. Since K can contain any number of elements (greater than 1), there are indefinitely many Boolean algebras. When we speak of a formula as a theorem of Boolean algebra *simpliciter*, we shall mean that it holds in every Boolean algebra, irrespective of the number of elements in K[383].

The operation symbolised by $\times$ is often called *multiplication*; and $(a \times b)$ is said to be the *product* or *intersection* of a and b. The operation symbolised by $+$ is often called *addition*; and $(a + b)$ is said to be the *sum* or *union* of a and b. $-a$ is said to be the *negation* or *complement* of a. $a \subset b$ can be read as 'a is included in b'[384].

If the elements are sets of objects, then $a \times b$ is the set of things which are both members of a and also members of b,

381 '$a = b$' means that a and b are the same element of the algebra.

382 1 can be called the *unit* (or *universal*) element of the algebra and 0 the *null* element.

383 As examples of Boolean algebras we may cite the matrices of Chapter 13 used to prove the distinctness of the Lewis systems. For most of them K has four members, the numbers 1–4 (i.e., $\mathrm{K} = \{1, 2, 3, 4\}$) and the tables for $\sim$ and . determine functions which satisfy the conditions for $-$ and $\times$ respectively.

384 Various alternative notations are common. Thus $a \times b$ is sometimes written as $a \cap b$, $a \,.\, b$ or simply ab; $a + b$ as $a \cup b$; $-a$ as $\bar{a}$ or a'; $a \subset b$ as $a \subseteq b$ or $a \leqslant b$; 0 as Z, Λ or ϕ; 1 as V.

$-a$ is the set of all those things (i.e. all those things in any member of K) which are not in a, and $a + b$ is the set of all those things which are either members of a or members of b (or both), 0 is the empty (or null) set, and 1 (the universal set) is the set containing everything which is in any member of K. On this interpretation an equation asserts identity of membership; i.e. we say that $a = b$ iff a and b have exactly the same members.

On a spatial interpretation, say where the elements are areas on a plane, we can speak of the points which go to make up those areas where in the previous interpretation we spoke of members of sets. But it is unnecessary to spend more time on the details of the various interpretations.

We now list (though without proofs [385]) a number of important theorems of Boolean algebra for future reference:

B1 $a = --a$
B2 $-(-a \times -b) = a + b$
B3 $-(-a + -b) = a \times b$
B4 $-a \times -b = -(a + b)$
B5 $-a + -b = -(a \times b)$
(B2–B5: De Morgan's Laws)
B6 $-a + b = 1$ iff $a \subset b$
B7 $-a + a = 1$
B8 $a \times a = a$
B9 $a + a = a$
B10 $-1 = 0$
B11 $0 + a = a$
B12 $a \subset 1$
B13 $((a + -b) \times (b + -a)) = 1$ iff $a = b$
B14 $(a \times b) \subset a$
B15 $a \subset (a + b)$
B16 $(a \times b) + c = (a + c) \times (b + c)$
B17 $(a + b) \subset c$ iff $a \subset c$ and $b \subset c$
B18 If $a \subset b$ and $b \subset c$ then $a \subset c$
B19 If $a \subset b$ and $b \subset a$ then $a = b$
B20 If $a \subset b$ and $c \subset d$ then $(a + c) \subset (b + d)$

In these theorems a, b, . . . are to be understood as members of K.

[385] Proofs of some of these theorems, and a more detailed account of the whole subject, will be found in any standard introduction to Boolean algebra. *Vide*, e.g., Stoll (op. cit.); Rasiowa and Sikorski [1963].

Where K has only a finite number of elements, $\langle K, -, \times \rangle$ is said to be a *finite Boolean algebra*, and where K has an infinite number of elements, $\langle K, -, \times \rangle$ is said to be an *infinite Boolean algebra*. We now define an *atom* in a finite Boolean algebra as follows[386]:

a is an atom iff (i) $a \neq 0$ and (ii) for every $b \in K$, if $b \subset a$ then either $b = a$ or $b = 0$.

(The easiest way to understand this definition is by thinking of the elements as sets. Then what the definition comes to is this: for a set to be an atom, (i) it must not be empty, and (ii) the only set that can be included in it (apart from the empty set) must be itself. On this interpretation atoms can be regarded as unit sets, i.e. sets with precisely one member. The complements of atoms are sometimes called *primes*.)

In any finite Boolean algebra $\langle K, -, \times \rangle$ there is a (finite) set $\{a^1, \ldots, a^n\}$ consisting of all the atoms in K, and the following principles hold:

B21 Each non-null element in K is the sum of a unique, determinate set of atoms.

B22 $(a^1 + \ldots + a^n) = 1$.

B23 For any $a, b \in K$, if b is the sum of all the atoms which are included in a, then $a = b$.

B24 Where n is the number of atoms in K, the number of elements in K is 2^n.

It is also possible for there to be an infinite number of atoms in K; in that case $\langle K, -, \times \rangle$ will of course be an infinite Boolean algebra. But not every infinite Boolean algebra is such that its elements are composed of atoms at all.

Boolean algebra and PC

We can establish a relation between a Boolean algebra $\langle K, -, \times \rangle$ and the Propositional Calculus in the following way. Given a wff of PC, we assign to each variable in it some member of K. (If, e.g. a is assigned to p, we write '$V(p) = a$'). We then evaluate the whole formula by the following rules:

1. If $V(\alpha) = a$, then $V({\sim}\,\alpha) = -a$;
2. If $V(\alpha) = a$ and $V(\beta) = b$, then $V(\alpha \,.\, \beta) = a \times b$.

[386] Cf. Stoll, op. cit., p. 188.

Since every wff of PC can be expressed in terms of $\sim$ and ., we can in this way assign some member of K to any wff[387]. We say that an algebra, $\langle K, -, \times \rangle$ *verifies* a wff, α, iff for every assignment from the members of K to the variables in α, $V(\alpha) = 1$; and we say that the algebra *falsifies* α iff it does not verify it, i.e. iff there is some assignment to the variables for which $V(\alpha) \neq 1$. It can be proved (though we shall not prove it here) that a wff of PC is valid iff it is verified by every Boolean algebra[388].

Boolean algebras with added operators

It is possible to augment Boolean algebras of the kind we have described by adding new operators, and some of the algebras so formed lead to mathematical models for various systems of modal logic. Following McKinsey[389], we add to a Boolean algebra $\langle K, -, \times \rangle$ the monadic operator $*$ with the following additional postulates:

2.1 If $a \in K$, then $*a \in K$.

2.2 If $a \in K$, then $a \subset *a$.

2.3 If $a, b \in K$, then $*(a + b) = (*a + *b)$.

2.4 $*0 = 0$.

In addition to these postulates we can also add:

3.1 If $a \in K$, then $*a = **a$.

or

3.2 If $a \in K$, then $*a = 1$ unless $a = 0$[390].

We shall call an algebra satisfying 1.1–2.4, a *T-algebra*, one which also satisfies 3.1, an *S4-algebra*, and one which satisfies 1.1–2.4 and 3.2, an *S5-algebra*[391].

[387] Note that in view of the definition of +, if $V(\alpha) = a$ and $V(\beta) = b$, then $V(\alpha \vee \beta) = a + b$; i.e. + represents $\vee$ as $\times$ represents ..

[388] *Vide*, e.g., Stoll (op. cit.), p. 194. We can in fact go further and say that a wff of PC is valid iff it is verified in the two-element Boolean algebra, viz. that in which $K = \{1, 0\}$, $-a$ is $1 - a$ (arithmetical subtraction), and $a \times b$ is the (arithmetical) product of a and b.

[389] McKinsey [1941].

[390] *Vide* Scroggs [1951].

[391] The proof given on p. 243 that the S5 axiom is independent of the basis of S4 proceeds by constructing a finite S4-algebra which is not an S5-algebra.

T-algebras have been called *extension algebras*[392], and S4-algebras *closure algebras*[393].

In this chapter we shall prove a number of results about T-algebras. These results can all be extended in a quite straightforward way to apply to S4- and S5-algebras, but we shall leave the task of so extending them to the reader. Many of the results also hold for algebras which can be linked in a similar way with most of the other modal systems considered in previous chapters[394].

Boolean algebras and the system T

For our present purposes it will be convenient to use a version of T in which the primitive operators are $\sim$, . and M, and to use the axioms of Von Wright's system M, whose deductive equivalence to T we proved on p. 125. We shall, however, refer to the system as T throughout.

We can establish a relation between a Boolean algebra $\langle K, -, \times, * \rangle$ and T in the same way as we did for PC, but with the additional rule that if $V(\alpha) = a$, then $V(M\alpha) = *a$. We then once more have the position that for every algebra $\langle K, -, \times, * \rangle$ and for every wff, α, of T, the assignment to each variable in α of some member of K determines the assignment to α of some member of K. We again say that the algebra verifies α iff for every such assignment to the variables, $V(\alpha) = 1$; and we define falsification by an algebra analogously.

THEOREM I

Every T-algebra $\langle K, -, \times, * \rangle$ *verifies every thesis of T*[395].

PROOF

The proof proceeds by showing that all the axioms of T are verified by every T-algebra and that the transformation rules preserve the property of being so verified.

392 *Vide* Lemmon [1960a].

393 *Vide* McKinsey and Tarski [1944]. The term 'closure algebra' is derived from the applicability of S4-algebras to certain problems in topology. When, as in topology, space is thought of in terms of sets of points, algebraic formulations of this kind become profitable.

394 *Vide* Lemmon [1966a], [1966b]. For an algebraic approach to the modal predicate calculus *vide* Rasiowa and Sikorski [1963], pp. 481–488.

395 This theorem is proved in Lemmon [1960a] for T using the procedure of McKinsey [1941] for S2. Although we are essentially using these authors' proofs, we do not always follow their terminology.

1. From the fact that $\langle K, -, \times \rangle$ is an ordinary Boolean algebra it follows that $\langle K, -, \times, * \rangle$ verifies the PC axioms.

2. The modal axioms are $p \supset Mp$ and $M(p \vee q) \equiv (Mp \vee Mq)$. We consider each in turn.

(i) Suppose that $p \supset Mp$ is falsified by a T-algebra $\langle K, -, \times, * \rangle$ when $V(p) = a$. Then we have:

$$-a + *a \neq 1$$

and hence, by B6:

$$a \not\subset *a$$

which contradicts 2.2.

(ii) Suppose $M(p \vee q) \equiv (Mp \vee Mq)$ is falsified when $V(p) = a$ and $V(q) = b$. Then, re-writing $M(p \vee q) \equiv (Mp \vee Mq)$ as $(\sim M(p \vee q) \vee (Mp \vee Mq)) . (\sim(Mp \vee Mq) \vee M(p \vee q))$, we have:

$$(-*(a+b) + (*a + *b)) \times (-(*a + *b) + *(a+b)) \neq 1$$

Hence, by 2.3:

$$(-*(a+b) + *(a+b)) \times (-(*a + *b) + (*a + *b)) \neq 1$$

and hence, by B7 $(-a + a = 1)$:

$$1 \times 1 \neq 1$$

which contradicts B8 $(a \times a = a)$.

3. The rule of substitution clearly preserves the property; for if $V(\alpha) = 1$ for *every* assignment from the members of K to a variable, say p, in α, then since every wff, β, is assigned some member of K, we shall also have $V(\alpha[\beta/p]) = 1$.

For Modus Ponens, suppose $V(\alpha) = a$ and $V(\beta) = b$; then if $a = 1$ and $-a + b = 1$, we have $-1 + b = 1$, hence (by B10) $0 + b = 1$, and hence (by B11) $b = 1$.

For the rule $\vdash(\alpha \equiv \beta) \rightarrow \vdash(M\alpha \equiv M\beta)$, suppose that $V(\alpha \equiv \beta) = 1$ (i.e., by Def $\equiv$, that $V((\sim\alpha \vee \beta) . (\sim\beta \vee \alpha)) = 1$); then where $V(\alpha) = a$ and $V(\beta) = b$, B13 gives $a = b$. Hence $*a = *b$; and hence by B13 and Def $\equiv$ again, $V(M\alpha \equiv M\beta) = 1$.

Finally, for the rule of Necessitation (N), suppose $V(\alpha) = 1$; then we have

$$\begin{aligned} V(L\alpha) &= -*-1 \\ &= -*0 && \text{(by B10)} \\ &= -0 && \text{(by 2.4)} \\ &= 1 && \text{(by Def 1)} \end{aligned}$$

Hence every T-algebra verifies every theorem of T. On the other hand, many T-algebras also verify many formulae which are not theorems of T. An algebra which verifies all formulae which are theorems of T *and no others* is known as a *characteristic*[396] T-algebra. It is not, of course, immediately obvious that for a given system there is a characteristic algebra. We can, however, prove the following theorem:

THEOREM II

There is a T-algebra which verifies only theses of T (and hence is a characteristic T-algebra).

To prove this theorem[397] we first show how to construct a certain algebra $\langle K, -, \times, * \rangle$, and then show (i) that it is a T-algebra, and (ii) that it falsifies every wff which is not a thesis of T.

In a Boolean algebra the elements of K can be any objects we like – taking them in the familiar way to be sets is quite inessential to their formal properties. In the algebra we are about to construct, we shall let the elements of K be certain *wffs of T*. More exactly, K will be an infinite proper subset of the wffs of T. We select which wffs are to be in K by Lindenbaum's method (*vide* p. 194n). We suppose that the wffs of T are in some standard ordering, and we put each wff, α, into K *unless* there is some wff, α', earlier in the ordering, which is in K and which is such that $\vdash_T(\alpha \equiv \alpha')$. To put this in another way: we could think of the wffs of T as grouped into sets, each set consisting of all and only those formulae which are equivalent (in T) to a given formula (and therefore to each other). Suppose that $\alpha_1, \ldots, \alpha_i, \ldots$ are the members of such a set, occurring in that order in the standard ordering; then of these we put α_1 into K and leave

[396] Cf. McKinsey, op. cit., p. 122.

[397] We follow the main lines of the proof (for S2) in McKinsey, op. cit., pp. 122–123. McKinsey distinguishes between an algebra and a matrix. In the case of T the only point in this distinction is that the matrix is regarded as specifying some element (of K) as the 'designated' element (i.e., the value taken by all the theorems). We take 1 as designated throughout, and have built this into our definition of 'verify'. Some authors use 0 for this purpose instead of 1, and accordingly modify their account in appropriate ways. McKinsey's S2-matrices (op. cit., p. 117) have a set of designated elements instead of a single one.

the others out. (We shall say that α_1 *represents* in K each of $\alpha_1, \ldots, \alpha_i, \ldots$). We treat all the other sets of this kind in the same way; and for the formula in K which represents a formula, α, we use the notation α'. As a result:

1. No two formulae in K are equivalent to each other.
2. For any wff, α, which is not in K, there is some wff, α', in K such that $\vdash_T(\alpha \equiv \alpha')$.
3. If $\vdash_T(\alpha \equiv \beta)$, then α and β will be represented by one unique formula in K; i.e. α' and β' will be the same formula.
4. Conversely, if α' and β' are the same formula, then $\vdash_T(\alpha \equiv \beta)$; and hence (since Eq holds in T), α and β are intersubstitutable in any theorem. Moreover, since $\vdash_T(\alpha \equiv \alpha')$, α and α' are intersubstitutable.

We have now defined K in our algebra. It remains to define $-$, $\times$ and $*$. Where a ($\in$ K) is the wff α, $-a$ is to be $(\sim\alpha)'$ – i.e. $-a$ is to be $\sim\alpha$ if it is also in K, or its representative in K if it is not. Where a and b ($\in$ K) are the wffs α and β, $(a \times b)$ is to be $(\alpha . \beta)'$; and $*a$ is to be $(M\alpha)'$. Since $0 =_{Df} (a \times -a)$, and since all wffs of the form $(\alpha . \sim\alpha)$ will be represented in K by the same formula, it is simplest to take 0 as $(p . \sim p)'$; and since $1 =_{Df} -0$, we can take 1 as $(p \vee \sim p)'$.

We now have an algebra $\langle K, -, \times, * \rangle$. We show that it is a T-algebra by showing that it satisfies all of conditions 1.1–2.4.

Conditions 1.1, 1.2 and 2.1 are easily seen to be satisfied, and call for no comment.

1.3 Suppose that for some a, $b \in K$, $(a \times b) \neq (b \times a)$. This means that, where a and b are the wffs α and β respectively, $(\alpha . \beta)'$ and $(\beta . \alpha)'$ are different formulae. But $\vdash_T((\alpha . \beta) \equiv (\beta . \alpha))$, and hence, by the construction of K, $(\alpha . \beta)'$ and $(\beta . \alpha)'$ are the *same* formula. Thus 1.3 is satisfied.

1.4 Suppose that for some $a, b, c \in K, a \times (b \times c) \neq (a \times b) \times c$. This means that, where a, b and c are the wffs α, β and γ respectively, $(\alpha . (\beta . \gamma)')'$ and $((\alpha . \beta)' . \gamma)'$ are not the same formula. But $\vdash_T((\alpha . (\beta . \gamma)) \equiv ((\alpha . \beta) . \gamma))$; and hence, since $\vdash_T((\beta . \gamma) \equiv (\beta . \gamma)')$ and $\vdash_T((\alpha . \beta) \equiv (\alpha . \beta)')$, Eq gives $\vdash_T((\alpha . (\beta . \gamma)') \equiv ((\alpha . \beta)' . \gamma)$. Therefore, by the construction of K, $(\alpha . (\beta . \gamma)')'$ and $((\alpha . \beta)' . \gamma)'$ *are* the same formula. Thus 1.4 is satisfied.

1.5 Suppose that for some a, b, $c \in K$, $(a \times -b) = (c \times -c)$.

This means that, where a, b and c are the wffs α, β and γ respectively, $(\alpha \,.\, (\sim\beta)')'$ and $(\gamma \,.\, (\sim\gamma)')'$ are the same formula; and hence $\vdash_T((\alpha \,.\, \sim\beta) \equiv (\gamma \,.\, \sim\gamma))$. By PC transformations this yields $\vdash_T((\alpha \,.\, \beta) \equiv \alpha)$, from which it follows that $(\alpha \,.\, \beta)'$ and α' are the same formula, and therefore (since α is in K) $(\alpha \,.\, \beta)'$ and α are the same formula – i.e. $(a \times b) = a$.

1.6 Since $\vdash_T((\alpha \,.\, \beta) \equiv \alpha) \rightarrow \vdash_T((\alpha \,.\, \sim\beta) \equiv (\gamma \,.\, \sim\gamma))$, the proof for 1.6 is similar to that for 1.5.

2.2 By Def $\subset$ this can be re-written as: if $a \in K$, then $(a \times *a) = a$. Hence we have to prove that if a is the wff α, then $(\alpha \,.\, (M\alpha)')'$ is the same formula as α. This follows, by the same method as before, from the fact that $\vdash_T((\alpha \,.\, M\alpha) \equiv \alpha)$.

2.3 The proof again proceeds similarly from $\vdash_T(M(\alpha \vee \beta) \equiv (M\alpha \vee M\beta))$.

2.4 As explained above, 0 is $(p \,.\, \sim p)'$. Hence $*0$ is $(M(p \,.\, \sim p)')'$. The proof then proceeds as in previous cases from $\vdash_T(M(p \,.\, \sim p) \equiv (p \,.\, \sim p))$.

This completes the proof that $\langle K, -, \times, * \rangle$ is a T-algebra. (Hence by Theorem I it verifies all theses of T.)

What remains to be proved is that $\langle K, -, \times, * \rangle$ falsifies all non-theorems of T. A wff, α, is falsified by the algebra iff there is some assignment from K to the variables in α for which $V(\alpha) \neq 1$. Suppose α is not a thesis of T. For each variable, p_i, in α we let $V(p_i) = p_i'$. As a result we have $V(\alpha) = \alpha'$. Hence it is sufficient to prove that $\alpha' \neq 1$, i.e. that α' is not the same formula as $(p \vee \sim p)'$; and this is so because if α' and $(p \vee \sim p)'$ were the same formula we should have $\vdash_T(\alpha \equiv (p \vee \sim p))$, and hence $\vdash_T\alpha$, contrary to the supposition that α is not a thesis of T.

This completes the proof of Theorem II.

We have shown that the algebra we have constructed is not merely a T-algebra but a characteristic T-algebra [398]. A feature of this algebra is, of course, that K is infinite. Indeed, it follows from some results of Dugundji [399] that there can be no finite characteristic T-algebra. We can, however, show that for any particular non-theorem of T there is a finite T-algebra which

[398] This does not prove that the algebra in question is the only characteristic T-algebra; in fact a system may have a number of distinct characteristic algebras. *Vide* Lemmon [1966a], p. 51.

[399] Dugundji [1940].

falsifies it[400]. (This parallels the case of the semantic diagrams in Part I, where for any wff which was not T-valid we could construct a finite falsifying T-diagram.) Essentially the proof of this will consist in taking the (infinite) characteristic T-algebra, which will of course falsify any given non-theorem, α, and showing how to construct a finite sub-algebra which will also falsify α. It turns out that the maximum number of elements which the sub-algebra must have depends on the number of well-formed parts of α: to be precise, if α has n well-formed parts, there will be a T-algebra, with at most 2^{2^n} elements, which falsifies α.

We shall prove all this a little later on, when we come to Theorem IV. As a preliminary we prove a theorem about the construction of sub-algebras:

THEOREM III

Given a T-algebra $\langle K, -, \times, *\rangle$ *and a finite subset of* K, $\{a_1, \ldots, a_n\}$, *there is a finite T-algebra* $\langle K', -', \times', *'\rangle$ *such that*

(*i*) *for every* a_i $(1 \leqslant i \leqslant n)$, $a_i \in K'$;

(*ii*) K′ *contains at most* 2^{2^n} *members*;

(*iii*) *for every* $a, b \in K'$, *if* $-a \in K'$ *then* $-a = -'a$, *if* $(a \times b) \in K'$ *then* $(a \times b) = (a \times' b)$, *and if* $*a \in K'$ *then* $*a = *'a$.

(*Note*: If $\langle K', -', \times', *'\rangle$ is a T-algebra, postulates 1.2 and 2.1 of course guarantee that if $a, b \in K'$ then $-'a$, $(a \times' b)$ and $*'a$ are in K′, but they do not guarantee that $-a$, $(a \times b)$ and $*a$ are in K′, since $-$, $\times$ and $*$ are operators of the original algebra, not of the sub-algebra. $-a$, $(a \times b)$ or $*a$, however, *could* be in K′, since we are entitled to select $\{a_1, \ldots, a_n\}$ from any of the members of K; or they might be in K′ because of some relation which we establish between the operators of the two algebras[401].)

We prove Theorem III by constructing an algebra satisfying conditions (i)–(iii) and then showing that it is a T-algebra.

[400] This characteristic of T is known as the *finite model property* (from Harrop [1958]). *Vide* also Lemmon [1960a], p. 7, and Drake [1962], p. 401. For a proof that S2 and S4 also have the finite model property, *vide* McKinsey [1941]; for a proof that S5 has it, *vide* Scroggs [1951] and Bergmann [1949a]. Scroggs (op. cit.) shows that any system properly containing S5 has a finite characteristic algebra.

[401] In fact we shall so define $-'$ and $\times'$ that (for any $a, b \in K'$) $-a \in K'$ and $(a \times b) \in K'$, but we shall not so define $*'$ that $*a \in K'$; indeed in the algebra we shall construct it will be quite possible to have $a \in K'$ (and therefore $*'a \in K'$) but $*a \notin K'$.

We let K′ be the set of all elements which can be made up from $a_1, \ldots, a_n$ by – and × (the operators, apart from *, in the original algebra). Clearly (i) is now satisfied. Moreover, $\langle K', -, \times \rangle$ is a Boolean algebra; and it is a standard result for Boolean algebras (without *) – though not one that we have proved here – that if we start with n members, $a_1, \ldots, a_n$, then the algebra will contain at most 2^n distinct atoms and 2^{2^n} elements[402]. Hence condition (ii) is satisfied. We next define –′ and ×′ by simply identifying them with – and × in so far as these operate on members of K′; i.e. for all a, $b \in K'$, we let $-'a$ be $-a$ and $(a \times' b)$ be $(a \times b)$. This ensures that since $\langle K', -, \times \rangle$ is a Boolean algebra, so is $\langle K', -', \times' \rangle$; and moreover that condition (iii) holds as far as –′ and ×′ are concerned.

We could not, however, identify *′ and * in the same way, for if we did we should either introduce new members into K′, and thus be in danger of failing to satisfy condition (ii), or else we should fail to make $\langle K', -', \times', *' \rangle$ a T-algebra, by not always having $*'a \in K'$ wherever $a \in K'$. What we have to do is to find a way of so defining *′ that whenever $a \in K'$, (1) $*'a$ will be some element which is in K′ already, whether $*a$ is in K′ or not, and (2) if $*a$ *is* in K′, $*'a$ will be the same as $*a$ (to comply with condition (iii)).

We can find such a definition of *′ in the following way. Whether $*a$ is in K′ or not, suppose there is some $b \in K'$ such that $a \subset b$ and $*b$ *is* in K′. In that case, b is said to *cover a*. We now define $*'a$ as the intersection of all the $*b$'s whose b's cover a. More exactly, where $b_1, \ldots, b_k$ are all the members of K′ which cover a,

$$*'a =_{\mathrm{Df}} (*b_1 \times \ldots \times *b_k)$$

We now show that this definition meets the requirements stated above for a definition of *′.

(1) For any $a \in K'$ there will always be at least one member of K′ which covers a: 1 will do this even if no other element will. For since $\langle K', -', \times' \rangle$ is a Boolean algebra, $1 \in K'$; since $1 = *1$[403], $*1 \in K'$; and by B12, $a \subset 1$. Moreover, if $*b_1, \ldots, *b_k \in K'$, then by the construction of K′, $(*b_1 \times \ldots \times *b_k) \in K'$. Hence if $a \in K'$, $*'a \in K'$.

[402] *Vidē* McKinsey (op. cit.), p. 124.

[403] By 2.2, $1 \subset *1$; by B12, $*1 \subset 1$; hence by B19, $1 = *1$.

(2) It is a theorem of T-algebras (though we omit the proof of it here) that if $a \subset b_1, \ldots, a \subset b_k$, then $*a \subset (*b_1 \times \ldots \times *b_k)$. Now if $b_1, \ldots, b_k$ are all the elements which cover a, then (by the definition of covering) $a \subset b_1, \ldots, a \subset b_k$; hence $*a \subset (*b_1 \times \ldots \times *b_k)$; and hence (by Def $*'$) $*a \subset *'a$.

Moreover, if $*a \in$ K$'$, then, since $a \subset a$, a will be covered by itself. In that case, where $b_1, \ldots, b_k$ are all the elements which cover a, a itself will be one of these; and hence $*'a$ will be $(*b_1 \times \ldots \times *a \times \ldots \times b_k)$. But by B14 $((a \times b) \subset a)$, $(*b_1 \times \ldots \times *a \times \ldots \times b_k) \subset *a$; i.e. $*'a \subset *a$.

Thus if $*a \in$ K$'$, $*a = *'a$; and this means that condition (iii) in Theorem III is satisfied for $*'$ as well as $-'$ and $\times'$.

We now have an algebra $\langle$K$', -', \times', *'\rangle$ which satisfies conditions (i)–(iii). To complete the proof of Theorem III we have to show that it is a T-algebra.

We have already shown that $\langle$K$', -', \times'\rangle$ is a Boolean algebra, and therefore 1.1–1.6 are satisfied. We proved a moment ago that if $a \in$ K$'$, then $*'a \in$ K$'$, i.e. that 2.1 is satisfied. We also showed that $*a \subset *'a$; but since by hypothesis $\langle$K$, -, \times, *\rangle$ is a T-algebra, $a \subset *a$; hence (by B18) $a \subset *'a$, and so 2.2 is satisfied. Since $\langle$K$', -', \times'\rangle$ is a Boolean algebra, $0 \in$ K$'$, and therefore since $*0 = 0$, $*0 \in$ K$'$. Hence by condition (iii), $*'0 = *0$, and so $*'0 = 0$ – i.e. 2.4 is satisfied. All that remains, therefore, is to show that 2.3 is also satisfied.

For this we have to prove that for all $a, b \in$ K$'$,

$$(1)\quad *'(a + b) = (*'a + *'b)$$

Let $a_1, \ldots, a_n$ be all the elements covering a; $b_1, \ldots, b_m$ all the elements covering b, and $c_1, \ldots, c_k$ all the elements covering $(a + b)$. Then by Def $*'$, (1) is

$$(2)\quad (*c_1 \times \ldots \times *c_k) = (*a_1 \times \ldots \times *a_n) + (*b_1 \times \ldots \times *b_m)$$

Since $\langle$K$, -, \times, *\rangle$ is a T-algebra, the right-hand side of (2) 'multiplies out' by repeated applications of B16 to:

$$(3)\quad (*a_1 + *b_1) \times \ldots \times (*a_n + *b_m)$$

where for every $i (1 \leqslant i \leqslant n)$ and every $j (1 \leqslant j \leqslant m)$, $(*a_i + *b_j)$ is a factor (i.e. one of the terms in the product) in (2); and by 2.3, (3) in turn becomes:

$$(4)\quad *(a_1 + b_1) \times \ldots \times *(a_n + b_m)$$

where for every i and j as before, $*(a_i + b_j)$ is a factor in (4). We therefore take as what is to be proved:

$$(5)\ (*c_1 \times \ldots \times *c_k) = *(a_1 + b_1) \times \ldots \times *(a_n + b_m)$$

It will be sufficient to show (i) that each of $c_1, \ldots, c_k$ is the same as some $(a_i + b_j)$ occurring in the right-hand side of (5), and (ii) that each such $(a_i + b_j)$ is the same as one of $c_1, \ldots, c_k$.

(i) Let c_l be any one of $c_1, \ldots, c_k$. Then since c_l covers $(a + b)$, $(a + b) \subset c_l$. Hence by B17, $a \subset c_l$ and $b \subset c_l$. Moreover, $*c_l \in \mathrm{K}'$. Hence c_l both covers a and covers b; i.e. c_l is some a_i and also some b_j, and therefore for this a_i and b_j, $c_l = (a_i + b_j)$ (by B9).

(ii) Since a_i covers a and b_j covers b, $a \subset a_i$ and $b \subset b_j$. Hence (by B20), $(a + b) \subset (a_i + b_j)$. Moreover, $*a_i$, $*b_j \in \mathrm{K}'$, and hence $(*a_i + *b_j) \in \mathrm{K}'$; but since $\langle \mathrm{K}, -, \times, * \rangle$ is a T-algebra, $(*a_i + *b_j) = *(a_i + b_j)$ (by 2.3), and therefore $*(a_i + b_j) \in \mathrm{K}'$. Hence $(a_i + b_j)$ covers $(a + b)$; i.e. $(a_i + b_j)$ is one of $c_1, \ldots, c_k$.

This completes the proof of Theorem III.

THEOREM IV

If α is a wff of T, containing n well-formed parts[404], *then α is a thesis of T iff it is verified by every T-algebra containing at most 2^{2^n} elements.*

PROOF

By Theorem I, if α is a thesis of T it is verified by every T-algebra whatsoever, and hence by every T-algebra containing at most 2^{2^n} elements. What remains therefore to be proved is that if α is not a thesis of T it is falsified by some T-algebra containing at most 2^{2^n} elements.

Suppose then that α is not a thesis of T. Let $\langle \mathrm{K}, -, \times, * \rangle$ be the characteristic T-algebra. By Theorem II, α will be falsified by this algebra: i.e. for some assignment from K to the variables in α, $\mathrm{V}(\alpha) \neq 1$. Let $\alpha_1, \ldots, \alpha_n$ be the well-formed parts of α, and let $a_1, \ldots, a_n$ be the members of K which (as a result of this assignment to the variables) are assigned to $\alpha_1, \ldots, \alpha_n$ respectively. Now let $\langle \mathrm{K}', -', \times', *' \rangle$ be a finite T-algebra constructed

404 McKinsey [1941] (p. 125) calls these *sub-sentences* of α. Note that in counting up the well-formed parts of a formula we assume it to be written in primitive notation.

in the manner explained in the proof of Theorem III, with $\{a_1, \ldots, a_n\}$ as the initial finite sub-set of K. In $\langle K', -', \times', *' \rangle$ we then make to the variables in α the same assignment as falsified α in $\langle K, -, \times, * \rangle$. As a result the elements thereby assigned to the well-formed parts of α will all be in K′; and since $\langle K', -', \times', *' \rangle$ satisfies condition (iii) of Theorem III, when we evaluate α using $-'$, $\times'$ and $*'$ in place of $-$, $\times$ and $*$, V(α) will be the same in this algebra as in $\langle K, -, \times, * \rangle$; i.e. $V(\alpha) \neq 1$ for this assignment in $\langle K', -', \times', *' \rangle$. Moreover, by Theorem III again, $\langle K', -', \times', *' \rangle$ has at most 2^{2^n} elements.

Theorem IV gives us – in theory at least – a solution to the decision problem for T. For any wff, α, has a finite number, n, of well-formed parts, and if n is finite so is 2^{2^n}. Moreover if any number is finite, there is only a finite number of T-algebras containing no more than that number of elements; and given a finite T-algebra it only requires a finite number of steps to determine whether a given formula is verified by it or not. Hence since by Theorem IV α is a thesis iff it is verified by every T-algebra containing not more than 2^{2^n} elements, its thesishood can be effectively determined in a finite number of steps.

This decision procedure, however, while it is of theoretical importance, is of no practical use (at least for those who do not have high-speed computers at their disposal), because of the size of the numbers involved. Consider the very simple formula $((p \vee q) \vee \sim Mp)$. This has six well-formed parts: p; q; Mp; $\sim Mp$; $(p \vee q)$; and $((p \vee q) \vee \sim Mp)$ itself. Hence it is a thesis iff it is verified by every T-algebra containing not more than 2^{2^6} elements. Now $2^{2^6} = 256^8$ – and we have not yet begun to calculate the number of T-algebras with not more than this number of elements or the number of steps required to test the formula in each of them. Fortunately we were able to devise a much more practical decision procedure for T in Chapter 5.

Boolean algebras and T-validity

In the previous section we have been concerned to match Boolean algebras with an axiomatic version of T, and thus to express in algebraic terms the conditions of thesishood in T. We shall now show how to match certain Boolean algebras with our earlier semantic models for T, and thus give an algebraic account of T-validity [405]. We shall then relate these new algebras to the old ones in such a way as to show that the criteria of thesishood and validity coincide; this will constitute an algebraic proof of the completeness of T, alternative to the proof we gave in Chapter 5. (As before, we leave it to the reader to extend these results to S4 and S5.)

We have already referred to the interpretation of the elements of a Boolean algebra as sets. The easiest way to get an intuitive grasp of the algebras we are about to construct is to think of their elements as *sets of worlds* (in a T-model), and of a proposition as the set of all those worlds in which it is true. It will then make good sense to assign to each propositional variable in a formula some member of K. Moreover, in any T-model $p \,.\, q$ is true in exactly those worlds in which both p is true and q is true, i.e. in the intersection of the set of worlds in which p is true and the set in which q is true. Hence if $V(p) = a$ and $V(q) = b$, we shall have $V(p \,.\, q) = (a \times b)$. Similarly, since $\sim p$ is true in exactly those worlds in which p is not true, if $V(p) = a$, $V(\sim p) = -a$. Thus the analogy with our earlier T-algebras is so far preserved. It only remains to show how M can be fitted into this pattern so as to give us an algebra which mirrors exactly the structure of a T-model.

For this purpose we make use of the notion of an *atom* in a Boolean algebra. We explained on p. 314 that every non-null element in a finite [406] Boolean algebra can be expressed as the sum of a number of atoms, and that when the elements are sets the atoms represent unit sets. Here we can think of the atoms as representing the individual worlds in a T-model. Now in any T-model $M\alpha$ is true in a world w_i iff α is true in some world w_j such that $w_i R w_j$, where R is a reflexive relation over the set of worlds in the model; and by introducing into a Boolean algebra

[405] One might of course say that the T-algebras we have already considered give us a definition of T-validity; but it is not one which is closely related to semantic definitions in Chapter 4.

[406] We do not lose any generality by confining ourselves to finite algebras, for the analogue of Theorem IV holds in the present context; i.e. although there is no finite algebra such that every wff of T is valid or invalid according as it is verified by it or not, still each particular wff is valid iff it is verified by every finite algebra (of the type we are constructing) with a certain maximum number of elements.

a relation R, reflexive over K, we can give the following analogous rule for evaluating $M\alpha$, given $V(\alpha)$:

If $V(\alpha) = a$, then where $a^1, \ldots, a^k$ are all the atoms of a (so that $a = (a^1 + \ldots + a^k)$), and $b^1, \ldots, b^m$ are all the atoms in K such that for each b^j $(1 \leqslant j \leqslant m)$, $b^j \mathrm{R} a^i$ (when a^i is one of $a^1, \ldots, a^k$), $V(M\alpha) = (b^1 + \ldots + b^m)$. (In other words if $V(\alpha) = a$, then $V(M\alpha)$ – i.e., $*a$ – is the sum of all the atoms related by R to any atom of a[407].)

We shall say that $\langle K, -, \times, R\rangle$ is a T′-algebra iff $\langle K, -, \times\rangle$ is a Boolean algebra and R is a reflexive relation over K. We say that such an algebra verifies a wff of T, α, iff for every assignment from K to the variables in α, $V(\alpha) = 1$[408] when α is evaluated by the rules just given for $\sim$, . and M.

It should be clear from the way we have defined a T′-algebra that given a T-model $\langle W, R, V\rangle$ we can construct a T′-algebra $\langle K, -, \times, R\rangle$ and a value-assignment to variables such that

[407] A slightly different definition would be simply to say that $*a$ is the sum of all the b's such that $b\mathrm{R}a$. Jónsson and Tarski [1951], p. 935, show that closure algebras (i.e. S4-algebras) can be characterized by this means when R is transitive. Another equivalent method (Drake [1962]) is to assume in place of R a set S of functions such that, for every $s \in S$ and $a, b \in K$, the following are satisfied:

a1. $s(a) \in K$
a2. $-s(a) = s(-a)$
a3. $(s(a) \times s(b)) = s(a \times b)$
a4. There is some $s_1 \in S$ such that $s_1(a) = a$.

We can then define $*a$ as the sum of all the $s(a)$'s for every $s \in S$, and get a T-algebra.

To get an S4-algebra we add:

a5. For every $s_i, s_j \in S$ there is some $s_k \in S$ such that $s_k(a) = s_i(s_j(a))$;

and for an S5-algebra:

a6. For every $s \in S$ there is some $s' \in S$ such that $s'(s(a)) = a$.

If the elements of K are interpreted as sentences of some language then the members of S can be interpreted as substitutions permitting certain transformations to be made which will yield other sentences. Such an interpretation enables the formulation of the 'syntactical' definition of possibility in McKinsey [1945] (p. 83). McKinsey defines a sentence as possible if it can be transformed into a true one by a set of substitutions which preserve its logical form. For him this amounts essentially to a substitution which satisfies a1–a5. He shows that the system of modal logic then obtained is at least as strong as S4. That it is precisely S4 was shown by Drake (op. cit.). A similar approach is taken in Makinson [1966b].

[408] 1 here of course must be understood as the unit element of the algebra (footnote 382) and not as a truth-value.

(i) to every $w_i \in W$ there corresponds some unique atom $a^i \in K$; (ii) when and only when $w_j R w_k$, for the corresponding atoms we have $a^j R a^k$; (iii) for any variable, p_m, $V(p_m)$ in the algebra is $(b^1 + \ldots + b^n)$ where $b^1, \ldots, b^n$ are all the atoms corresponding to the worlds in the model in which $V(p_m) = 1$. In such a case we shall say that the algebra, together with the value-assignment, *corresponds with* the model.

It follows from all this that iff a wff, α, has the value 1 in every world in a given T-model, then for the corresponding T′-algebra and value-assignment, $V(\alpha)$ will be the sum of all the atoms in K. Now the sum of all the atoms in K is 1; hence in this case $V(\alpha) = 1$. Moreover, to each T′-algebra together with a value-assignment there will correspond a T-model; hence iff $V(\alpha) = 1$ in every world in every T-model – i.e. iff α is T-valid – α is verified by every T′-algebra. We have thus arrived at an algebraic definition of validity in T.

Our remaining task is to prove that for each T′-algebra $\langle K, -, \times, R\rangle$ there is a T-algebra $\langle K, -, \times, *\rangle$ which verifies exactly the same formulae, and that for every T-algebra there is a T′-algebra which verifies exactly the same formulae. We shall then have the result that if a wff, α, is verified by every T′-algebra (and is therefore valid), then it is verified by every T-algebra (and is therefore, by Theorem IV, a thesis of T); i.e. we shall have an algebraic proof of the completeness of T.

We therefore prove the following two theorems:

THEOREM V

If $\langle K, -, \times, R\rangle$ *is a T′-algebra, there is a T-algebra* $\langle K, -, \times, *\rangle$ *which verifies exactly the same wffs as* $\langle K, -, \times, R\rangle$ *does.*

PROOF

Given $\langle K, -, \times, R\rangle$ as a T′-algebra, to construct an algebra $\langle K, -, \times, *\rangle$ we have only to define $*$. We do so by letting $*0$ be 0, and for every non-null $a \in K$ letting $*a$ be $(b^1 + \ldots + b^m)$, where $b^1, \ldots, b^m$ are all the atoms related by R (in the T′-algebra) to any atom of a as explained above. We now show that $\langle K, -, \times, *\rangle$ as thus defined is a T-algebra.

Since $\langle K, -, \times\rangle$ is by hypothesis a Boolean algebra, we have only to show that 2.1–2.4 hold.

2.1 If $a = 0$, then $*a = *0$. But, by definition, $*0 = 0$; hence $*a = 0$, and therefore $*a \in$ K. If a is non-null, a is the sum of a number of atoms, each in K. Since R is reflexive, each such atom is related by R to itself. Hence there will always be some atom related to an atom of a, and therefore once more $*a \in$ K.

2.2 Let $a^1, \ldots, a^k$ be all the atoms of a. Then $a = (a^1 + \ldots + a^k)$. But since R is reflexive, each of $a^1, \ldots, a^k$ is one of $b^1, \ldots, b^m$. Hence by B15 $(a \subset (a + b))$, $a \subset *a$.

2.3 The atoms of $(a + b)$ are precisely the atoms of a together with the atoms of b. Hence the atoms related by R to any atom of $(a + b)$ are precisely those related to any atom of a together with those so related to any atom of b; therefore

$$*a + *b = *(a + b)$$

2.4 holds by the definition of $*$.

Hence $\langle$K, $-$, $\times$, $*\rangle$ is a T-algebra.

Moreover it is easy to see that $\langle$K, $-$, $\times$, R$\rangle$ and $\langle$K, $-$, $\times$, $*\rangle$ verify exactly the same formulae. For since K, $-$ and $\times$ are identical in the two algebras, $\sim$ and . will be represented in the same way in each. And given that $V(\alpha) = a$, we represent $M\alpha$ in a T-algebra by $*a$, i.e. in the present case by $(b^1 + \ldots + b^m)$ – which is precisely the way in which $M\alpha$ is evaluated in the T$'$-algebra.

THEOREM VI

If $\langle$K, $-$, $\times$, $*\rangle$ *is a T-algebra, there is a T$'$-algebra* $\langle$K, $-$, $\times$, R$\rangle$ *which verifies exactly the same wffs as* $\langle$K, $-$, $\times$, $*\rangle$ *does.*

PROOF

Given $\langle$K,$-$,$\times$,$*\rangle$ as a T-algebra, to construct an algebra $\langle$K,$-$,$\times$,R$\rangle$ we have only to define R. We do so by letting R be a relation such that, for any $a, b \in$ K, bRa iff for each atom b^j of b there is some atom a^i of a such that $b^j \subset *a^i$. It will follow that R is reflexive over K; for by the definition of R we shall have aRa iff for each atom a^j of a there is some atom a^i of a such that $a^j \subset *a^i$, and since a^i can be a^j itself, 2.2 guarantees that this condition will always hold. Since R is reflexive, $\langle$K,$-$,$\times$,R$\rangle$ is a T$'$-algebra.

It remains to prove that the two algebras verify exactly the same formulae. Since K, $-$ and $\times$ are identical in each, we have only to show that expressions containing M are evaluated in the same way in each. Let $V(\alpha)$ be a for a given wff α; then $V(M\alpha)$ in the T-algebra is $*a$, and $V(M\alpha)$ in the T′-algebra is $(b^1 + \ldots + b^m)$, where $b^1, \ldots, b^m$ are all the atoms in K related by R to any atom of a. By the definition of R, what this means is that (i) every b^j among $b^1, \ldots, b^m$ is included in one of the $*a^i$'s, where the a^i's are the atoms of a, and (ii) every atom included in any of the $*a^i$'s is one of the b^j's; i.e., where $a^1, \ldots, a^k$ are all the atoms of a,

$$(b^1 + \ldots + b^m) = (*a^1 + \ldots + *a^k)$$

Hence by 2.3:

$$(b^1 + \ldots + b^m) = *(a^1 + \ldots + a^k)$$

and hence by B23:

$$(b^1 + \ldots + b^m) = *a$$ **Q.E.D.**

As we explained earlier, with the proofs of Theorems V and VI we have an algebraic proof of the completeness of T.

APPENDIX ONE

Natural Deduction and Modal Systems

For the sake of readers who are accustomed to doing logic by what is known as the method of *natural deduction*[409], we give here a brief outline of how this method can be extended to deal with certain modal systems. We confine ourselves to propositional systems. We follow in the main the account given by Fitch[410], and we assume that the reader is familiar with the notions of hypothesis and subordinate proof, and with the various rules of Introduction, Elimination and Reiteration (for non-modal logic), which form the main machinery of the method.

Given a set of rules sufficient for the derivation of all valid formulae of PC, we introduce three rules for formulae containing L[411]. The second of these rules has three alternative forms, and according to which of these we adopt we shall obtain T, S4 or S5.

1. *Necessity Elimination*

If $L\alpha$ occurs at any stage in a proof, then α may occur at a later stage.

For the remaining rules we need the notion of a *strict subordinate* proof. In a strict subordinate proof we can proceed according to the usual rules for subordinate proofs, except that in place of the ordinary rule of Reiteration we have a more restrictive

[409] This method was first developed by Gentzen in [1934], and is found in a good many logic text-books. Essentially it provides an alternative to the axiomatic presentation of a system. For an introductory account of the topic *vide* Fitch [1952] or Anderson and Johnstone [1962].

[410] Fitch (op. cit.), pp. 64–80. Fitch also considers modal LPC briefly on pp. 164–166.

[411] Fitch (op. cit.) pp. 64–66.

rule of Strict Reiteration, which can take any of the following forms:

2. *Strict Reiteration*
A. α may occur in a strict subordinate proof if $L\alpha$ occurs earlier in the proof to which it is subordinate.
B. $L\alpha$ may occur in a strict subordinate proof if $L\alpha$ occurs earlier in the proof to which it is subordinate.
C. $\sim L\alpha$ may occur in a strict subordinate proof if $\sim L\alpha$ occurs earlier in the proof to which it is subordinate[412].

We have finally:

3. *Necessity Introduction*
If α occurs in a *categorical* strict subordinate proof (i.e. if α is proved from no hypotheses in a strict subordinate proof), then $L\alpha$ may occur later in the proof to which it is subordinate.

Analogous rules for M, $\prec$ and $=$ can be derived from the above rules (together with the standard PC rules) in a quite straightforward manner.

We now show that rules 1, 2A and 3 yield T[413], 1, 2B and 3 yield S4, and 1, 2C and 3 yield S5. We assume, of course, that these rules are added to a set of rules sufficient to derive PC. We first derive the modal axioms of T. We indicate a strict subordinate proof by Fitch's notation of a square placed to the left of the top of the vertical line covering the proof.

A5 $Lp \supset p$

(1)	$\mid Lp$	Hypothesis
(2)	$\mid p$	(1), Necessity elimination
(3)	$Lp \supset p$	(1), (2), Implication introduction

412 Only forms A and B occur in Fitch. Wisdome [1964], p. 298, credits C to Robert Price. An equivalent formulation would be: 'α may occur in a strict subordinate proof if α occurs earlier in the proof to which it is subordinate and α is fully modalized.'

413 Fitch (op. cit., p. 66) says that these rules yield a system 'almost the same' as S2.

A6 $L(p \supset q) \supset (Lp \supset Lq)$

(1)	$\mid L(p \supset q)$	Hypothesis
(2)	$\mid \quad \mid Lp$	Hypothesis
(3)	$\mid \quad \mid \quad \Box \mid p$	(2), Strict reiteration A
(4)	$\mid \quad \mid \quad \mid p \supset q$	(1), Strict reiteration A
(5)	$\mid \quad \mid \quad \mid q$	(3), (4), Implication elimination
(6)	$\mid \quad \mid Lq$	(5), Necessity introduction
(7)	$\mid Lp \supset Lq$	(2), (6), Implication introduction
(8)	$L(p \supset q) \supset (Lp \supset Lq)$	(1), (7), Implication introduction

For the rule of Necessitation we have to show that whenever there is a proof of α there is also a proof of $L\alpha$. Now to say that there is a proof of α is to say that there is such a proof which is not subordinate to any other proof. Hence no item in it is obtained by reiteration from any proof to which it is subordinate; and therefore it could be treated as a strict subordinate proof. So we could always obtain a proof of $L\alpha$ as follows:

$$\left.\begin{array}{l|l} \Box & - \\ & - \\ & - \\ & \alpha \end{array}\right\} \text{Proof of } \alpha$$

$L\alpha$ Necessity introduction

We have thus derived the whole basis of T. By using rule 2B we can in addition very simply prove the characteristic S4 axiom A7:

A7 $Lp \supset LLp$

(1)	$\mid Lp$	Hypothesis
(2)	$\mid \quad \Box \mid Lp$	(1), Strict reiteration B
(3)	$\mid LLp$	(2), Necessity introduction
(4)	$Lp \supset LLp$	(1), (3), Implication introduction

And by using rule 2C we can give an exactly analogous proof of the S5 axiom A8, in the form $\sim Lp \supset L\sim Lp$.

We have thus shown that the systems obtained by the new rules are at least as strong as T, S4 and S5 respectively. It can in fact be proved (though we shall not prove it here) that they are exactly T, S4 and S5 [414].

[414] The first attempt to use natural deduction methods for modal systems seems to have been made by Moh Shaw-Kwei [1950]; but, except for one which was equivalent to S4, his systems turned out not to be equivalent to any of the Lewis systems. Other natural deduction systems for T, S4 or S5 are to be found in Curry [1950, 1952], Kanger [1957a], Ohnishi and Matsumoto [1957, 1959], Matsumoto [1960], Ohnishi [1961a, b], and Canty [1964]. Ohnishi and Matsumoto have also developed natural deduction formulations of S2 and S3. For a summary of natural deduction systems of modal PC, *vide* Feys [1965], pp. 106–109 and 173–185.

APPENDIX TWO

Entailment and Strict Implication

To say that one proposition entails another is simply an alternative, and often more convenient, way of saying that the second logically follows from, or is logically deducible from, the first[415]. Now as we said back in Chapter 2 (p. 24), we have intended the symbol $\prec$ to be interpreted as 'entails'. Some writers have, however, objected that certain formulae which are theses even of the weakest modal system we have considered, S0.5, are unplausible as principles of entailment or deducibility. These formulae are often called the 'paradoxes of strict implication'[416], and the most important of them are the following[417]:

$$(1)\ (p \,.\, {\sim}p) \prec q$$
$$(2)\ q \prec (p \vee {\sim}p)$$
$$(3)\ {\sim}Mp \supset (p \prec q)$$
$$(4)\ Lq \supset (p \prec q)$$

When $\prec$ is interpreted as 'entails', (1) means that from any proposition of the form $(p \,.\, {\sim}p)$ any proposition whatever can be deduced, and (2) means that from any proposition whatever

415 This use of 'entails' has for some time been standard in philosophy. It derives from Moore [1919] (reprinted in his *Philosophical Studies*; *vide* esp. p. 291).

416 Tendentiously; for those on the other side in the controversy regard the formulae as expressing perfectly sound principles of deducibility, and on anyone's account they express sound and quite unparadoxical truths about strict implication. The 'paradoxes' seem to have been first stated (and incidentally accepted as unparadoxical) in modern logic by MacColl [1906b], p. 513. For some information about mediaeval anticipations of them *vide* Kneale [1962b] pp. 281 ff.

417 In S2 and stronger systems we can also prove ${\sim}Mp \prec (p \prec q)$ and $Lq \prec (p \prec q)$. Two early attempts to formalize a relation which does not lead to the 'paradoxes' (Emch [1936] and Vredenduin [1939]) avoided them in the S2 forms but contained our (1) and (2) as theorems.

there can be deduced any proposition of the form $(p \vee \sim p)$. (3) and (4) are more general: (3) means that from any logically impossible proposition (whether of the form $(p \,.\, \sim p)$ or not) any proposition whatever can be deduced, and (4) means that every necessary proposition (whether of the form $(p \vee \sim p)$ or not) can be deduced from any proposition whatever.

If these are not sound principles of deducibility, that would of course tell against the claim of the standard modal systems to be correct logics of entailment. But in order to decide whether they are sound principles of deducibility or not we have to look into what we take ourselves to be asserting when we assert that one proposition is deducible from another.

Now one plausible account – and perhaps the only one which approaches real clarity – is that to say that q is deducible from p is to say that it is logically impossible for p to be true but q false. Deducibility is after all the relation which obtains between the conclusion and the premiss(es) of a valid deductive inference, and what we require in a valid inference is the logical guarantee that we shall not have the premiss(es) true but the conclusion false. Now by this account the 'paradoxes' *are* sound principles of deducibility; and hence it is not their presence in but their absence from a system which would tell against its claim to be a correct logic of entailment. To take the case of (1): to say that $(p \,.\, \sim p)$ entails q is on this account to say that it is logically impossible for $(p \,.\, \sim p)$ to be true but q false, i.e. it will amount to saying that $(p \,.\, \sim p \,.\, \sim q)$ is logically impossible; but since $(p \,.\, \sim p)$ is itself impossible, so is $(p \,.\, \sim p \,.\, \sim q)$. Similar comments will apply to the other 'paradoxes'. Moreover, this account will guarantee that $\prec$ *can* be interpreted as 'entails'; for in all the standard systems $(\alpha \prec \beta)$ is defined as $\sim M(\alpha \,.\, \sim\beta)$ (or, what comes to the same thing, as $L(\alpha \supset \beta)$), where M is interpreted as 'it is logically possible that'.

No one is likely to deny that the logical impossibility of $(p \,.\, \sim q)$ is a *necessary* condition of q's deducibility from p, but it has been suggested that it is not a *sufficient* condition on the ground that a further condition of q's deducibility from p is that there should be some connection of 'content' or 'meaning' between p and q. It is, however, extremely difficult, if not impossible, to state this additional requirement in precise terms; and to insist on it seems to introduce into an otherwise clear and workable account of deducibility a gratuitously vague element which will make it impossible to determine whether a given formal system is a correct logic of entailment or not.

Even those who are inclined to accept this further requirement for deducibility, however, have to face the following argument. On any account we shall have to regard q as deducible from p when it can be derived from p by some valid principle or principles of deductive inference[418]. Now the following principles seem intuitively to be valid:

A. Any conjunction entails each of its conjuncts.

B. Any proposition, p, entails $(p \vee q)$, no matter what q may be.

C. The premisses $(p \vee q)$ and $\sim p$ together entail the conclusion q (the principle of the *disjunctive syllogism*).

D. Whenever p entails q and q entails r, then p entails r (the principle of the *transitivity of entailment*).

C. I. Lewis has shown that by using these principles we can always derive any arbitrary proposition, q, from any proposition of the form $(p . \sim p)$, in the following way[419]:

	(i) $p . \sim p$
From (i), by A:	(ii) p
From (ii), by B:	(iii) $p \vee q$
From (i), by A:	(iv) $\sim p$
From (iii) and (iv), by C:	(v) q

[418] To say that q may be derived from p by *some valid principle(s) of inference* is (as noted in Lewis and Langford [1932], pp. 252–255) not the same as saying that it may be derived by the principles of a given system, or by principles we have already established up to a given point in the development of a system. Rather it is to say that the principles which enable us to pass from p to q are *sound* ones, whether they occur in any particular system or not. *Vide* Pollock [1966], pp. 184–185, for a discussion of this confusion in writers later than Lewis.

[419] Cf. Lewis and Langford [1932], pp. 250–251, where there is also found an analogous derivation, relevant in a similar way to our (2), of $\sim q \vee q$ from p. For a mediaeval anticipation of Lewis' derivation of q from $p . \sim p$ *vide* Kneale, loc. cit. and [1956], pp. 239–240. It is also possible to derive the result that $(p . \sim p)$ entails $\sim q$ (a simple and equally general variant of the 'paradox' in question) by starting from the principle that $(p . q)$ entails p and applying to it the principle of *antilogism*, viz. that if $(p . q)$ entails r then $(p . \sim r)$ entails $\sim q$ (*vide* Lewis [1914a], p. 246n, and Moh Shaw-Kwei [1950], p. 70).

By D we then have the result that $(p \,.\, {\sim}p)$ entails q.

This derivation shows that the price which has to be paid for denying that $(p \,.\, {\sim}p)$ entails q is the abandonment of at least one of A–D. Frankly, this price seems to us exorbitantly high, since all of A–D seem intuitively sound and the principle that $(p \,.\, {\sim}p)$ entails q is at worst an innocent one: it could never lead us astray in practice by taking us from a true premiss to a false conclusion, since no proposition of the form $(p \,.\, {\sim}p)$ can ever be true.

This is not, of course, to deny that we can have a formal logic of a relation which does not hold between $(p \,.\, {\sim}p)$ and q but does hold in the great majority of cases where we should want to say, intuitively, that entailment holds. Anderson and Belnap's System E, which we outlined on pp. 298 ff, is one such logic. In it none of the 'paradoxes' are theorems (when $\strictif$ is replaced everywhere by the symbol $\rightarrow$ which they read as 'entails'). However, as Lewis' derivation of q from $(p \,.\, {\sim}p)$ shows, any such system (provided it contains the usual rule of uniform substitution) must also omit at least one of the principles we labelled A–D. In the system E in fact the disjunctive syllogism fails, in the sense that $((p \vee q) \,.\, {\sim}p) \rightarrow q$ is not a theorem (and indeed could not be added to the system without bringing in its train the 'paradoxes' which the system was expressly designed to exclude).

The mere construction of such an axiomatic system, however, does not give us a precise concept of entailment which could be a rival to the one mentioned above which is so adequately expressed by $\strictif$; and in our judgement the absence of even a single one of the principles A–D is a far better intuitive reason for rejecting a system as a correct logic of entailment than the presence of the 'paradoxes' could be. In fact we are inclined to go further: the 'paradoxes' seem to us on reflection not to be tiresome (though harmless) eccentricities which we have to put up with in order to have the disjunctive syllogism, transitivity of entailment and the rest, but sound principles in their own right: a logic of entailment *ought*, for example, to contain some principle which reflects our inclination to say to someone who has asserted something self-contradictory, 'If one were to accept *that*, one could prove anything at all' – and the principle that

$(p \,.\, {\sim}p)$ entails q expresses this in just the way that a formal system might be expected to [420].

[420] For further discussion of issues raised in this Appendix and references to the literature *vide* Bennett [1954, 1965], Duncan-Jones [1935], Geach [1958], Lewy [1950, 1958], Nelson [1930], Pollock [1966], Smiley [1959], Strawson [1948], Von Wright [1957].

APPENDIX THREE

Axiomatic Bases for Propositional Modal Systems

We summarize here the axiomatic bases of all the systems of propositional modal logic which we have considered, except for a few mentioned in Chapter 16 which stand somewhat apart from the rest and whose bases are conveniently set out in that chapter in any case.

We first give lists of definitions, rules and axioms, with page references and (where appropriate) the names used for them in the text. We then specify the axiomatic bases with reference to these lists.

Definitions

Def M:	$M\alpha =_{\mathrm{Df}} \sim L \sim \alpha$	[p. 30]
Def M':	$M\alpha =_{\mathrm{Df}} (\alpha \prec L\alpha) \prec L\alpha$	[p. 295]
Def L:	$L\alpha =_{\mathrm{Df}} \sim M \sim \alpha$	[p. 125]
Def L':	$L\alpha =_{\mathrm{Df}} (\alpha \prec \alpha) \prec \alpha$	[p. 295]
Def L'':	$L\alpha =_{\mathrm{Df}} (\alpha = 1)$	[p. 297]
Def $\prec$:	$(\alpha \prec \beta) =_{\mathrm{Df}} L(\alpha \supset \beta)$	[p. 30]
Def $\prec'$:	$(\alpha \prec \beta) =_{\mathrm{Df}} \sim M(\alpha \,.\, \sim\beta)$	[p. 217]
Def $=$:	$(\alpha = \beta) =_{\mathrm{Df}} (\alpha \prec \beta) \,.\, (\beta \prec \alpha)$	[p. 30]

We do not list definitions of non-modal operators.

Rules

PC:	If α is a valid wff of PC, then $\vdash\alpha$ [421]	[p. 255]
MP:	$\vdash\alpha$, $\vdash(\alpha \supset \beta) \rightarrow \vdash\beta$	[Modus Ponens, p. 18]
N:	$\vdash\alpha \rightarrow \vdash L\alpha$	[Necessitation, p. 31]

[421] We shall also say that a basis contains the rule PC if it contains a set of non-modal axioms and rules sufficient for the Propositional Calculus.

L≡: $\vdash(\alpha \equiv \beta) \rightarrow \vdash(L\alpha \equiv L\beta)$ [TR4, p. 124]

M≡: $\vdash(\alpha \equiv \beta) \rightarrow \vdash(M\alpha \equiv M\beta)$ [TR5, p. 125]

LA: $\vdash(\alpha \supset \beta) \rightarrow \vdash(L\alpha \supset \beta)$ [TR6, p. 127]

LC: $\vdash(\alpha \supset \beta) \rightarrow \vdash(\alpha \supset L\beta)$, if α is fully modalized [TR7, p. 127]

Eq: $\vdash\alpha$, $\vdash(\gamma = \delta) \rightarrow \vdash\beta$, where β differs from α only in having δ in some of the places where α has γ [Eq, p. 217; RP1.2, p. 247]

Adj: $\vdash\alpha$, $\vdash\beta \rightarrow \vdash(\alpha \,.\, \beta)$ [p. 217]

MP′: $\vdash\alpha$, $\vdash(\alpha ⥽ \beta) \rightarrow \vdash\beta$ [p. 217]

PCL′: If α is a valid wff of PC or an axiom, then $\vdash L\alpha$ [RP1.1, p. 247]

L⥽: $\vdash(\alpha ⥽ \beta) \rightarrow \vdash(L\alpha ⥽ L\beta)$ [RP2.1, p. 248]

PCL: If α is a valid wff of PC then $\vdash L\alpha$ [p. 256]

N⊃: $\vdash(\alpha \supset \beta) \rightarrow \vdash(\alpha ⥽ \beta)$ [R2, p. 296]

N≡: $\vdash(\alpha \equiv \beta) \rightarrow \vdash(\alpha = \beta)$ [R, p. 297]

⊃/⥽: $\vdash\alpha \rightarrow \vdash\beta$, where β differs from α only in having $\supset$ wherever α has ⥽ [R4, p. 296]

Modal axioms

A. *Axioms for systems in Part I*

1. $Lp \supset p$ [A5, p. 31; AP1.2, p. 246]
1a. $p \supset Mp$ [A11, p. 125]
2. $L(p \supset q) \supset (Lp \supset Lq)$ [A6, p. 30; AP2.1, p. 248]
3. $Lp \supset LLp$ [A7, p. 46]
3a. $MMp \supset Mp$ [A14, p. 126]
4. $Mp \supset LMp$ [A8, p. 49; AP5.1, p. 253]
4a. $M{\sim}Mp \supset {\sim}Mp$ [A15, p. 130]
5. $p \supset LMp$ [B, p. 57]
6. $(Lp \,.\, Lq) \equiv L(p \,.\, q)$ [A9, p. 123]
6a. $M(p \vee q) \equiv (Mp \vee Mq)$ [A12, p. 125]
7. $L(p \supset p)$ [A10, p. 123]
8. $L(p \supset q) \supset L(Lp \supset Lq)$ [A13, p. 126; AP3.1, p. 252]

B. *Axioms for Lewis systems* (*if not given above*)

(i) Original Lewis axioms:

9. $(p \,.\, q) ⥽ (q \,.\, p)$ [AS1.1, p. 217]
10. $(p \,.\, q) ⥽ p$ [AS1.2, p. 217; H2, p. 244]
11. $p ⥽ (p \,.\, p)$ [AS1.3, p. 217; H1, p. 244]

12. $((p \cdot q) \cdot r) \prec (p \cdot (q \cdot r))$ [AS1.4, p. 217]
13. $((p \prec q) \cdot (q \prec r)) \prec (p \prec r)$ [AS1.5, p. 217]
14. $(p \cdot (p \prec q)) \prec q$ [AS1.6, p. 217]
15. $M(p \cdot q) \prec Mp$ [AS2.1, p. 230]
16. $(p \prec q) \prec ({\sim}Mq \prec {\sim}Mp)$ [AS3.1, p. 233; H6, p. 244]
17. $Lp \prec LLp$ [AS4.1, p. 236]
18. $Mp \prec LMp$ [AS5.1, p. 237]

(ii) For alternative axiomatizations:

19. $LL(p \supset p)$ [p. 271]
20. $((r \cdot p) \cdot {\sim}(q \cdot r)) \prec (p \cdot {\sim}q)$ [H3, p. 244]
21. ${\sim}Mp \supset {\sim}p$ [H4, p. 244]
22. $p \prec Mp$ [H5, p. 244]
23. $({\sim}M{\sim}p \prec Mp) \prec (p \prec {\sim}M{\sim}Mp)$ [p. 245]
24. $((p \prec q) \cdot (q \prec r)) \supset (p \prec r)$ [AP1.1, p. 246]

C. *Axioms for other systems based on L or M (if not given above)*

25. $p \prec LMp$ [B′, p. 257]
26. $p \prec LLMp$ [B″, p. 259]
27. $L_n p \supset L_{n+1} p$ [p. 259]
28. $MLp \supset LMp$ [G1, p. 261]
29. $L(Lp \supset Lq) \vee L(Lq \supset Lp)$ [D2, p. 261]
30. $MLp \supset (((p \prec Lp) \prec Lp) \prec Lp)$ [p. 263n]
31. $((p \prec Lp) \prec Lp) \supset (MLp \supset Lp)$ [M1, p. 263]
32. $((p \prec Lp) \prec p) \supset (MLp \supset p)$ [N1, p. 263]
33. $p \supset (MLp \supset Lp)$ [R1, p. 263]
34. $LMLp \supset Lp$ [p. 264]
35. $M(Lp \prec LLp)$ [p. 264]
36. $LMMp \prec LMp$ [p. 264]
37. $LMLp \prec Lp$ [p. 265]
38. ${\sim}Lp \supset L{\sim}Lp$ [p. 265]
39a. $(LMp \cdot LMq) \supset M(p \cdot q)$ [Ka, p. 265]
39b. $LMp \supset MLp$ [Kb, p. 265]
39c. $LM(Mp \supset Lp)$ [Kc, p. 265]
39d. $LML(p \supset Lp)$ [Kd, p. 265]
39e. $LML(Mp \supset p)$ [Ke, p. 265]
40. $p \supset L(Mp \supset p)$ [H1, p. 266]
41. $((p \prec Lp) \prec p) \supset p$ [J1, p. 266]
42. MMp [C13, p. 267]
43. $LMMp$ [p. 269]

D. *Axioms for systems based on* $\prec$ *or* $=$

44. $q \prec (p \prec p)$ [M1, p. 295]
45. $(((p \prec q) \prec r) \prec q) \prec ((q \prec s) \prec (p \prec s))$ [M2, p. 295]
46. $p \prec p$ [p. 295]
47. $(p \prec (q \prec r)) \prec ((p \prec q) \prec (p \prec r))$ [p. 295]
48. $(q \prec r) \prec ((p \prec q) \prec (p \prec r))$ [C3.1, p. 295]
49. $\alpha \prec (\beta \prec \alpha)$, if α and β are strict [C3.2, p. 295]
50. $\alpha \prec (p \prec \alpha)$, if α is strict [C4.1, p. 295]
51. $((\alpha \prec p) \prec \alpha) \prec \alpha$, if α is strict [C5.1, p. 295]
52. $(p \prec q) \prec (s \prec ((q \prec r) \prec (p \prec r)))$ [C4′.1, p. 296]
53. $((p \supset q) \supset p) \prec p$ [C4′.2, p. 296]
54. $(((p \prec q) \prec r) \prec s) \prec ((q \prec s) \prec (p \prec s))$ [C5′.1, p. 296]
55. $p \prec (q \supset p)$ [C5′.2, p. 296]
56. $(p \prec (p \supset q)) \prec (p \prec q)$ [C5′.3, p. 296]
57. $(p \prec q) \prec (p \supset q)$ [C5′.4, p. 296]
58. $0 \prec p$ [C5′.5, p. 296]
59. $(p = q) \supset (\alpha \supset \beta)$, where β differs from α only in having q in some of the places where α has p [I, p. 297]

Axiomatic bases

In specifying axiomatic bases we mention first the modal operator taken as primitive, next the definitions, then the rules, and finally the axioms, in each case by quoting the appropriate number or symbol from the above lists. When, however, a system, S′, is obtained by adding some item, *m*, to an already specified system, S, we specify S′ as S + *m*. Bases containing PC are to be assumed to have the rule MP. The rule of uniform substitution for propositional variables is assumed throughout.

We use the symbol 'S4*' for axiomatizations of S4, such as those given in Part I, in which the rule of Necessitation can be guaranteed to hold for all extensions, and reserve 'S4' for axiomatizations in the style of Lewis, in which this cannot be guaranteed. Even though we have no extensions of S5 to consider, we distinguish analogously between S5* and S5.

A. *Systems in Part I*

T = {*L*, Defs *M*, $\prec$, =, PC, N, 1, 2} (pp. 30f)
or {*L*, Defs *M*, $\prec$, =, PC, L≡, 1, 6, 7} (pp. 123f)
or {*M*, Def *L*, PC, N, M≡, 1a, 6a} (p. 125)

S4* = T + 3 (or 3a) (pp. 46, 126)
    or {*L*, Defs *M*, ⥽, =, N, 1, 8} (pp. 125f, 253)
    or {=, Def *L*″, PC, N≡, 59} (p. 297)

S5* = T + 4 (or 4a) (pp. 49, 129f)
    or {*L*, Defs *M*, ⥽, =, PC, LA, LC} (p. 127)
    or the same with LA replaced by 1 (p. 127)
    or S4* + 5 (p. 58)

The Brouwerian system = T + 5 (p. 58)

B. *The Lewis systems*

S1 = {*M*, Defs *L*, ⥽′, =, Eq, Adj, MP′, 9–14} (p. 216f)
    or the same with 14 replaced by 22 (p. 217n)
    or {*L*, Defs *M*, ⥽, =, MP, PCL′, Eq, 1, 24} (p. 246)

S2 = S1 + 15 (p. 230)
    or {*L*, Defs *M*, ⥽, =, MP, PCL′, L⥽, 1, 2} (p. 248)

S3 = S1 + 16 (p. 233)
    or {*M*, Defs *L*, ⥽′, =, MP, 11, 10, 20, 21, 22, 16} (p. 244)
    or {*L*, Defs *M*, ⥽, =, MP, PCL′, 1, 8} (p. 252)

S4 = S1 + 17 (p. 236)
    or S3 + N (p. 235)
    or S3 + 19 (p. 271)

(See also below, after C5′)

S5 = S1 + 18 (p. 237)
    or S4 + 25 (p. 258)
    or S1 + 26 (p. 258)
    or S3 + 23 (p. 245)

(See also below, after C5′)

The following relations hold among the Lewis systems and the systems in A above:

S4* = S4; S5* = S5

S1 + N = S1 + 19 = T

S1 + 25 = the Brouwerian system (T + 5)

S4* + 5 = S4 + 25 = S5

C. *Other systems based on L or M*

S0.5 = {*L*, Defs *M*, ⥽, =, PCL, MP, 1, 2} (pp. 256f)

$S4_n$ = T + 27 (pp. 259f)

T^+_n = $S4_n$ + 5 (p. 260)

S4.2 = S4* + 28 (p. 261)
S4.3 = S4* + 29 (p. 261)

S4.1 = S4* + 32 S4.1.1 = S4* + 31 S4.2.1 = S4.2 + 32 S4.3.1(D) = S4.3 + 32 *or* S4.3 + 30 S4.4 = S4* + 33	(p. 263f)
S4.5 = S4* + 34 (= S5)	(p. 264)
S3.1 = S3 + 35 S3.2 = S3 + 36 S3.3 = S3 + 37 S3.5 = S3 + 38	(p. 264f)
K1 = S4* + any one of 39a–e	(p. 265)
K1.1 = S4* + 41 K1.2 = S4* + 40 K2 = S4* + 28 + 39b K2.1 = S4.2 + 41 K3 = K1 + 29 K3.1 = S4.3 + 41 K4 = S4.4 + 39b	(p. 266)
S6 = S2 + 42 S7 = S3 + 42 S8 = S3 + 43 S9 (S7.5) = S3.5 + 42	(p. 269)

D. *Systems with ⥽ as primitive*

C3 = {⥽, MP′, 46–49} C4 = {⥽, MP′, 46, 47, 50} C5 = C4 + 51 *or* {⥽, MP′, 44, 45}	(p. 295)

(Def L' can be added to C4, and Def L' and Def M' to C5)

C4′ = {⥽, MP′, N⊃, ⊃/⥽, 52, 53, 56, 57} C5′ = {⥽, MP′, N⊃, ⊃/⥽, 54–57} C4′ + 58 + Def L' + Def M = S4 C5′ + 58 + Def L' + Def M = S5	(p. 296)

The following diagram illustrates the relations holding among most of the systems in the above list. The notation S → S′ means that S contains S′.

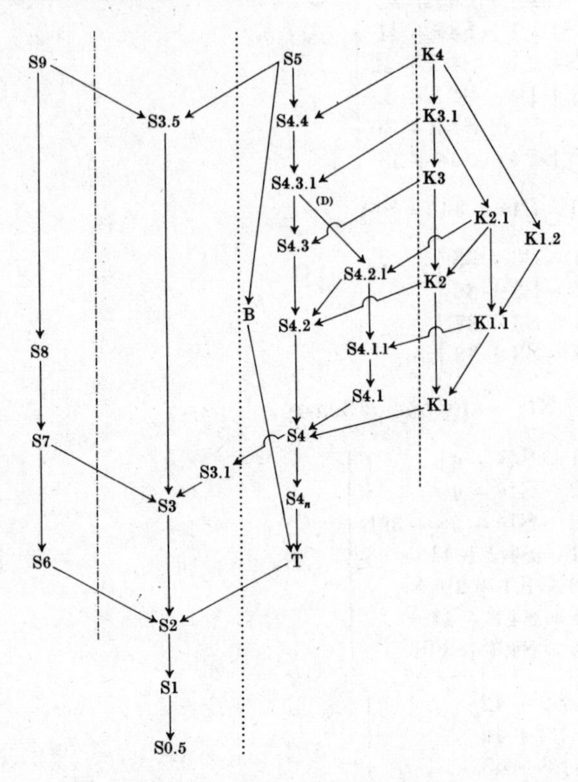

Systems to the left of the line –·–·– and those to the right of the line ------ neither contain nor are contained in S5; all the other systems are contained in S5. Systems to the right of the line ···· (those containing T) have the rule $\vdash\alpha \rightarrow \vdash L\alpha$; those to the left of this line do not, and those to the left of the line –·–·– are inconsistent with the rule.

APPENDIX FOUR

Notation

The notation we have used throughout this book is only one of many in standard use. For non-modal logic, in both propositional and predicate calculi, we use the notation of *Principia Mathematica*, except that we use 'literary' bracketing instead of the system of dots found there. Our modal notation derives from various sources. M as a possibility operator was used by Oskar Becker in 1930[422], and L as a necessity operator by Robert Feys in 1950[423], while ⥽ and $=$ as strict implication and strict equivalence signs derive from Lewis[424]. All four are in common use nowadays, though there is probably a tendency for authors who use ⥽ and $=$ to use $\Diamond$ and $\Box$[425] instead of our M and L respectively. Our main reason for preferring M and L is that they are easier to write. As a strict equivalence operator, $\equiv$[426] has come to be a strong rival to $=$. It has much to commend it, since it minimizes the chance of confusion with other uses of $=$ (e.g. as an identity sign), and we hesitated a long time about whether to adopt it.

For greater ease of comprehension we have, when citing the results of other authors, regularly translated their notation into

[422] Becker [1930]. M is also used as an operator in the three-valued logic of Łukasiewicz and Tarski [1930] and was quite probably in use in pre-war Poland.

[423] The earliest use of L as a monadic operator for necessity that we have been able to find is in Feys [1950]. Feys claims to be using Łukasiewicz' notation and so the symbol may well date from much earlier.

[424] ⥽ dates from Lewis [1918] and $=$ from Lewis [1912].

[425] $\Diamond$ first occurs in Lewis and Langford [1932]. The first published use of $\Box$ is in Barcan [1946], but the symbol itself was devised shortly before that by F. B. Fitch to provide a necessity symbol in the same typographical style as Lewis' diamond. (We are grateful to Professor Barcan Marcus and Professor Fitch for supplying us with this information by letter.)

[426] This symbol occurs in McKinsey [1941].

our own, and the reader who consults the originals must expect to find many formulae which will look strange to him if this is the only book on modal logic he has studied. We therefore list here, for at least the more important of the works we have referred to, the main ways in which their notation differs from ours. We shall not, however, always note divergences in non-modal notation. We arrange the list in alphabetical order of authors. Full references will be found in the Bibliography.

BARCAN [1946] (and later) uses $\Box$ for L, $\Diamond$ for M and $\equiv$ for $=$.

CARNAP [1946] uses N for L.

DUMMETT AND LEMMON [1959]: see under LEMMON [1960a].

FEYS [1937] uses A for L, E for M, and (as non-modal operators) N for $\sim$, V for $\vee$, & for ., F for $\supset$ and H for $\equiv$. Feys here obtains the effect of our Def M by laying down as a law (p. 534) the equivalence 24.1: Ap H NENp. Our A5 is given in the form (23.11, p. 533) p F Ep, and A6 in the imported form (25.3, p. 535) A(p F q). &. Ap : F : Aq. The rule of Necessitation is 25.2 (p. 534): 'D'une loi logique p on peut conclure à une loi logique Ap'. The system thus obtained he calls 'logique t' (p. 536). With the addition of his 26.5, Ep . F . AEp (our A8) it gives S5, and with the addition of a formula (28.43) which he writes as Arp . F . Ar^2p it gives S4 (p. 537). Feys calls S4 'logique r'.

FEYS [1965] uses $\Box$ for L, $\Diamond$ for M, $\Rightarrow$ for $\prec$ and $\Leftrightarrow$ for $=$. Among non-modal operators he uses $\wedge$ for ., $\rightarrow$ for $\supset$ and $\leftrightarrow$ for $\equiv$; and in LPC (p. 147), **A**x for (x), **E**x for $(\exists x)$ and a, b, . . . for $\phi, \psi, \ldots$.

FITCH [1952] uses $\Box$ for L, $\Diamond$ for M and $\equiv$ for $=$.

HACKING [1963] uses $\supset$ for $\prec$.

KRIPKE [1963a, b] uses $\Box$ for L, $\Diamond$ for M, $\wedge$ for ., and P^n, Q^n, . . . as n-adic predicate variables. He states all axioms as schemata, with A, B, . . . as variables for wffs. Our A5, A6, A7, A8 are his A1, A3, A4, A2 respectively; he refers to MP as R1 and to N as R2.

LEMMON [1957] uses the Polish notation for PC, with L and M as necessity and possibility operators and C' for strict implication. His numbering of axioms and rules (p. 177) corresponds with ours (pp. 246–252) as follows: our AP1.1 = his (3); AP1.2 = (2); RP1.1 = (a′); RP1.2 = (b′); AP2.1 = (1′); RP2.1 = (b); AP3.1 = (1).

LEMMON [1960a] uses $\Box$ for L and $\Diamond$ for M; $\cup$, $\cap$ and $-$ for

$+$, $\times$ and $-$ respectively; M for K; $x, y, \ldots$ for members of M; and **E** for $*$. Lemmon calls T-algebras *extension algebras* (hence his use of **E** for $*$). In DUMMETT AND LEMMON [1959] similar notation and terminology are used. In closure (S4) algebras **C** is used instead of **E** for $*$, and an *interior*, I, is defined as $-\mathbf{C}-$.

LEWIS AND LANGFORD [1932] uses $\Diamond$ for M, $\prec$ and $=$ as they are used in this book, and p $\bigcirc$ q for 'p is consistent with q' (i.e. $M(p \,.\, q)$). These authors have no single necessity symbol, but write $\sim\Diamond\sim$ throughout[427].

ŁUKASIEWICZ [1953] uses Polish notation, with Γ for L and Δ for M.

MCKINSEY [1941] uses $x, y, \ldots$ for members of K and $<$ for inclusion. He uses $\rightarrow$ and $\leftrightarrow$ so that $x \rightarrow y = -*(x \times -y)$ and $x \leftrightarrow y = (x \rightarrow y) \times (y \rightarrow x)$. Otherwise his notation (for Boolean algebra) is the same as ours. For propositional operators he follows Lewis and Langford [1932] except that he uses $\equiv$ for $=$.

PARRY [1939] follows Lewis and Langford [1932].

PRIOR [1955] (and elsewhere) uses the Polish notation, with L and M as necessity and possibility operators and C' and E' for strict implication and strict equivalence. In LPC, Πx is used for (x), $\sum x$ for $(\exists x)$ and Ixy for $x = y$.

In PRIOR [1963a] Fpq is used for $p \prec q$.

SCROGGS [1951] follows McKinsey [1941].

SIMONS [1953] uses the notation of Lewis and Langford [1932], except that he uses $\wedge$ for conjunction and states axioms as schemata.

SOBOCIŃSKI [1953] uses the Polish notation, with L and M as necessity and possibility operators. (In later articles he uses $\mathfrak{C}$ for strict implication (LC) and $\mathfrak{E}$ for strict equivalence (LE)).

VON WRIGHT [1951] uses M as we do but N for our L; $a, b, c, \ldots$ as propositional variables; and &, $\rightarrow$ and $\leftrightarrow$ for ., $\supset$ and $\equiv$ respectively.

427 Lewis in his early articles used the notation of *Principia Mathematica* but interpreted it in a strict sense. (In [1914b], p. 590n, he explained that this was for typographical reasons.) He interpreted $\sim p$ as 'p is impossible', and for 'p is false' wrote $-p$; as a material implication sign he used $<$. In Lewis [1918] $\prec$ replaced $\supset$ as a strict implication sign. The first appearance of $\Diamond$ was in Lewis and Langford [1932].

APPENDIX FIVE

Kripke's Model Structures and Hintikka's Model Sets

Model structures and models in Kripke's semantics

The device of semantic models which we have used extensively in this book is derived essentially from the work of Kripke. We have, however, deviated in several ways from his notation, terminology and general method of setting out his material, and we give a brief account of these here.

Kripke defines a *normal model structure*[428] as an ordered triple (**G**, **K**, ***R***) where **K** is a non-empty set, **G** is a member of **K**, and ***R*** is a reflexive relation over **K**. This gives an M model structure (the system we call T he calls M); S4, S5 and *Brouwersche* model structures are obtained by requiring in addition that ***R*** be transitive, transitive and symmetrical, and symmetrical, respectively. **K** corresponds to our W, **G** to some $w_i \in$ W (Kripke uses the letters **G**, **H**, . . . as we use w_i, w_j, . . .), and ***R*** is of course our R.

Given a model structure (**G**, **K**, ***R***) we obtain a *model* for a wff, A, by adding a function $\Phi(P, \mathbf{H})$ (our V), whose first argument, P, ranges over the atomic formulae (propositional variables) in A, whose second argument, **H**, ranges over members of **K**, and whose values are the members of the set {**T**, **F**} (our {1, 0}). I.e., where P is a propositional variable in A and **H** is a member of **K**, in each case either $\Phi(P, \mathbf{H}) = \mathbf{T}$ or $\Phi(P, \mathbf{H}) = \mathbf{F}$. Given $\Phi(P, \mathbf{H})$ as thus defined for each atomic formula in A and each $\mathbf{H} \in \mathbf{K}$, then where B is any sub-formula (well-formed part) of A, $\Phi(B, \mathbf{H})$ is defined inductively as follows. (Kripke uses $\wedge$ for . and $\Box$ for L.)

If B is an atomic formula, $\Phi(B, \mathbf{H})$ is already defined. If $\Phi(B, \mathbf{H}) = \Phi(C, \mathbf{H}) = \mathrm{T}$, then $\Phi(B \wedge C, \mathbf{H}) = \mathbf{T}$; otherwise $\Phi(B \wedge C, \mathbf{H}) = \mathbf{F}$. If $\Phi(B, \mathbf{H}) = \mathbf{T}$, then $\Phi({\sim}B, \mathbf{H}) = \mathbf{F}$; otherwise

[428] Kripke [1963a], p. 68.

$\Phi(\sim B, \mathbf{H}) = \mathbf{T}$. If $\Phi(B, \mathbf{H}') = \mathrm{T}$ for every $\mathbf{H}'$ such that $\mathbf{H}\boldsymbol{R}\mathbf{H}'$, then $\Phi(\Box B, \mathbf{H}) = \mathbf{T}$; otherwise $\Phi(\Box B, \mathbf{H}) = \mathbf{F}$[429].

We can therefore evaluate $\Phi(A, \mathbf{H})$ as $\mathbf{T}$ or $\mathbf{F}$ for each $\mathbf{H} \in \mathbf{K}$. A is said to be *true* in a model Φ associated with a model structure $(\mathbf{G}, \mathbf{K}, \boldsymbol{R})$ if $\Phi(A, \mathbf{G}) = \mathbf{T}$; *false* in that model if $\Phi(A, \mathbf{G}) = \mathbf{F}$. A is said to be *valid* iff it is true in all its models (for every model structure).

What we have in this book called a model is something wider than what Kripke calls a model. A model in his sense is a model in our sense but with some particular 'world' singled out for special consideration. Truth in a model (in his sense) is truth in that designated world in the corresponding model (in our sense). Hence the definition of validity as truth in every model (in his sense) is precisely equivalent to its definition as truth in every world in every model (in our sense); that is, his definition and ours give exactly the same formulae as valid.

For other modal systems the differences between Kripke's treatment and ours are analogous to the ones we have just described.

Hintikka's 'model sets'

An alternative approach to the definition of validity in modal systems has been developed by K. J. J. Hintikka[430]. Although we have not used his methods in this book, his work has shown that they can be employed with profit in the study of modal logic, and we therefore give a brief account of them here.

Hintikka defines a *model system*, Ω, as a set of *model sets*, $\mu, \nu, \ldots$, related by what he calls the relation of *alternativeness* (a dyadic relation corresponding to the $\boldsymbol{R}$ of Kripke's semantics). Model sets so related to a model set, μ, are said to be the *alternatives* to μ. A model set is a set, μ, of formulae which satisfies the following conditions[431]:

(C . ~) If μ contains an atomic formula or an identity, it does not contain its negation.

[429] These conditions are clearly the same as those stated for V on p. 73, with the trivial difference that Kripke takes conjunction as primitive instead of disjunction.

[430] Hintikka [1961], [1963].

[431] Hintikka uses F and G for formulae, & for ., (Ex) for $(\exists x)$, (Ux) for (x), and N for L. He uses 'atomic formula' to cover both propositional variables and atomic formulae of LPC. The conditions as stated here are found in his [1963], pp. 66–67.

(C . &) If $(\boldsymbol{F} \,\&\, \boldsymbol{G}) \in \mu$, then $\boldsymbol{F} \in \mu$ and $\boldsymbol{G} \in \mu$.

(C . v) If $(\boldsymbol{F} \vee \boldsymbol{G}) \in \mu$, then $\boldsymbol{F} \in \mu$ or $\boldsymbol{G} \in \mu$.

(C . E) If $(Ex)\boldsymbol{F} \in \mu$, then $\boldsymbol{F}(a/x) \in \mu$ for at least one free individual symbol a. (Here $\boldsymbol{F}(a/x)$ is the result of replacing x everywhere by a in $\boldsymbol{F}$.)

(C . U) If $(Ux)\boldsymbol{F} \in \mu$ and b is a free individual symbol which occurs in at least one formula of μ, then $\boldsymbol{F}(b/x) \in \mu$.

(C . self ≠) μ does not contain any formulae of the form $\sim(a = a)$.

(C . =) If $\boldsymbol{F} \in \mu$, $(a = b) \in \mu$, and if $\boldsymbol{G}$ is like $\boldsymbol{F}$ except for the interchange of a and b at some (or all) of their occurrences, then $\boldsymbol{G} \in \mu$ provided that $\boldsymbol{F}$ and $\boldsymbol{G}$ are atomic formulae or identities.

(C . N) If $N\boldsymbol{F} \in \mu \in \Omega$, then $\boldsymbol{F} \in \mu$.

(C . M*) If $M\boldsymbol{F} \in \mu \in \Omega$, then there is in Ω at least one alternative to μ which contains $\boldsymbol{F}$.

(C . N⁺) If $N\boldsymbol{F} \in \mu \in \Omega$ and if ν is an alternative to μ in Ω, then $\boldsymbol{F} \in \nu$.

(The statement of the conditions in this form assumes that we have rules for reducing negated non-atomic formulae to formulae in which negation signs are prefixed only to atomic formulae or identities; but the standard systems do have such rules.)

A formula, $\boldsymbol{F}$, is said to be *satisfiable* iff it could be a member of some model set in a model system, and *valid* iff its negation is unsatisfiable – i.e. iff its negation could not be a member of any model set in any model system.

The system determined by these conditions is of course only one of the many possible modal systems. It is in fact LPC + T without the Barcan formula and with contingent identity. Other systems can, however, be obtained either by imposing the familiar requirements of transitivity, symmetry, etc. on the alternativeness relation or by modifying in various ways the conditions for being a model set [432].

[432] There have been several other semantical studies of modal logic. Among the earliest were those in McKinsey [1945], which is discussed briefly in footnote 407, and in Carnap [1946]. Carnap, however, dealt only with S5. Kanger's account (in [1957a] and elsewhere) is very like Kripke's, except in terminology. Most of the recent treatments have either followed Kripke or developed in a very similar manner.

Solutions to Exercises

In most cases these solutions are little more than hints about how to proceed and in some cases no solutions at all are given.

Exercises – 2 (p. 42)

2.1. (a) PC21 (Syll) × DR1 × A6 × Def ⥽ × Imp (PC23).

(b) $(p \supset q) \supset ((p \, . \, r) \supset (q \, . \, r))$ × DR1 × T8 × Imp (PC23).

(c) T15 (Implicational form) × PC26.

(d) T7 × Def $\supset$ and LMI.

(e) T7, T1.

(f) Exercise 2.1(d), T10, PC22 $[Lp/p, M(p \, . \, q)/q, Mp \, . \, Mq/r]$, $(p \supset (q \, . \, r)) \supset ((p \supset q) \, . \, (p \supset r))$.

2.2. Proof by induction on the construction of A (cf. p. 19n) using DR1, DR3, PC10.

2.3. Let α be $p \supset Lp$.

2.4. (a) Proof as for the consistency of T (pp. 41f), noting that the PC-transform of $Lp \vee L{\sim}p$ is also PC-valid.

(b) Use $Lp \vee L{\sim}p$ and the rest of the basis of T to derive $p \supset Lp$.

Exercises – 3 (p. 60)

3.1. (a) PC14, A2 $[Lp/p, Lq/q]$ × DR1 × A7 × Syll × PC27.

(b) A13 (p. 126) $[p \equiv q/p, r/q]$, × R4 (T19) × T4.

3.2. By induction as in 2.2 but using A13 (p. 126) in place of DR1.

3.3. Prove $((p \supset p)$ ⥽ $q) \equiv ((p \supset p) = q)$; $Lq \supset ((p \supset p) = q)$; $(L(p \supset p) = Lq) \supset LLq$; and use them and $((p = q) \supset (Lp = Lq)$ to obtain A7. Use the proof of A13 to obtain $(p = q) \supset (Lp = Lq)$ in S4.

3.4. (a) Prove (Lp ⥽ $Lq) \equiv (Lp \supset Lq)$ (from T30) and use $(p \supset q) \vee (q \supset p)$.

(b) Prove first $L(Mp \supset Lq) \supset L(p \supset Lq)$ (T1, DR1, PC) and then $L(Mp \supset q) \supset L(Mp \supset Lq)$ (T29, Def $\supset$). For the converse implication prove $L(p \supset Lq) \supset L(Mp \supset Lq)$ and $L(Mp \supset Lq) \supset L(Mp \supset q)$.

3.5. (a) $L(p \vee q) . (L(p \vee r) \vee Ls)$

(b) $(Lp \vee L{\sim}(p . q) \vee Mq) . (M{\sim}q \vee L{\sim}(p . q) \vee Mq)$

(c) $(M(p . q) \vee L({\sim}p \vee q) \vee L({\sim}p \vee {\sim}q)) . (M(p . {\sim}q) \vee L({\sim}p \vee q) \vee L({\sim}p \vee {\sim}q))$

(d) $(M(p \vee q) \vee M{\sim}(p \vee q) \vee Lr) . (M(p \vee q) \vee {\sim}r \vee L(p \supset p))$

(e) $(Mq \vee M(p . {\sim}q) \vee L{\sim}p \vee Lp) . (L{\sim}p \vee Lp \vee M(p . {\sim}q) \vee L{\sim}p \vee Lp))$

Exercises – 4 (p. 80)

4.1. (a) (b) *vide* p. 65 (where $n = 2$).

(c) Let A and B be any two distinct modalities in the list of fourteen. In each case give an S4-setting which shows $Ap \equiv Bp$ to be invalid.

4.2. (b) Show that a setting which is both a Brouwerian setting and an S4 setting is also an S5 setting by using footnote 39.

4.3. *Vide* pp. 289f (S4.3).

4.4. *Vide* pp. 274–278 (S2).

4.5. *Vide* pp. 276–278 (S3).

4.6. *Vide* pp. 281f (S6, S8).

Exercises – 5 (p. 104)

5.2. Let the invalid formula be $p \supset L((p \supset p) \supset p)$.

5.3. (a) invalid; (b) valid; (c) valid; (d) invalid.

5.5. *Vide* pp. 99f.

5.6. *Vide* p. 101.

Exercises – 6 (p. 122)

6.1. (a) invalid; (b) invalid; (c) invalid; (d) invalid; (e) valid; (f) valid.

6.3. (a) invalid; (b) valid; (c) valid; (d) valid; (e) invalid.

6.4. (a) Ordered MCNF is $({\sim}q \vee L(p \supset r) \vee Mp) . ({\sim}q \vee L(p \supset r) \vee M{\sim}(q \supset r)$. Invalid (second conjunct is invalid).

(b) Ordered MCNF is $(L(p \supset q) \vee Lr \vee M({\sim}(p \supset q) \vee {\sim}(q \supset p))) . (L(q \supset p) \vee Lr \vee M({\sim}(p \supset q) \vee {\sim}(q \supset p))) . (L{\sim}r \vee Lr \vee M({\sim}(p \supset q) \vee {\sim}(q \supset p)))$. Invalid (last conjunct is invalid).

(c) Ordered MCNF is $(L{\sim}p \vee Lp \vee Lp \vee M{\sim}p) . (Lp \vee M({\sim}p \vee {\sim}p))$. Valid.

Exercises – 7 (p. 130)

7.1. (a) Prove TR4 in T″ (using DR1 and PC15, PC16).

(b) Interpret Lp as $(p \,.\, {\sim}p)$ and show that in that case every formula obtainable without A10 is valid but that A10 is not.

7.2. (a) Show that M contains T′ (use LMI and De Morgan).

(b) Use LMI to derive TR4 in M* and then prove TR3 from it as if in T′.

7.3. (a) For Necessitation let α and β both be $(p \supset p)$ and use A10 and PC. For A6 let α be p, β be $(p \supset q)$ and γ be q, and use PC.

(b) Necessitation is proved as in T*. For A6 first prove $(Lp \,.\, L(p \supset q)) \supset q$ (using A5 and PC) and proceed as for T*. For A7 let α, β, be p and let γ be Lp and use PC. (The proof that T contains T* and that S4 contains S4* should be obvious.)

7.4. (a) *Vide* p. 258 (B is $p \supset LMp$ and follows from B″ by A5).

(b) Use A11 and A8 to obtain $LM(p \supset p)$, which may be used as A10 was to prove TR3 in M*.

Bibliography

After each item will be found a list of the numbers of the footnotes in which reference to the item in question is made. All references to this bibliography are in the footnotes.

Abbreviations

JSL *The Journal of Symbolic Logic*, Association for Symbolic Logic, Providence, Rhode Island.

LA *Logique et Analyse*, E. Nauwelearts, Louvain.

NDJ *Notre Dame Journal of Formal Logic*, University of Notre Dame, Notre Dame, Indiana.

ZML *Zeitschrift für mathematische Logik und Grundlagen der Mathematik*, VEB Deutscher Verlag der Wissenschaften, Berlin.

ACKERMANN, W.

[1956] Begrundung einer strengen Implikation. *JSL* Vol. 21 (1956) pp. 113–128 (**337, 341**).

ALBAN, M. J.

[1943] Independence of the primitive symbols of Lewis' calculi of propositions. *JSL* Vol. 8 (1943) pp. 24–26 (**293, 320**).

AMBROSE, A. and LAZEROWITZ, M.

[1948] *Fundamentals of symbolic logic*, New York, Rinehart & Co., 1948 (**1**).

ANDERSON, A. R.

[1954] Improved decision procedures for Lewis's calculus S4 and Von Wright's calculus M. *JSL* Vol. 19 (1954) pp. 201–214 (**48, 56**). (Correction in *ibid.* Vol. 20 (1955) p. 150.)

[1956] Independent axiom schemata for S5. *JSL* Vol. 21 (1956) pp. 255–256 (**234**).

[1957] Independent axiom schemata for Von Wright's M. *JSL* Vol. 22 (1957) pp. 241–244 (**232**).

[1960] Completeness theorems for the system E of entailment and EQ of entailment with quantification. *ZML* Vol. 16 (1960) pp. 201–216 (**348**).

[1963] Some open problems concerning the system E of entailment. *Acta Philosophica Fennica* (1963) *Modal and Many-valued Logics* pp. 7–18 (**342**).

ANDERSON, A. R. and BELNAP, N. D.

[1959a] Modalities in Ackermann's 'Rigorous Implication'. *JSL* Vol. 24 (1959) pp. 107–111 (**341, 347**).

[1959b] A simple treatment of truth-functions. *JSL* Vol. 24 (1959) pp. 301, 302 (**344**).

[1962a] Tautological entailments. *Philosophical Studies* Vol. 13 (1962) pp. 9–24 (**342**).

[1962b] The pure calculus of entailment. *JSL* Vol. 27 (1962) pp. 19–52 (**339, 345**).

ANDERSON, A. R., BELNAP, N. D. and WALLACE, J. R.

[1960] Independent axiom schemata for the pure theory of entailment. *ZML* Vol. 6 (1960) pp. 93–95 (**340**).

ANDERSON, J. M. and JOHNSTONE, H. W.

[1962] *Natural deduction.* Wadsworth Publishing Co. Inc., Belmont, California, 1962 (**409**).

ÅQVIST, L.

[1964] Results concerning some modal systems that contain S2. *JSL* Vol. 29 (1964) pp. 79–87 (**201, 242, 278, 301, 302, 312**).

ARISTOTLE

[B.C. 350] *Prior analytics.* (Tr. A. J. Jenkinson, Oxford University Press, 1928) (**168**).

BARCAN, R. C. (Mrs J. A. Marcus)

[1946] A functional calculus of first order based on strict implication. *JSL* Vol. 11 (1946) pp. 1–16 (**87, 304, 425**).

[1947] The identity of individuals in a strict functional calculus of second order. *JSL* Vol. 12 (1947) pp. 12–15 (**141, 304**).

[1950] The elimination of contextually defined predicates in a modal system. *JSL* Vol. 15 (1950) p. 92 (**304**).

[1962] Interpreting quantification. *Inquiry* Vol. 5 (1962) pp. 252–259 (**121**).

[1963] Classes and attributes in extended modal systems. *Acta Philosophica Fennica* (1963) *Modal and Many-valued Logics* pp. 123–135 (**167**).

[1967] Essentialism in modal logic. *Noûs* Vol. 1 (1967) pp. 91–96 (**166**).

BAYART, A.

[1958] La correction de la logique modale du premier et second ordre S5. *LA* Vol. 1 (1958) pp. 28–44 (**42, 92**).

[1959] Quasi-adéquation de la logique modale de second ordre S5 et adéquation de la logique modale de premier ordre S5. *LA* Vol. 2 (1959) pp. 99–121 (**117**).

BECKER, O.

[1930] Zur Logik der Modalitäten. *Jahrbuch für Philosophie und Phänomenologische Forschung*, Vol. 11 (1930) pp. 497–548 (**25, 29, 37, 186, 210, 215, 217, 249, 254, 364, 422**).

[1952] *Untersuchungen über den Modalkalkül.* Westkulturverlag Anton Heim, Meisenheim am Glan, 1952.

BELNAP, N. D.

Vide Anderson and Belnap.

Vide also Anderson, Belnap and Wallace.

BENNETT, J. F.

[1954] Meaning and implication. *Mind* Vol. 63 (1954) pp. 451–463 (**420**).

[1965] Review of Anderson and Belnap [1962b] and Smiley [1959] *JSL* Vol. 30 (1965) pp. 240–241 (**420**).

BERGMANN, G.

[1948] Contextual definitions in non-extensional languages. *JSL* Vol. 13 (1948), p. 140.

[1949a] The finite representations of S5. *Methodos* Vol. 1 (1949) pp. 217–219 (**400**).

[1949b] A syntactical characterization of S5. *JSL* Vol. 14 (1949) pp. 173–174.

BETH, E. W. and NIELAND, J. J. F.

[1965] Semantic construction of Lewis's systems S4 and S5. *The Theory of Models* (ed J. W. Addison, L. Henkin, A. Tarski) Amsterdam, North Holland Publishing Co., 1965, pp. 17–24 (**327**).

BOCHEŃSKI, I. M.

[1961] *A history of formal logic.* (Translated and edited by Ivo Thomas) Notre Dame, University of Notre Dame Press, 1961 (**168**).

BRONSTEIN, D. J. and TARTER, H.

[1934] Review of Lewis and Langford's *Symbolic logic. The Philosophical Review* Vol. 43 (1934) pp. 305–309 (**334**).

BULL, R. A.

[1964a] A note on the modal calculi S4.2 and S4.3. *ZML* Vol. 10 (1964) pp. 53–55 (**263**).

[1964b] An axiomatization of Prior's modal calculus Q. *NDJ* Vol. 5 (1964) pp. 211–214 (**360**).

[1965a] An algebraic study of Diodorean modal systems. *JSL* Vol. 30 (1965) pp. 58–64 (**268, 316**).

[1965b] A modal extension of intuitionist logic. *NDJ* Vol. 6 (1965) pp. 142–146 (**363**).

[1965c] A class of extensions of the modal system S4 with the finite model property. *ZML* Vol. 11 (1965) pp. 127–132 (**263**).

[1966a] That all normal extensions of S4.3 have the finite model property. *ZML* Vol. 12 (1966) pp. 341–344 (**263**).

[1966b] MIPC as the formalization of an intuitionist concept of modality. *JSL* Vol. 31 (1966) pp. 609–616 (**363**).

CANTY, J. T.

[1964] A natural deduction system for modal logic. *NDJ* Vol. 5 (1964) pp. 199–210 (**414**).

[1965] Systems classically axiomatized and properly contained in Lewis's S3. *NDJ* Vol. 6 pp. 309–318 (**232**).

CARNAP, R.

[1946] Modalities and quantification. *JSL* Vol. 11 (1946) pp. 33–64 (**35, 42, 63, 432**).

[1947] *Meaning and necessity*, Chicago, University of Chicago Press, 1947 (**150**).

CHISHOLM, R.

[1967] Identity through possible worlds: some questions. *Noûs* Vol. 1 (1967) pp. 1–8 (**122**).

CHURCH, A.

[1936] A note on the *Entscheidungsproblem*. *JSL* Vol. 1 (1936) pp. 40–41 (correction in *ibid*. pp. 101–102) (**77, 97**).

[1951] A formulation of the logic of sense and denotation. *Structure, Method and Meaning* (ed. P. Henle, H. M. Kallen, S. K. Langer), Liberal Arts Press, New York, 1957 (**150**).

[1956] *Introduction to mathematical logic Vol. I*. Princeton, Princeton University Press, 1956 (**1, 7, 8, 9, 74, 77, 95, 318**).

CHURCHMAN, C. W.

[1938] On finite and infinite modal systems. *JSL* Vol. 3 (1938) pp. 77–82 (**210, 218, 255**).

COHEN, L. J.

[1960] A formalization of referentially opaque contexts. *JSL* Vol. 25 (1960) pp. 193–202 (**146**).

CRESSWELL, M. J.

[1960] *Investigations in modal logic*. Thesis submitted in partial requirement for the M.A. degree in the University of New Zealand, Victoria University of Wellington Library (**202**).

[1965] Another basis for S4. *LA* Vol. 8 (No. 31) (1965) pp. 191–195 (**333**).

[1966a] The completeness of S0.5. *LA* Vol. 9 (No. 34) (1966) pp. 263–266 (**314**).

[1966b] Functions of propositions. *JSL* Vol. 31 (1966) pp. 545–560 (**318**).

[1967a] Note on a system of Åqvist. *JSL* Vol. 32 (1967) pp. 58–60 (**312**).

[1967b] The interpretation of some Lewis systems of modal logic. *Australasian Journal of Philosophy* Vol. 45 (1967) pp. 198–206 (**305**).

[1967c] Propositional identity. *LA* Vol. 10 (no. 39–40) (1967) pp. 283–292 (**318**).

CURRY, H. B.

[1950] *A theory of formal deducibility*. Notre Dame, University of Notre Dame Press, 1950 (**414**).

[1952] The elimination theorem when modality is present. *JSL* Vol. 17 (1952) pp. 249–265 (**414**).

DONNELLAN, K. S.

[1966] Reference and definite descriptions. *The Philosophical Review* Vol. 75 (1966) pp. 281–304 (**143**).

DRAKE, F. R.

[1962] On McKinsey's syntactical characterization of systems of modal logic. *JSL* Vol. 27 (1962) pp. 400–406 (**400, 407**).

DUGUNDJI, J.

[1940] Note on a property of matrices for Lewis and Langford's calculi of propositions. *JSL* Vol. 5 (1940) pp. 150–151 (**370**, **399**).

DUMMETT, M. A. E. and LEMMON, E. J.

[1959] Modal logics between S4 and S5. *ZML* Vol. 5 (1959) pp. 250–264 (**261**, **267**, **274**, **365**, **366**).

DUNCAN-JONES, A. E.

[1935] Is strict implication the same as entailment? *Analysis* Vol. 2 (1935) pp. 70–78 (**420**).

EMCH, A. F.

[1936] Implication and deducibility. *JSL* Vol. 1 (1936) pp. 26–35 and p. 58 (**417**).

FARIS, J. A.

[1962] *Truth-functional logic*, London, Routledge and Kegan Paul, 1962 (**1**).

FEYS, R.

[1937] Les logiques nouvelles des modalités. *Revue Néoscholastique de Philosophie* Vol. 40 (1937) pp. 517–553 and Vol. 41 (1938) pp. 217–252 (**20**).

[1950] Les systèmes formalisés des modalités Aristotéliciennes. *Revue Philosophique de Louvain* Vol. 48 (1950) pp. 478–509 (**28**, **351**, **423**).

[1965] *Modal logics*. Louvain, E. Nauwelaerts, 1965 (**187**, **233**, **257**, **274**, **276**, **277**, **284**, **335**, **351**, **414**).

FITCH, F. B.

[1937] Modal functions in two-valued logic. *JSL* Vol. 2 (1937) pp. 125–128 (**318**).

[1939] Note on modal functions. *JSL* Vol. 4 (1939) pp. 115–116 (**318**).

[1948] Corrections to two papers on modal logic. *JSL* Vol. 13 (1948) pp. 38–39 (**318**).

[1952] *Symbolic logic; an introduction*, New York, Ronald Press Co., 1952 (**409**, **410**, **411**, **413**).

[1966] A consistent modal set theory. (Abstract.) *JSL* Vol. 31 (1966) p. 701 (**164**).

[1967] A complete and consistent modal set theory. *JSL* Vol. 32 (1967) pp. 93–103 (**164**).

FRAASSEN, B. C. VAN

[1966] The completeness of free logic. *ZML* Vol. 12 (1966) pp. 219–234 (**126**).

FRAENKEL, A. A. and BAR-HILLEL, Y.

[1958] *Foundations of set theory*, Amsterdam, North-Holland Publishing Co., 1958 (**163**).

FREGE, G.

[1892] Über Sinn und Bedeutung. *Zeitschrift für Philosophie und Philosophische Kritik* (N.S.) Vol. 100 (1892) pp. 25–50 (English translation: On sense and reference, *translations from the writings of Gottlob Frege*, Peter Geach and Max Black, Oxford, Basil Blackwell, 1952) (**143, 150**).

GEACH, P. T.

[1958] Entailment. *Aristotelian Society Supplementary Volume XXXII* (1958) pp. 157–172 (**420**).

[1963] The problem of identifying objects of reference. *Acta Philosophica Fennica* (1963) *Modal and Many-valued Logics* pp. 41–52 (**143**).

GENTZEN, G.

[1934] Untersuchungen über das logische Schliessen. *Mathematische Zeitschrift*, Vol. 39 (1934) pp. 176–210, 405–413. (**409**).

GOBLE, L. F.

[1966] The iteration of deontic modalities, *LA* Vol. 9 (no. 34) (1966) pp. 197–209 (**347**).

GÖDEL, K.

[1933] Eine Interpretation des intuitionistischen Aussagenkalküls. *Ergebnisse eines mathematischen Kolloquiums* Vol. 4 (1933) pp. 34–40 (**20, 192, 364**).

GROVER, D. L.

[1966] *Entailment: a critical study of system E and a critical bibliography of the subject in general.* Thesis submitted in partial requirement for the M.A. degree in the Victoria University of Wellington, 1966, Victoria University of Wellington Library (**338, 346**).

HACKING, I.

[1963] What is strict implication? *JSL*, Vol. 28 (1963) pp. 51–71 (**327, 330, 331**).

HALLDÉN, S.

[1948a] A note concerning the paradoxes of strict implication and Lewis's system S1. *JSL* Vol. 13 (1948) pp. 138–139 (**197a**).

[1948b] A question concerning a logical calculus related to Lewis's system of strict implication which is of special interest for the study of entailment. *Theoria* Vol. 14 (1948) pp. 265–269.

[1949a] Results concerning the decision problem of Lewis's calculi S3 and S6. *JSL* Vol. 14 (1949) pp. 230–236 (**275, 294, 297, 300**).

[1949b] A reduction of the primitive symbols of the Lewis calculi. *Portugaliae Mathematica* Vol. 8 (1949) pp. 85–88 (**320**).

[1951] On the semantic non-completeness of certain Lewis calculi. *JSL* Vol. 16 (1951) pp. 127–129 (**289, 290**).

HAMBLIN, C. L.

[1959] The modal 'probably'. *Mind* Vol. 58 N.S. (1959) pp. 234–240 (**245**).

HANSON, W. H.

[1966] On some alleged decision procedures for S4. *JSL* Vol. 31 (1966) pp. 641–643 (**56**).

HARROP, R.

[1958] On the existence of finite models and decision procedures for propositional calculi. *Proceedings of the Cambridge Philosophical Society* Vol. 54 (1958) pp. 1–13 (**400**).

HENKIN, L.

[1949] The completeness of the first-order functional calculus. *JSL* Vol. 14 (1949) pp. 159–166 (**96, 97, 109**).

[1950] Completeness in the theory of types. *JSL* Vol. 15 (1950) pp. 81–91 (**117**).

HEYTING, A.

[1930] Die formalen Regeln der intuitionistischen Logik. *Sitzungsberichte der Preussischen Akademie der Wissenschaften* Physikalische-mathematische Klasse 1930, pp. 42–56 (**362**).

HINTIKKA, K. J. J.

[1961] Modality and quantification. *Theoria* Vol. 27 (1961) pp. 110–128 (**42, 92, 121, 143, 146, 430**).

[1963] The modes of modality. *Acta Philosophica Fennica* (1963) *Modal and Many-valued Logics* pp. 65–81 (**126, 143, 146, 430, 431**).

[1967] Individuals, possible worlds, and epistemic logic. *Noûs* Vol. 1 (1967) pp. 33–62 (**122**).

HUGHES, G. E. and LONDEY, D. G.

[1965] *The elements of formal logic*. London, Methuen, 1965 (**1, 5**).

HUNTINGTON, E. V.

[1904] Sets of independent postulates for the algebra of logic. *Transactions of the American Mathematical Society* Vol. 5 (1904) pp. 288–309 (**379**).

[1937] Postulates for assertion, conjunction, negation and equality. *Proceedings of the American Academy of Arts and Sciences* Vol. 72 (1937) pp. 1–44 (**332**).

ISHIMOTO, A.

[1954] A set of axioms of the modal propositional calculus equivalent to S3. *The Science of Thought* Vol. 1 pp. 1–11 (**335**).

[1956a] A note on the paper 'A set of axioms of the modal propositional calculus equivalent to S3'. *Ibid.* Vol. 2 (1956) pp. 69–72 (**335**).

[1956b] A formulation of the modal propositional calculus equivalent to S4, *ibid.* Vol. 2 (1956) pp. 73–82 (**335**).

JÓNSSON, B. and TARSKI, A.

[1951] Boolean algebras with operators. *American Journal of Mathematics* Vol. 73 (1951) pp. 891–939 (**407**).

KALMAR, L.

[1936] Zurückführung des Entscheidungsproblems auf den Fall von Formeln mit einer einzigenbinaren Funktionsvariablen. *Compositio Mathematica* Vol. 4 (1936) pp. 137–144 (**95**).

KANGER, S.

[1957a] *Provability in logic*, Stockholm, Almqvist & Wiksell, 1957 (**42, 414, 432**).

[1957b] The morning star paradox. *Theoria* Vol. 23 (1957) pp. 1–11 (**143, 146, 147, 148**).

[1957c] A note on quantification and modalities, *ibid*. Vol. 23 (1957) pp. 133–134 (**148**).

KAPLAN, D.

[1966] Review of Kripke [1963a]. *JSL* Vol. 31 pp. 120–122 (**105**).

KENNY, A.

[1963] Oratio obliqua. *Aristotelian Society Supplementary Volume XXXVII* (1963) pp. 127–146 (**143**).

KLEENE, S. C.

[1952] *Introduction to metamathematics*, Amsterdam, North Holland Publishing Co., 1952 (**152, 154, 156**).

KNEALE, W. C.

[1956] The province of logic, in *Contemporary British Philosophy* (ed. H. D. Lewis), London, George Allen and Unwin, 1956 pp. 237–261 (**419**).

[1962a] Modality *de dicto* and *de re*. *Logic, Methodology and Philosophy of Science*, Stanford 1962, pp. 622–633 (**130**).

KNEALE, W. C. and KNEALE, M.

[1962b] *The development of logic*, Oxford University Press, 1962 (**168, 416**).

KRIPKE, S. A.

[1959] A completeness theorem in modal logic. *JSL* Vol. 24 (1959) pp. 1–14 (**117**).

[1962] The undecidability of monadic modal quantification theory. *ZML* Vol. 8 (1962) pp. 113–116 (**94**).

[1963a] Semantical analysis of modal logic I, normal propositional calculi. *ZML* Vol. 9 (1963) pp. 67–96 (**42, 48, 52, 59, 62, 248, 316, 349, 428**).

[1963b] Semantical considerations on modal logics. *Acta Philosophica Fennica* (1963) *Modal and Many-valued Logics* pp. 83–94 (**92, 124, 125, 126, 127, 128**).

[1965a] Semantical analysis of intuitionistic logic I. *Formal Systems and Recursive Functions* (ed. J. N. Crossley, M. A. E. Dummett) Amsterdam, North Holland Publishing Co., 1965, pp. 92–129 (**364**).

[1965b] Semantical analysis of modal logic II, non-normal modal propositional calculi. *The Theory of Models* (ed. J. W. Addison, L. Henkin, A. Tarski) Amsterdam, North Holland Publishing Co., 1965, pp. 206–220 (**232, 305, 308, 309, 310, 311, 357, 376**).

LAMBERT, K.

[1963] Existential import revisited. *NDJ* Vol. 4 (1963) pp. 288–292 (**126**).

LEMMON, E. J.

[1956] Alternative postulate sets for Lewis's S5. *JSL* Vol. 21 (1956) pp. 347–349 (**72**).

[1957] New foundations for Lewis modal systems. *JSL* Vol. 22 (1957) pp. 176–186 (**69, 236, 237, 241, 245, 246, 349, 354**).

[1959] Is there only one correct system of modal logic? *Aristotelian*

Society Supplementary Volume XXXIII (1959) pp. 23–40 (**47a, 247**).

[1960a] An extension algebra and the modal system T. *NDJ* Vol. 1 (1960) pp. 2–12 (**392, 395, 400**).

[1960b] Quantified S4 and the Barcan formula. (Abstract.) *JSL* Vol. 24 (1960) pp. 391–392 (**85**).

[1963] A theory of attributes based on modal logic. *Acta Philosophica Fennica* (1963) *Modal and Many-valued Logics* pp. 95–122 (**166, 167**).

[1965a] Deontic logic and the logic of imperatives. *LA* No. 29 (1965) pp. 39–70 (**350**).

[1965b] Some results on finite axiomatizability in modal logic. *NDJ* Vol. 6 (1965) pp. 301–307 (**232**).

[1966a] Algebraic semantics for modal logics I. *JSL* Vol. 31 (1966) pp. 46–65 (**377, 394, 398**).

[1966b] Algebraic semantics for modal logics II, *ibid.* pp. 191–218 (**377, 394, 398**).

LEMMON, E. J., MEREDITH, C. A., MEREDITH, D., PRIOR, A. N.

[1957] and THOMAS, I. *Calculi of pure strict implication* (cyclostyled). Christchurch, Canterbury University College, 1957 (**323, 324**).

LEWIS, C. I.

[1912] Implication and the algebra of logic. *Mind* N.S. Vol. 21 (1912) pp. 522–531 (**179, 180, 181, 182, 321, 424**).

[1913] Interesting theorems in symbolic logic. *Journal of Philosophy* Vol. 10 (1913) pp. 239–242 (**179, 181, 183**).

[1914a] A new algebra of strict implication. *Mind* N.S. Vol. 23 (1914) pp. 240–247 (**179, 181, 182, 419**).

[1914b] The matrix algebra for implication. *Journal of Philosophy* Vol. 11 (1914) pp. 589–600 (**179, 181, 184, 427**).

[1918] *A survey of symbolic logic*. Berkeley, University of California, 1918 (**179, 185, 211, 424, 427**). (N.B. The chapter on strict implication is not included in the 1961 Dover reprint.)

[1920] Strict implication. An emendation. *Journal of Philosophy* Vol. 17 (1920) pp. 300–302 (**211**).

LEWIS, C. I. and LANGFORD, C. H.

[1932] *Symbolic logic*, New York, Dover publications, 1932. Second Edition, 1959 (**23, 27, 37, 187, 191, 193, 194, 203, 204, 205, 206, 207, 208, 211, 215, 219, 221, 222, 223, 224, 225, 226, 227, 228, 249, 253, 287, 288, 294, 295, 317, 418, 419, 425, 427**).

LEWIS, D. K.

[1968] Counterpart theory and quantified modal logic. *The Journal of Philosophy* Vol. 65 (1968) pp. 113–126 (**318**).

LEWY, C.

[1950] Entailment and necessary propositions, in *Philosophical Analysis* (ed. Max Black) Ithaca, Cornell University Press, 1950, pp. 195–210 (**420**).

[1958] Entailment. *Aristotelian Society Supplementary Volume XXXII* (1958) pp. 123–142 (**420**).

LINSKY, L.

[1966] Substitutivity and descriptions (APA symposium paper) *The Journal of Philosophy* Vol. 63 (1966) pp. 673–683 (**143**).

LÖB, M. H.

[1966] Extensional interpretations of modal logics. *JSL* Vol. 31 (1966) pp. 23–45 (**318**).

ŁUKASIEWICZ, J.

[1951] On variable functors of propositional arguments. *Proceedings of the Royal Irish Academy* Vol. 54 A2 pp. 25–35 (**367, 374**).

[1953] A system of modal logic. *The Journal of Computing Systems* Vol. 1 (1953) pp. 111–149 (**244, 369, 370, 374, 375**).

ŁUKASIEWICZ, J. and TARSKI, A.

[1930] Untersuchungen über den Aussagenkalkül. *Comptes Rendus des Séances de la Société des Sciences et des Lettres de Varsovie* Vol. 23 (1930) (**422**).

MCCALL, S.

[1963] *Aristotle's modal syllogisms*. Amsterdam, North Holland Publishing Co., 1963 (**168**).

MACCOLL, H.

[1880] Symbolical reasoning. *Mind* Vol. 5 (1880) pp. 45–60 (**169, 170, 171, 172**).

[1903] Symbolic reasoning V. *Mind* (N.S.) Vol. 12 (1903) pp. 355–364 (**173, 174, 176**).

[1906a] *Symbolic logic and its applications*. London (1906) (**173, 177**).

[1906b] Symbolic reasoning VIII. *Mind* (N.S.) Vol. 15 (1906) pp. 504–518 (**173, 175, 416**).

MCKINSEY, J. C. C.

[1934a] A note on Bronstein's and Tarter's definition of strict implication. *Philosophical Review* Vol. 43 (1934) pp. 518–520 (**334**).

[1934b] A reduction in the number of postulates for C. I. Lewis' system of strict implication. *Bulletin of the American Mathematical Society* Vol. 40 (1934) pp. 425–427 (**191**).

[1940] Proof that there are infinitely many modalities in Lewis's system S2. *JSL* Vol. 5 (1940) pp. 110–112 (**40**).

[1941] A solution of the decision problem for the Lewis systems S2 and S4 with an application to topology. *JSL* Vol. 6 (1941) pp. 117–134 (**389, 395, 396, 397, 400, 402, 404, 426**).

[1944] On the number of complete extensions of the Lewis systems of sentential calculus. *JSL* Vol. 9 (1944) pp. 41–45.

[1945] On the syntactical construction of systems of modal logic. *JSL* Vol. 10 (1945) pp. 83–96 (**42, 271, 281, 407, 432**).

[1953] Systems of modal logic which are not unreasonable in the sense of Halldén. *JSL* Vol. 18 (1953) pp. 109–113 (**292**).

MCKINSEY, J. C. C. and TARSKI, A.

[1944] The algebra of topology. *Annals of Mathematics* Vol. 45 (1944) pp. 141–191 (**393**).

[1948] Some theorems about the sentential calculi of Lewis and Heyting. *JSL* Vol. 13 (1948) pp. 1–15 (**216, 364**).

MAKINSON, D. C.

[1966a] There are infinitely many Diodorean modal functions. *JSL* Vol. 31 (1966) pp. 406–408 (**36**).

[1966b] How meaningful are modal operators? *Australasian Journal of Philosophy* Vol. 44 (1966) pp. 331–337 (**407**).

[1966c] On some completeness theorems in modal logic. *ZML* Vol. 12 (1966) pp. 379–384 (**105**).

MARCUS, MRS J. A. *vide* Barcan, R. C.

MASSEY, G. J.

[1966] The theory of truth-tabular connectives both truth-functional and modal. *JSL* Vol. 31 (1966) pp. 593–608 (**320**).

[1967] Binary connectives functionally complete by themselves in S5 modal logic. *JSL* Vol. 32 (1967) pp. 91–92 (**320**).

MATSUMOTO, K.

[1960] Decision procedure for modal sentential calculus S3. *Osaka Mathematical Journal* (1960) pp. 167–175 (**309, 414**).

Vide also Ohnishi and Matsumoto

MENDELSON, E.

[1964] *Introduction to mathematical logic*. Princeton, Van Nostrand, 1964 (**154, 156**).

MEREDITH, C. A.

[1956a] *Contributions to the investigation of pure strict implication.* (Cyclostyled) Christchurch, Philosophy Department, Canterbury University College, 1956 (**324, 326**).

[1956b] *Interpretations of different modal logics in the 'Property calculus'*. 1 p. (cyclostyled) Christchurch, Philosophy Department, Canterbury University College, 1956 (recorded and expanded by A. N. Prior) (**42**).

Vide also Lemmon, Meredith, Prior and Thomas [1957].

MEREDITH, C. A. and PRIOR, A. N.

[1964] Investigations into implicational S5. *ZML* Vol. 10 (1964) pp. 203–220 (**324, 326**).

MOH SHAW-KWEI

[1950] The deduction theorems and two new logical systems. *Methodos* Vol. 2 (1950) pp. 56–75 (**414, 419**).

MONTAGUE, R.

[1963] Syntactical treatments of modality. *Acta Philosophica Fennica* (1963) *Modal and Many-valued Logics* pp. 153–166 (**318**).

MONTGOMERY, H. A. and ROUTLEY, F. R.

[1966] Contingency and non-contingency bases for normal modal logics. *LA* Vol. 9 (no. 35–36) (1966) pp. 318–328 (**335a**).

MOORE, G. E.

[1919] External and internal relations. *Proceedings of the Aristotelian Society* (1919–20). (Reprinted in *Philosophical Studies*, London, Kegan Paul, Trench, Trubner & Co., 1922 (subsequent reprints, Routledge and Kegan Paul) pp. 276–309) (**415**).

MYHILL, J. R.

[1958] Problems arising in the formalization of intensional logic. *LA* Vol. 1 (No. 2) (1958) pp. 74–83 (**121**).

NELSON, E. J.

[1930] Intensional relations. *Mind* Vol. 39 (1930) pp. 440–453 (**420**).

OHNISHI, M.

[1961a] Gentzen decision procedures for Lewis's systems S2 and S3. *Osaka Mathematical Journal* Vol. 13 (1961) pp. 125–137 (**309**).

[1961b] Von Wright-Anderson's decision procedures for Lewis's systems S2–S3. *Ibid.* Vol. 13 (1961) pp. 139–142 (**309**).

OHNISHI, M. and MATSUMOTO, K.

[1957] Gentzen method in modal calculi I. *Ibid.* Vol. 9 (1957) pp. 113–130 (**48, 56, 414**).

[1959] Gentzen method in modal calculi II. *Ibid.* Vol. 11 (1959) pp. 115–120 (**414**).

PARRY, W. T.

[1934] The postulates for 'strict implication'. *Mind* Vol. 43 (1934) pp. 78–80 (**222**).

[1939] Modalities in the *Survey* system of strict implication. *JSL* Vol. 4 (1939) pp. 131–154 (**29, 69, 176, 197, 211, 212, 213, 252, 262, 273, 276, 282**).

[1953] Review of Becker [1952], *JSL* Vol. 18 (1953) pp. 327–329 (**245**).

PARSONS, T.

[1967] Grades of essentialism in quantified modal logic. *Noûs* Vol. 1 (1967) pp. 181–191 (**166**).

POLLOCK, J. L.

[1966] The paradoxes of strict implication. *LA* No. 34 (Vol. 9) pp. 180–196 (**418, 420**).

[1967a] Logical validity in modal logic. *The Monist* Vol. 51 (1967) pp. 128–135 (**47**).

[1967b] Basic modal logic. *JSL* Vol. 32 (1967) pp. 355–365 (**245**).

POLIFERNO, M. J.

[1961] Decision algorithms for some functional calculi with modality. *LA* Vol. 4 (1961) pp. 138–153 (**94**).

[1964] Correction to a paper on modal logic. *LA* Vol. 7 (1964) pp. 32–33.

PRIOR, A. N.

[1954] The interpretation of two systems of modal logic. *The Journal of Computing Systems*, Vol. 4 (1954) pp. 626–630 (**376**).

[1955] *Formal logic*. Oxford University Press, 1955. Second Edition, 1962 (**1**, **71**, **130**, **135**, **137**, **141**).

[1955b] Diodoran modalities. *The Philosophical Quarterly* Vol. 5 (1955) pp. 205–213 (**265**).

[1956] Modality and quantification in S5. *JSL* Vol. 21 (1956) pp. 60–62 (**90**).

[1957] *Time and modality*. Oxford University Press, 1957 (**29**, **121**, **243**, **264**, **265**, **358**, **359**, **363**, **371**, **372**, **373**, **376**).

[1958] Diodorus and modal logic, a correction. *The Philosophical Quarterly* Vol. 8 (1958) pp. 226–230 (a correction to [1955b]) (**266**).

[1959] Notes on a group of modal systems. *LA* Vol. 2 (1959) pp. 122–127 (**361**).

[1961] Some axiom-pairs for material and strict implication. *ZML* Vol. 7 (1961) pp. 61–65 (**325**, **328**).

[1962] Possible worlds. *The Philosophical Quarterly* Vol. 12 (1962) pp. 36–43 (**316**).

[1963a] The theory of implication. *ZML* Vol. 9 (1963) pp. 1–6 (**328**, **329**).

[1963b] Is the concept of referential opacity really necessary? *Acta Philosophica Fennica* (1963) *Modal and Many-valued Logics* pp. 189–198 (**143**).

[1963c] Oratio obliqua. *Aristotelian Society Supplementary Volume XXXVII* (1963) pp. 115–126 (**143**).

[1964a] Axiomatizations of the modal calculus Q. *NDJ* Vol. 5 (1964) pp. 215–217 (**361**).

[1964b] K1, K2, and related modal systems. *NDJ* Vol. 5 (1964) pp. 299–304 (**282**).

[1967] *Past, present and future*. Oxford University Press, 1967 (**91**, **264**, **268**, **269**).

Vide also Lemmon, Meredith, Prior and Thomas.

Vide also Meredith and Prior.

PURTILL, R. L.

[1968] About identity through possible worlds. *Noûs* Vol. 2 (1968) pp. 87–89 (**122**).

QUINE, W. V.

[1940] *Mathematical logic*. Harvard University Press, New York, 1940. Second edition, 1957 (**1**, **74**).

[1947] The problem of interpreting modal logic. *JSL* Vol. 12 pp. 43–48 (**143**).

[1950] *Methods of logic*. New York, Holt 1950, Revised edition Holt Rinehart and Winston, 1959 (**1**, **74**).

[1953] Reference and modality, in *From a logical point of view*, Cambridge, Mass., Harvard University Press, 1953, pp. 139–159 (Second edition 1961) (**143**, **149**).

[1960] *Word and object*. Cambridge, Mass., MIT Press, 1960 (**143**).

[1963] *Set theory and its logic*. Cambridge, Mass., Belknap Press, Harvard University Press, 1963 (**163**).

[1966] *The ways of paradox and other essays*. New York, Random House, 1966 (**143**).

RASIOWA, H. and SIKORSKI, R.

[1963] *The mathematics of metamathematics*. Warsaw, Pánstwowe Wydawnictwo Naukowe, 1963 (**385, 394**).

RESCHER, N.

[1959] On the logic of existence and denotation. *The Philosophical Review*, Vol. 58 (1959) pp. 157–180 (**126**).

[1966] On modal renderings of intuitionistic propositional logic. *NDJ* Vol. 7 (1966) pp. 277–280 (**364**).

ROSSER, J. B.

[1953] *Logic for mathematicians*. New York, McGraw-Hill Book Co., 1953 (**154, 155, 195**).

RUNDLE, B. B.

[1965] Modality and quantification, in *Analytical Philosophy*, Second Series (ed. R. J. Butler), Oxford, Basil Blackwell, 1956, pp. 27–39 (**143**).

SCROGGS, S. J.

[1951] Extensions of the Lewis system S5. *JSL* Vol. 16 (1951) pp. 112–120 (**390, 400**).

SEGERBERG, K.

[1967] Some modal logics based on a three-valued logic. *Theoria* Vol. 33 (1965) pp. 53–71 (**363**).

SHEFFER, H. M.

[1913] A set of five independent postulates for Boolean algebras. *Transactions of the American Mathematical Society* Vol. 14 (1913) pp. 481–488 (**319**).

SIMONS, L.

[1953] New axiomatizations of S3 and S4. *JSL* Vol. 18 (1953) pp. 309–316 (**229, 298**).

[1962] A reduction in the number of independent axiom schemata for S4. *NDJ* Vol. 3 (1962) pp. 256–258 (**232**).

SLATER, B. A.

[1963] Talking about something. *Analysis* Vol. 23 (1963–64) pp. 49–53 (**143**).

SMILEY, T. J.

[1959] Entailment and deducibility. *Proceedings of the Aristotelian Society*, Vol. 59 (1958–59) pp. 223–254 (**420**).

[1961] On Łukasiewicz's Ł-modal systems. *NDJ* Vol. 2 (1961) pp. 149–153 (**370**).

[1963] Relative necessity. *JSL* Vol. 28 (1963) pp. 113–134 (**368**).

SMULLYAN, A. F.

[1948] Modality and description. *JSL* Vol. 13 (1948) pp. 31–37 (**157, 160**).

SOBOCIŃSKI, B.

[1953] Note on a modal system of Feys–Von Wright. *The Journal of Computing Systems* Vol. 1 (1953) pp. 171–178 (**20, 27, 40, 67, 68**).

[1962a] A contribution to the axiomatization of Lewis's system S5. *NDJ* Vol. 3 (1962) pp. 51–60 (**218, 235, 251, 352**).

[1962b] A note on the regular and irregular modal systems of Lewis, *NDJ* Vol. 3 (1962) pp. 109–113 (**287, 352**).

[1962c] On the generalized Brouwerian axioms. *NDJ* Vol. 3 (1962) pp. 123–128 (**255, 352**).

[1963] A note on modal systems. *NDJ* Vol. 4 (1963) pp. 155–157 (**352**).

[1964a] Remarks about the axiomatizations of certain modal systems. *NDJ* Vol. 5 (1964) pp. 71–80 (**274, 280, 282**).

[1964b] Modal system S4.4. *NDJ* Vol. 5 (1964) pp. 305–312 (**262, 270, 272, 283**).

[1964c] Family $\mathscr{K}$ of the non-Lewis modal systems. *NDJ* Vol. 5 (1964) pp. 313–318 (**279, 283**).

STOLL, R. R.

[1961] *Sets, logic and axiomatic theories.* San Francisco and London, Freeman, 1961 (**163, 379, 385, 386, 388**).

STRAWSON, P. F.

[1948] Necessary propositions and entailment-statements. *Mind* Vol. 57, 1948, pp. 184–200 (**420**).

SUGIHARA, T.

[1962] The number of modalities in T supplemented by the axiom CL^2pL^3p. *JSL* Vol. 27 (1962) pp. 407–408 (**258**).

THOMAS, I.

[1962] Solutions of five modal problems of Sobociński. *NDJ* Vol. 3 (1962) pp. 199–200 (**42, 333**).

[1963a] $S1^0$ and Brouwerian axioms. *NDJ* Vol. 4 (1963) pp. 151–152 (**255, 353**).

[1963b] $S1^0$ and generalized S5-axioms. *NDJ* Vol. 4 (1963) pp. 153–154 (**353**).

[1963c] A final note on $S1^0$ and the Brouwerian axioms. *NDJ* Vol. 4 (1963) pp. 231–232 (**255, 353**).

[1964a] Modal systems in the neighbourhood of T. *NDJ* Vol. 5 (1964) pp. 59–61 (**256, 259**).

[1964b] Ten modal models. *JSL* Vol. 29 (1964) pp. 125–128 (**248, 255, 353**).

Vide also Lemmon, Meredith, Prior and Thomas.

VREDENDUIN, P. G. J.

[1939] A system of strict implication. *JSL* Vol. 4 (1939) pp. 73–76 (**417**).

WAJSBERG, M.

[1933] Ein erweiteter Klassenkalkül. *Monatshefte für Mathematik und Physik* Vol. 40 (1933) pp. 113–126 (**35, 42, 63, 197**).

WHITEHEAD, A. N. and RUSSELL, B. A. W.

[1910] *Principia mathematica.* Cambridge, Cambridge University Press, 3 Vols., First edition 1910–1913, Second edition 1923–1927 (**6, 152, 154, 160, 162, 178**).

WISDOME, W. A.

[1964] Possibility elimination in natural deduction. *NDJ* Vol. 5 (1964) pp. 295–298 (**412**).

WRIGHT, G. H. VON

[1951] *An essay in modal logic*. Amsterdam, North Holland Publishing Co., 1951 (**20, 27, 48, 56, 68, 70, 73, 130, 133, 134, 135**).

[1957] The concept of entailment, in *Logical Studies*. London, Routledge and Kegan Paul, 1957, pp. 166–191 (**420**).

YONEMITZU, N.

[1955] A note on modal systems, Von Wright's and Lewis's S1. *Memoirs of the Osaka University of the Liberal Arts and Education Bulletin of Natural Science*, No. 4 (1955) p. 45 (**199**).

Index of Authors

Index of Subjects

List of Symbols and Important Rules

(For other notations *vide* Appendix 4)